Social Inclusion and Usability of ICT-Enabled Services

Social Inclusion and Usability of ICT-Enabled Services is a cutting-edge research book written for researchers, students, academics, technology experts, activists, and policy makers. The book explores a wide range of issues concerning innovative ICT-enabled digital services, their usability, and their consequent role in social inclusion. It includes the impacts of the use of ICT-enabled digital services on individuals, organizations, governments, and society, and offers a theoretically informed and empirically rich account of the socio-technical, management, and policy aspects of social inclusion and innovative ICT-enabled digital services.

This publication offers insights from the perspectives of IS, media, and communications, management, and social policy, drawing on research from these disciplines to inform readers on diverse aspects of social inclusion and usability of ICT-enabled digital services. The originality of this book lies in the combination of socio-technical, management and policy perspectives offered by the contributors, and integrated by the editors, as well as in the interdisciplinary and both theoretically framed and empirically rich features of the various chapters of the book. While providing a timely account of existing evidence and debates in the field of social inclusion and technology usability, this book will also offer some original insights into what practitioners, experts, and researchers are to expect in the near future to be the emerging issues and agendas concerning the role of technology usability in social inclusion and the emerging forms and attributes of the latter.

Through a collection of high quality, peer-reviewed papers, *Social Inclusion and Usability of ICT-enabled Services* will enhance knowledge of social inclusion and usability of ICT-enabled digital services and applications at a diverse level.

Jyoti Choudrie holds the position of Professor of Information Systems at University of Hertfordshire, United Kingdom.

Panayiota Tsatsou is an Associate Professor of Media and Communication at the University of Leicester, United Kingdom.

Sherah Kurnia is a Senior Lecturer at the School of Computing and Information Systems, the University of Melbourne, Australia.

Routledge Studies in Innovation, Organization and Technology

For a full list of titles in this series, please visit www.routledge.com

37 **The Entrepreneurial University**
Context and Institutional Change
Edited by Lene Foss and David V. Gibson

38 **Dynamics of Knowledge-Intensive Entrepreneurship**
Business Strategy and Public Policy
Edited by Franco Malerba, Yannis Caloghirou, Maureen McKelvey and Slavo Radosevic

39 **Collaborative Innovation**
Developing Health Support Ecosystems
Edited by Mitsuru Kodama

40 **Innovation Drivers and Regional Innovation Strategies**
Edited by Mario Davide Parrilli, Rune Dahl-Fitjar, and Andrés Rodríguez-Pose

41 **Information Theft Prevention**
Theory and Practice
Romanus Izuchukwu Okeke and Mahmood Hussain Shah

42 **The New Production of Users**
Changing Innovation Collectives and Involvement Strategies
Edited by Sampsa Hyysalo, Torben Elgaard Jensen, and Nelly Oudshoorn

43 **Foundations of Information Systems**
Research and Practice
Andrew Basden

44 **Social Inclusion and Usability of ICT-Enabled Services**
Edited by Jyoti Choudrie, Panayiota Tsatsou and Sherah Kurnia

Social Inclusion and Usability of ICT-Enabled Services

Edited by Jyoti Choudrie, Panayiota Tsatsou and Sherah Kurnia

Routledge
Taylor & Francis Group
NEW YORK AND LONDON

First published 2018
by Routledge
711 Third Avenue, New York, NY 10017

and by Routledge
2 Park Square, Milton Park, Abingdon, Oxon OX14 4RN

Routledge is an imprint of the Taylor & Francis Group, an informa business

Library of Congress Cataloging-in-Publication Data
A catalog record for this book has been requested

ISBN: 978-1-138-93555-6 (hbk)
ISBN: 978-1-315-67731-6 (ebk)

Typeset in Sabon
by Apex CoVantage, LLC

Contents

List of Figures

List of Tables

Preface

Information and Communications Technology (ICT) has much potential to improve people's lives and make the world a better place. The promise is that ICT can make people more productive, improve health outcomes, provide better government services, and help people to be more connected to their family and friends. This book focuses on innovative ICT-enabled services for social inclusion. Designed and used appropriately, ICT-enabled services can enable people to more effectively communicate, contribute, and participate at work, at home, and in the wider society more generally.

However, as the chapters in this book illustrate, achieving these benefits is not always straightforward. There are many challenges that need to be addressed. For example, how can we ensure that people are not digitally excluded? If they do not have access (e.g., they cannot afford it), it is simply not possible for them to take advantage of ICT-enabled services. This is the case in many underdeveloped countries where ICT4D has become an important topic. Similarly, if they do not have the skills to use ICT appropriately, then there is no way for them to effectively communicate, contribute, and participate. Digital literacy, or as I prefer to call it nowadays, digital fluency, has become an important prerequisite for participation in today's digital world. Another important consideration might be various technological barriers themselves—sometimes the software or advice can prevent people from doing things, particularly if it is not designed well.

This book looks at all these issues from a variety of perspectives. There are conceptual and theoretical chapters looking at what it means to be socially inclusive and there are empirical chapters that look at social inclusion in various countries. Some chapters look at the use of ICT at work, in particular mobile knowledge work and telemetry work, whereas others look at ICT for entertainment; some chapters look at ICT services for health, particularly in rural areas (sometimes called e-health), whereas others look at ICT services for government (often called e-government). The development of smart urban spaces in smart cities is something that many local government organizations are investigating if not already introducing. In contrast to the use of ICT-enabled services for government, there are various groups of people or social movements that use ICT to protest against current government policies and services. One chapter in this book looks at a group of people, older adults, who for one reason or another feel excluded or disadvantaged in some way, and thus ICT becomes a way of them to more effectively voice their grievances.

In summary, the opportunities for the use of ICT-enabled services for social inclusion are great, but so are the challenges. I trust this book will provide those who are interested in this topic with a deeper understanding of the issues. Using ICT to improve the welfare of disadvantaged people, to improve health and government services, and, more generally, to make the world a better place is something that we should all be working toward. I trust that this book will contribute in a both intellectual and practical way to the achievement of these goals.

Michael D. Myers
Professor of Information Systems
University of Auckland
New Zealand

Acknowledgments

I would like to thank my wonderful peers in the form of my co-editors, Sherah and Panayiota who worked tirelessly with me and the authors who contributed toward the completion of this project. I would also like to thank Jerry Forrester, and Keith Randle who supported and encouraged me when submitting the proposal, my head of department, Associate Dean of Research and Enterprise and Dean of Hertfordshire Business School for affording me the time to work on this manuscript. My best friend, Razwana, and dearest neighbor, Barbara, who would from time to time ask about the book's progress and encourage me when I was worried. Above all, I dedicate this manuscript to my parents and Brother Bobby who stand by me no matter what and work tirelessly with me when I need to meet deadlines, or face off any challenges. Without all of you, this manuscript would not be complete. Thank you.

Jyoti Choudrie

I would like to thank the authors for their great contributions. Above all, though, I would like to thank my daughter, Marie-Christine, who gives me all sorts of reasons to work hard as a parent, person, and researcher. I dedicate this book to her with love and gratitude.

Panayiota Tsatsou

The vision of this book would not have been achieved without the valuable contributions from all the authors and the support from the School of Computing and Information Systems at the University Melbourne. Most importantly, this book project would not have been completed without the excellent work of and collaboration from Jyoti and Panayiota. I dedicate this book to my parents who believe in me and give me freedom to pursue my dreams, to my brothers and sister who always care for and protect me, to my husband who always gives me strengths and encouragements when facing challenges, and to my wonderful children, Shelin and Sheldon, for making me live my life to the fullest.

Sherah Kurnia

Notes

1 We would like to express our gratitude to the Oxford Internet Institute, University of Oxford, for allowing us to use the latest data from the Oxford Internet Surveys (OxIS) for the empirical analyses in Chapter 11.

2 This work was supported by the British Academy Grant SG102167 "Discovering Digital Me: Forging links Across Digital Identity, Digital Literacy and Digital Economy" led by Gillian Youngs and Panayiota Tsatsou with research assistance from Carolyn Watt.

Introduction

Panayiota Tsatsou, Sherah Kurnia, and Jyoti Choudrie

The history of Information and Communication Technologies (ICTs) is full of developments, surprises, debates, and as many certainties as many uncertainties. Landmarks in the history of ICTs that stand out, including the first packet-switching network, the Transmission Control Protocol and the Internet Protocol (IP) in the late 1960s, the development of hypertext language in 1989, the release of the World Wide Web by CERN for general use in 1993, and the spectacular developments around broadband, Web 2.0 and Web 3.0 from the mid-2000s onward, cannot portray sufficiently the scale, magnitude, and wealth of breakthroughs in the domain of ICTs and ICT-enabled services.

At the same time, ICTs and their histories have been marked by dichotomies, dualisms, and binary tensions, the most important of which is that concerning the role of ICTs in the society. On the one hand, acknowledging the significance of rapidly developing technologies such as Internet-based technologies resulted in strongly argumentative, normative and even emotional approaches that either glorified or cursed the Internet for driving a fast changing and increasingly challenging and globalizing mode of living. On the other hand, claims about the "Internet revolution" have been treated with particular caution, resulting in the downplaying of the Internet as technology, the advocacy of the importance of social contexts within which technology is designed, and approaches in support of the social shaping of technology (Tsatsou 2014)

At the core of such dualisms and associated debates is the question of the role of ICTs in social inclusion. Social inclusion is of immense interest to technology experts, activists, and policy makers due to the way information and communication technological innovations such as those based on the Internet have been rapidly and largely unexpectedly emerged, affecting people's positioning in the society. Questions regarding the role that technology might play in people's positioning in the society began to be examined by researchers and scholars mostly from the turn of the 21st century (e.g., Warschauer 2003), with the emphasis ranging from the study of the role of media technologies and ICTs in enhancing people's citizenship (e.g., Coleman and Blumler, 2009; Mossberger, Tolbert, and McNeal 2008) to the study of the barriers to equal opportunities for everyone to access and use such technologies (e.g., Ferro et al., 2009; Norris 2001; Tsatsou 2011, van Dijk 2005). However, existing research hardly offers a wide-ranging and inclusive account on the role of innovative ICT-enabled services in social inclusion. Instead, it either focuses on single techno-centric issues, such as that of technology diffusion and adoption (e.g., Rogers 1995), or it approaches questions concerning social inclusion in a rather vague and fragmented way. Even research that focuses on human interaction via technology does not seem to delve sufficiently into issues relating to technology's management, the diversity of society's responses to technological development and the policy strategies that mediate the intercourse between society and technology.

There is a continuing need to make sense of complex socio-technical systems that are associated with the adoption, management, and policy of new technological services in order to unpack the intercourse of social inclusion and usability of ICT-enabled services. It is our hope and belief that this book can make some contribution toward fulfilling this need.

The vision of this book is to bring together latest thinking and groundbreaking research in the area of innovative ICT-enabled services and social inclusion, in order to dig deeper into the socio-technical, management, and policy dimensions related to our increased reliance upon ICT-enabled services for reinforcement and enhancement of social inclusion. Specifically, the book focuses on aspects and benefits of social inclusion that are highly driven by the usability and employment of ICT-enabled services, rather than on the features of the technology itself or just the sociological aspects of the issue. The case studies in this book delve into, explore further, and critically assess whether ICT-enabled services reinforce social inclusion and how they impact the delivery of public or community related services to individuals and whole populations who previously did not take advantage of such services due to demographic, personal, or broadly social conditions. Also, the case studies in this book shed light on whether ICT-enabled services and products can lead to enhanced social capital and whether individuals or groups become more aware of macro-space, non-personal issues due to more engagement with technologically distributed information and the provision of new technological means for exchange, sharing, and collective action. The book offers analysis and evidence that inform the reader on the potential of improved teleworking capacity and how ICTs can lead to a new world of entertainment, occupation, social networking, political communication, and e-governance possibilities. The contents of this volume also touch upon diverse groups of the society, such as young people, older adults, farmers, professionals, and students.

In the midst of ICT-enabled innovations that are being introduced and spread rapidly and in all different contexts, this book provides valuable insights on critical aspects of social inclusion that arise or could emerge as innovative ICT-enabled products and services proliferate. The book offers original insights into emerging issues and agendas concerning the role of usability of ICT-enabled services in social inclusion and maps out the implications for the emergence of new forms and attributes of social inclusion. It contains 15 original and timely contributions to the critical thinking and knowledge enhancement about social inclusion and usability of ICT-enabled services, which run at both the theoretical and empirical case study levels, involve the study of various population groups, and draw concepts and insights from different disciplines. It is a cutting-edge research book written not only for researchers, students, and academics but also for stakeholders and policy makers. The contributing authors are scholars and researchers with expertise in IS, management, ICT, new/digital media, and communications policy, offering a collection of chapters that provides original and powerful insight into the value of ICT-enabled services for the various facets and evolving features of social inclusion.

Hence, the contribution of this book can be summarized as mostly lying in the following areas:

1. Shedding light on long-standing and ongoing debates and themes related to the theme of social inclusion and usability of ICT-enabled services.
2. Examining a range of countries and contexts (e.g., Australia, India, United Kingdom, USA), while also shedding light on a range of sectors (e.g., e-health, telework, e-government, e-entertainment, farming) where the presence and influence of ICT-enabled services are broadly perceived as crucial.
3. Adopting an interdisciplinary and reflective approach to developing new knowledge on phenomena, topics, and issues that are pertinent to the theme of social inclusion and usability of ICT-enabled services.

This book is structured into three parts. Part 1 focuses on theoretical concepts related to social inclusion. Part 2 captures a number of cutting-edge ICT-enabled services used within organizations and society, while Part 3 assess the adoption, use, and management of ICT-enabled services to achieve social inclusion. Details of each chapter are outlined next.

Part 1: Examining the Theoretical Foundations of Social Inclusion and Usability

The chapter by Efpraxia D. Zamani, titled "Social Inclusion and ICTs: A Literature Review Through the Lens of the Capability Approach," argues that in the context of today's hyper-connected society, adopting ICTs, which are considered to be a way toward achieving social inclusion is no longer adequate to ensure participation of all. ICTs need to specifically address the needs of disadvantaged individuals to empower them and support them in improving their welfare and meeting their goals. Drawing upon the Capability Approach, this chapter looks into the abilities, needs, and personal wants of disadvantaged individuals in relation to ICTs.

Christoph F. Breidbach contributes a chapter titled "Connectivity: A Socio-technical Construct to Examine ICT-Enabled Service" that explains vividly how advances in ICT transformed interactions between service providers and their customers. It shows that ICT-enabled customer-firm interfaces are increasingly the norm in service contexts such as health care or consulting. It argues that academic research has not fully caught-up with this new reality. The chapter introduces the "connectivity" construct as a socio-technical lens to advance the current understanding of ICT-enabled service.

The chapter by H. Patricia McKenna titled "Re-conceptualizing Social Inclusion in the Context of 21st-Century Smart Cities" explores and develops the social inclusion concept in the context of smart cities with the aim of re-conceptualizing inclusion for 21st-century urban environments. Aspects and benefits of innovative ICT-enabled services and social inclusion are examined in contemporary urban spaces, along with emerging forms, and attributes. This chapter introduces and theorizes ambient inclusion, and operationalizes an ambient inclusion framework for the exploration of innovative ICT-enabled spaces, services, and designs in smart cities.

Finally, the chapter by Arthur Glenn Maail, Sherah Kurnia, and Shanton Chang titled "Enhancing Social Inclusion Through Optimal Community Participation Levels in ICT4D Projects" acknowledges that user participation in Information and Communication Technology for Development (ICT4D) projects is critical to promote social inclusion for the local community. The authors develop a conceptual framework that identifies the optimal level of user participation based on the approach deployed and the relevant conditional factors affecting user participation for each approach. The study enhances the current understanding in managing user participation in the development of ICT4D projects.

Part 2: ICT-Enabled Services of Value to Society and Organizations

The chapter by David J. Yates, Girish J. "Jeff" Gulati, and Christine B. Williams titled "Understanding the Impact of Politication Structure, Governance, and Public Policy on E-Government" examines two distinct but related measures of e-government effectiveness—namely, the online service index and the e-participation index reported in the 2014 e-government survey conducted by the United Nations. The study analyses the impact of political structure, public sector performance, and policy initiatives on both indices in 175 countries. The study suggests that the path to e-government leverages different strategies depending on a nation's political structure and processes, and that authoritarian countries may be utilizing e-government to maintain the status quo.

Avijit Sarkar, James Pick, and Jessica Rosales contribute a chapter titled "ICT-Enabled E-Entertainment Services in U.S. Counties: Socio-economic

Determinants and Geographic Patterns" that examines spatial patterns and determinants of the use of the Internet for entertainment purposes in the counties of the United States. Spatial patterns of e-entertainment diffusion in U.S. counties indicate an urban-rural divide. High use of e-entertainment amidst low e-entertainment clusters is often found in counties that are home to large public universities, military reservations, or government labs. This chapter fills an important void in the digital-divide literature as research and related discourse shifts from measuring and examining differences in access of ICTs to differences in actual use of the Internet.

The chapter titled "E-Health as an Enabler of Social Inclusion" by Ken Clarke, Adam Lodders, Robyn Garnett, Anne Holland, Rodrigo Mariño, and Zaher Joukhadar shows how e-health can improve the access and reach of health services to include those sections of the community who can be otherwise excluded due to factors such as advanced age, having English as a second language, or remote location. The practical examples in this chapter demonstrate how well designed e-health initiatives that have simple user interfaces and "ambient" sensing technologies can drive and obtain both better health outcomes and improved levels of social engagement. Society can also benefit from efficiency gains that allow health-care professionals to address the needs of more people, particularly in rural and remote areas with clinical skill shortages.

The chapter by Victoria Carty titled "Challenging the Cost of Higher Education With the Assistance of Digital Tools: Case Studies of Protest Activity in Canada and the United States" presents two case studies involving Canada and the United States that focus on how activists use digital tools to challenge the cost of higher cost of education using digital tools. It shows that new communication technologies allow social movement actors to operate in a new political terrain that enhances not only virtual but also face-to face forms of interaction, debate, and critical thinking. Social movement theories are applied to show the relevance of peer-to-peer networks that allow activist to circumvent state- and corporate-controlled and owned media, allowing them to take ownership of the narrative of their protest activity.

Rachelle Bosua, Sherah Kurnia, Marianne Gloet, and Antonette Mendoza's chapter is titled "Telework Impact on Productivity and Well-Being: An Australian Study." They argue that forms and locations of work have changed dramatically over the last few decades. Facilitated by ICTs, it is expected that mobility in work (telework) will increase. Using Australia as the study context, this chapter shows that telework improves team productivity and individual well-being and highlights key elements that contribute to effective telework: appropriate IT support, trust, management support, a supportive telework culture, and access to hybrid telework arrangements. A focus on one or more of these elements would improve social inclusion of workers from a productivity and individual well-being perspective.

The chapter by Sherah Kurnia, Md Mahbubur Rahim, Serenity Hill, Kirsten Larsen, Patrice Braun, Danny Samson, and Prakash Singh titled "Supporting Regional Food Supply Chains With an E-Commerce Application" evaluates the effectiveness of the Open Food Network (OFN) in connecting and supporting regional food supply chain communities in Australia. In this chapter, the authors show how an innovative e-commerce application improves farmer's access to local markets and consumers' access to fresh local produce, as well as optimizes the regional food distribution and improves the local community welfare. Thus, they demonstrate the potential of ICT-enabled innovations to include disadvantaged local food players in order to establish a more socially sustainable and fair trading environment.

Part 3: Adoption, Usage, and Management Aspects Surrounding Social Inclusion and Usability of ICT-Enabled Services

The chapter by Bianca C. Reisdorf and Darja Groselj titled "Digital Divides, Usability, and Social Inclusion: Evidence From the Field of E-Services in the United Kingdom" examines how digital divides and usability affect engagement with e-services. Using the context of British Internet users and non-users, the chapter investigates whether Internet users are more likely to use any government services as compared to non-users and what factors shape use of e-government services. The study shows how that apart from socio-demographic factors, digital skills, and high quality, ubiquitous Internet access are crucial for use of e-government services. This has implications for the necessity for high levels of usability in the design of e-services and the acknowledgment that Internet users are not one homogeneous group.

The next chapter by Mohammad Hossein Jarrahi and Luke Williamson titled "Mobility of Work: Usability of Digital Infrastructures and Technological Divide" focuses on the technological barriers of mobile knowledge work. The study examines technological barriers that diminish the usability of systems for mobile knowledge workers and plague communication, information management, spatial mobility, and navigation of organizational boundaries. These barriers can lead to a potential digital divide between mobile workers and those who work from traditional work settings. This study further offers an overview of adaptive strategies pursued by mobile knowledge workers in addressing these technological barriers. Findings from this work inform the design of more inclusive technological infrastructures that cater to the needs of mobile knowledge workers.

Deana A. Rohlinger and Shawn Gaulden contribute a chapter titled "Overcoming Obstacles to Activism With ICTs: An Analysis of MoveOn.Org and the Florida Tea Party Movement." Drawing on participant observation data and 52 interviews with supporters of MoveOn.org and the Florida Tea Party Movement, the study identifies three obstacles to political participation: motivational obstacles, organizational obstacles, and feelings of efficacy. The findings show that ICTs can help individuals overcome motivational obstacles by connecting information to political activities and by creating opportunities for individuals to learn new political skills. Organizational obstacles can be overcome by cultivating identities focused on supporters' roles as citizens and make them feel efficacious by linking them to a larger political community.

The chapter by Ranjan Vaidya titled "Social Inclusion, Farmer Resignation, and the Challenges of Information Technology Implementation" explores the role of information technology in achieving social inclusion of farmers in India using stakeholder theory and Bourdieu's theory of practices. For this purpose, a qualitative approach is used to study the case of a state government organization in India that implemented an information technology project to provide fair prices to the poor farmers by interconnecting the agricultural market yards of the state. The study highlights stakeholder resignation as a type of practice that is exercised when the powerless stakeholders have lost all hopes of their social inclusion. The study has important implications for academics, policy makers, and the agricultural industry.

The next chapter by Jyoti Choudrie, Sutee Pheeraphuttharangkoon, and Uchenna Ojiako titled "Smartphones Adoption and Usage of 50+ Adults in the United Kingdom" explains that smartphones are devices providing advanced computing capabilities and connectivity that offer immense benefits and convenience to diverse users. The chapter investigates the adoption, usage, and diffusion of smartphones within the United Kingdom's older adults through an online survey of North London households. The study shows that Perceived Enjoyment was the strongest adoption factor followed by performance expectancy, compatibility, facilitating conditions, and effort expectancy. Moreover, education and experience were identified as the moderating adoption variables. The identification and understanding of the factors that encourage or inhibit smartphone use within the older adult population can assist future researchers to identify ICTs that can assist in reducing social isolation within the older adult population.

The final chapter in this section is by Panayiota Tsatsou, Gillian Youngs, and Carolyn Watt and it is titled "Literacy and Identity Links Forging Digital Inclusion? Critical Reflections and Signposts From a Qualitative Study." It argues that the concept of "interactivity" lies at the core of digital literacy and suggests a micro-perspective on people's identity in general, and age identity in particular, in order to conceptually shed light on the joint role that literacy and identity play in digital inclusion. It also reports on qualitative research that operationalizes and empirically tests these conceptual suggestions. The findings suggest that literacy and age identity affect digital inclusion separately as well as jointly, while age influences factual and perceptual literacy gaps. The findings challenge the assumptions about "digital natives" and "digital immigrants" and invite further exploration of "intra-generational" nuances in the realm of digital inclusion for a better understanding of the importance of digital inclusion for social inclusion.

References

Coleman, Stephen and Jay G. Blumler. *The Internet and Democratic Citizenship: Theory, Practice and Policy*. Cambridge and New York: Cambridge University Press, 2009.

Ferro, Enrico, Yogesh Kumar Dwivedi, J. Ramón Gil-Garcia and Michael D. Williams, eds. *Handbook of Research on Overcoming Digital Divides: Constructing an Equitable and Competitive Information Society*. Hershey, PA: IGI Global, 2009.

Mossberger, Karen, J. Caroline Tolbert and Ramona S. McNeal. *Digital Citizenship: The Internet, Society, and Participation*. Cambridge, MA: MIT Press, 2008.

Norris, Pipa. *Digital Divide: Civic Engagement, Information Poverty, and the Internet Worldwide*. Cambridge: Cambridge University Press, 2001.

Rogers, Everett M. *Diffusion of Innovations*, Vol. 4. New York: Free Press, 1995.

Tsatsou, Panayiota. *Digital Divides in Europe: Culture, Politics and the Western-Southern Divide*. Berlin: Peter Lang, 2011.

Tsatsou, Panayiota. *Internet Studies: Past, Present and Future Directions*. Surrey: Ashgate Publishing, 2014.

Van Dijk, Jan A. G. M. *The Deepening Divide: Inequality in the Information Society*. Thousand Oaks CA, London and New Delhi: Sage, 2005.

Warschauer, Mark. *Technology and Social Inclusion: Rethinking the Digital Divide*. Cambridge, MA: MIT Press, 2003.

Part I

Examining the Theoretical Foundations of Social Inclusion and ICT-Enabled Services

1 Social Inclusion and ICTs

A Literature Review Through the Lens of the Capability Approach

Efpraxia D. Zamani

1. Introduction

Social inclusion is considered to lead to equal participation in society, and to people's welfare (Gingrich and Lightman 2015). Policy makers have put forth the adoption of ICTs, arguing that these can deliver reforms and support disadvantaged individuals to overcome long-standing inequalities (Sourbati 2012). However, mere adoption of ICTs is not enough to achieve social inclusion. Rather, it is necessary that the design and the functionality of ICTs meet the needs and expectations of disadvantaged individuals (Zheng 2009), as these can help them meet their personal goals.

Disadvantaged individuals exhibit quite different characteristics, which form their needs and expectations. ICTs that may support one type of individuals and help them overcome some impediments may inhibit others' access. At the same time, desirability is important for the successful use of ICTs. Therefore, ICTs need to account both for what individuals are able and wish to achieve.

This chapter investigates the abilities, needs, and desires of individuals who experience social exclusion by focusing on the role of ICTs. It is based on a literature review of previous studies at the crossroad of social inclusion and ICTs. Drawing upon Sen's Capability Approach (Sen 2000), it identifies the individual abilities, the opportunities and the expectations a disadvantaged person has so that s/he can lead a life s/he values (Zheng 2009). In this sense, it illustrates the needs and requirements of disadvantaged individuals, with the aim to facilitate the discussion on social inclusion through ICTs.

2. Background

2.1 Social Inclusion and ICTs

Social inclusion is much more than challenging financial inequalities (Notley 2009). It entails that individuals participate equally in structures, activities, and their communities, because they are able and permitted to partake in society (Guildford 2000). Along these lines, the availability of ICTs is a necessary condition for one to be able to conduct a certain type of life in today's interconnected society (e.g., Jaeger and Xie 2009). Specifically, there is a linkage between ICTs and social development (Helsper 2008), and ICTs can help in bridging social inequalities, while, alongside other things, they may support development and lead to improved education, government, and health care (e.g., Zheng 2009). In addition, ICTs can strengthen existing social ties and support the development of new ones, because mobile phones, online social networks and the likes enable two-way communication (Zinnbauer 2007). Equally so, ICTs may lead to social exclusion when there is a lack of basic ICTs skills, infrastructures (Zheng and Walsham 2008) or due to increased costs (Khorshed and Imran 2015). It is thus important, on the one hand, to ensure the necessary conditions, the financial resources and often the political support for the deployment of technical infrastructures and other practical arrangements (Fernback 2005). On the other hand, governmental and private initiatives need to safeguard that ICTs allow and support the inclusion of individuals in everyday activities (Eisma et al. 2004).

2.2 *Capabilities and Deprivation*

The Capability Approach builds upon two concepts: functionings and capabilities. Capabilities relate to the notion of freedom and "what real opportunities you have regarding the life you may lead" (Sen 1987: 36). Functionings refer to the well-being of a person, and to associated achievements and expectations (Sen 1987; Zheng 2009). Functionings and capabilities are related to each other in the sense that capabilities describe one's freedom to achieve a set of potential functionings—i.e., the possibility to achieve something. Based on one's choice, the individual will then achieve one functioning from the many potential ones.

Another aspect of the Capability Approach is human diversity (Zheng and Walsham 2008), which is introduced into the theoretical framework in the form of conversion factors, and can be personal, social and environmental (Robeyns 2005). The resources one has at their disposal, such as the Internet or food, contribute toward them achieving a set of functionings. For example, a smartphone can contribute toward one's participation in their community by allowing for communication channels with the local authority and friends. Based on one's conversion factors, we understand whether and to what extent they can convert the available resources into desired functionings. An individual with basic ICT skills can take advantage of the smartphone's features, whereas someone with no ICT skills will not be able to operate the device. Therefore, personal conversion factors are internal—i.e., physical condition, literacy, and ICTs skills, etc. Social conversion factors relate to social norms, policies, and power structures among others. Environmental conversion factors then relate to the geographical location, features of the natural, and human made environment, such as infrastructures (Robeyns 2005; Zheng and Walsham 2008).

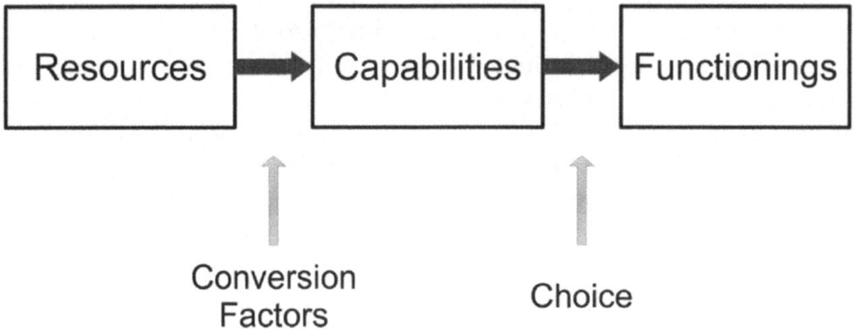

Figure 1.1 An Illustration of the Capability Approach (Adapted from Verd and Lopez Andreu (2011)).

Figure 1.1 presents the relationship between resources, conversion factors, capabilities, and functionings.

As a result, individual differences, the opportunities, and the variations across societies need to be taken together, toward pinpointing the needs of individuals who are considered to be at risk or already affected by social exclusion. In this light, ICTs can strengthen one's capabilities by bringing down the barriers that prohibit people from further cultivating them. Yet, in order to convert capabilities into functionings, there need to exist strategies and policies that mediate the use of ICTs (Alhareth, Alhareth, and Dighrir 2015).

Examining individuals at the verge of social exclusion through the Lens of the Capability Approach can be useful specifically because of the theory's central tenet; it is very clear that its objective is to provide deprived individuals with the freedom to enjoy what they themselves consider valuable.

Following the method section, this chapter identifies groups at the risk of social exclusion, or those who are already deprived of their capabilities, based on a literature review of earlier works and pinpoints their particular requirements. The use of secondary rather than primary material has been chosen because social inclusion has received significant attention, and therefore there is already ample empirical material regarding the requirements of various groups; however, despite the availability of abundant information, this is rather fragmented, and most often, a single study focuses exclusively on one single group of individuals (e.g., older adults). Therefore, this chapter examines collectively the requirements of several groups so as to investigate possible similarities and differences, and facilitate the comparison of common themes across them. As a result, it is posited that it can enhance the practical use of previous findings by policy makers and help in forming policy actions that can boost social inclusion.

3. Method

A primary literature review was undertaken in order to identify relevant studies. This was done through Google Scholar, because it includes gray literature and citation metrics that help in quickly appreciating the article's impact. The keywords used were "social inclusion," "social exclusion," and "ICT." This set of keywords isn't very restrictive, and therefore relevant literature was also identified through Scopus and snowballing through citation tracking, so as to refine the pool of articles. Publication time was the second criterion; the search covered the 2000–2015 period, which allowed the saturation of the findings. Studies being more focused on the digital divide and less on social inclusion were excluded.

Finally, much of the social inclusion literature deals with the use of ICTs by individuals with disabilities. According to the World Health Organization, disability "is an umbrella term, covering impairments, activity limitations, and participation restrictions"[1] and can thus be physical, sensory, intellectual, mental, developmental, or a combination of these. Therefore, social inclusion of the hard of hearing would entail, for instance, actions for increased accessibility (Harris and Bamford 2001), whereas the case of mental health disabilities would suggest actions to reduce stigma, support recovery, ensure anonymity, etc. (Stephens-Reicher et al. 2011). In other words, each disability entails different resources and conversion factors, and leads to different functionings and examining the needs of disabled individuals collectively as a single group would inadvertently lead to reducing the importance of each subgroup's individual differences. For this reason, studies examining ICT-facilitated social inclusion of disabled individuals in general have also been excluded.

The final pool consisted of 46 research papers. The analysis entailed first their classification according to the examined group, and the collated findings led to the identification of six main groups. Next, the analysis continued by examining the studies along the main concepts of the Capability Approach—namely, resources, conversion factors, capabilities, and functionings.

4. Identifying Groups at the Verge of Social Exclusion

This section discusses the needs and requirements of the six identified groups and examines their capabilities that are essential to take advantage of the opportunities offered by ICTs. It is that this chapter investigates the achievable functionings without examining the purposeful rejection of or the lack of interest in ICTs (Haddon 2000).

4.1 Older Adults

Being elderly suggests that, most likely, the individual is to some extent already isolated with a constrained social network, which "reduces the likelihood of benefiting from social applications of the Internet" (Helsper 2008: 12). At the same time, user interfaces and interaction modalities (e.g., keyboard) used by laptops, tablets, and the like, tend to be confusing for the elderly. This reduces older adults' accessibility to services such as online social networks, the Internet and text messaging among others (Teixeira et al. 2012), and adds additional constraints to their communications. However, communication technologies are integral for the elderly's successful inclusion in their communities.

Older adults tend to experience difficulties in relation to their motor skills (Teixeira et al. 2012), they are less mobile, experience health problems, cognitive impairment, or may be affected by more than one of the aforementioned issues. As a result, not all technologies will allow their equal access to ICTs (Jaeger 2006). Their characteristics may work against them and cause their deprivation of their capabilities, and prohibit them from pursuing the life they would like to lead.

At the same time, it is these personal characteristics that will act as the conversion factors based on which older adults will convert their capabilities into the functionings they choose. Older individuals today grow old under much different conditions compared to the past; they retire later and continue working part-time after retirement, they desire to remain independent and active, and generally do not consider themselves "old" (Roberts 2010). This entails that they want and need to remain stimulated. In addition, while they tend to live alone, they need to live embedded in a caring support network, so that they can have a good quality of life as they age. Through this network, they have the opportunity to be in touch through a two-way communication with their carers, who can be family members or professionals, and send/receive information, keep updated with important information and so forth (Waterworth et al. 2009).

As a result, it is necessary that ICTs support them in overcoming whatever difficulties arise due to aging. Older adults seek to have adequate and easy-to-use communication channels and quite often turn to ICTs in order to satisfy this need (Choudrie, Grey, and Tsitsianis 2010). Conversely, the products and services they use need to be appealing, easy to use and accommodating of their situation (Pirkl 2009), so that they can live independently, age successfully, and remain active members of today's society.

4.2 Single Parents

Quite often, single parents find themselves at risk of social isolation when they create their own household, and experience increased levels of loneliness (Haddon 2000). Previous studies (e.g., Hardey 1989) have underlined that female-headed single parent households in particular are faced with an increased risk to be near poverty levels, quite often because of a diminished income due to divorce or separation (Russo Lemor 2005). Therefore, for this type of individuals, Internet access, and even more traditional forms of ICTs like the telephone, can be of particular importance. On the one hand,

ICTs can help them in overcoming their difficulties by facilitating communication, allowing connectivity with others, organizing daily life (Haddon and Silverstone 2000) and even telework when the circumstances require them to stay at home (Schofield Clark 2009). Notwithstanding, their dependent members (i.e., children) can also benefit, as they can take advantage of online opportunities (Tripp 2011), and receive their parents' support in their educational goals (Schofield Clark 2009). On the other hand, over time, the mode of access may be more important; even if single parent households do not have an Internet connection, it may be the case that their members (i.e., the parent and the children) have access through their work or their school, even for personal use (Haddon 2000).

However, ICTs are merely a commodity, and can only become a valuable resource when the individual has the ability to convert them into functionings. Conversion factors such as financial issues and the single parent's status govern potential capabilities, which may take the form of communicating with others through ICTs, working from home when needed, and supporting children with schoolwork. Their ability to convert these capabilities into achievable functionings is bounded by their ICTs skills, time constraints, and financial situations, among other things. In any case, the various valuable functionings for single parents would include overcoming their loneliness and sense of isolation, because, as findings illustrate, these individuals would be in a position to communicate more with their social networks and experience a sense of belonging with their closed ones. In addition, access to ICTs would equip them to conduct a more flexible and secure, financially wise life, since they can telework and receive information regarding possible opportunities.

4.3 Ethnic Minorities

The social inclusion of immigrants, refugees, and individuals of immigrant descent or of ethnic minorities is of particular interest because their participation in society is important for both them and the hosting country (Diaz Andrade and Doolin 2016). It has been found that ethnic minorities in general actively seek ways to maintain their social ties with the origin country (Khorshed and Imran 2015), but that they also try to integrate themselves in the hosting country (Lupiañez, Codagnone, and Dalet 2015). Further, they experience difficulties in accessing the labor market due to a lack of basic literacy and numeracy skills, ICTs skills and various obstacles in accessing relevant information (Broadbent and Papadopoulos 2013).

Along these lines, ICTs and digital media can be valuable instruments. Through the Internet, online social networking, and instant messaging applications, ethnic minorities can maintain connections with loved ones (Lupiañez, Codagnone, and Dalet 2015), and in doing so they can identify the positive aspects of technology, and further take advantage of its merits (Broadbent and Papadopoulos 2013; Ling et al. 2012). In addition, ICTs help them in developing links with the host society, because they can access information regarding opportunities for participation and social activities, as well as local information about their community and the available local services (e.g., health services) (Ling et al. 2012).

Furthermore, ICTs support the social inclusion of these groups by facilitating their access to the labor market and strengthening their language skills. While ICTs are less used for learning and educational purposes, it has been posited that this is a case of weaker confidence at the initial stages of appropriation (Lupiañez, Codagnone, and Dalet 2015), whereas when such resources are indeed available and successfully used, these groups are in a position to take full advantage of them, develop their language skills, seek employment opportunities (Broadbent and Papadopoulos 2013), and even support the children of the family with homework (Codagnone and Kluzer 2011).

While the social exclusion of ethnic minorities may have nothing to do with the deprivation of ICTs (Webb 2006), often expensive computing devices and technologies that are difficult to access do function as an impediment (Khorshed and Imran 2015). However, if such resources are available and accessible in the form of multilingual digital resources (Zinnbauer 2007) and digital literary courses, focused on basic language skills and basic education (literacy and numeracy) (Codagnone and Kluzer 2011), ICT-based learning can encourage ethnic minority groups to speak more within the hosting country (Webb 2006). In other words, resources such as an Internet connection and multilingual resources, controlling for personal, and social issues, such as literacy and the minorities' status, can help these individuals in obtaining capabilities, such as the development and improvement of their language and digital skills, continuous access to local information, and the use of ICT-based communication tools. In turn, these capabilities can help them in achieving valuable functionings, such as access to the labor market, communication channels with their country of origin and their integration and participation in the local community.

4.4 Offenders

A common problem that policy makers are faced with is the reduction of recidivism and reintegration of released offenders within the society (Fox 2014). Within this context, it is posited that acquiring relevant skill sets while incarcerated can prepare them for their future lives (Bedford, Dearden, and Dorman 2005). However, the provision of educational opportunities within prison is quite complex because these need to abide to the regulations of the penal system, while satisfying the needs of those inmates who wish to improve their prospects (Hurkmans et al. 2013). A way to reconcile these two may be the delivery of ICT-enabled programs (Eikeland, Manger, and Asbjornsen 2009). Studies show (e.g., Bedford, Dearden, and Dorman 2005) that ICTs can help the rehabilitation process, because indirectly help offenders find ways to improve themselves, envisage a life outside their cells, and ultimately have a productive role in their community following incarceration (Pike and Adams 2012).

Some necessary resources need to be at the disposal of offenders. Offenders require hardware and software, such as computers, word processors, spreadsheet programs, audio-visual recordings, interactive media and offline lectures (Pike and Adams 2012; Barkan et al. 2011). Tutoring sessions (distant or on site) and secure online messaging with tutors can be of particular value (Pike and Adams 2012). However for educational programs to be successful, offenders need to be allowed and encouraged to use these resources (Pike and Adams 2012). Therefore, there needs to be administrative support to secure the approval and participation of staff (Monteiro, Barros, and Leite 2015; Barkan et al. 2011). This way, these resources can become actual capabilities, because offenders can then take advantage of ICTs and address possible educational and technological-related deficits and develop their skill set for future employment (Hurkmans et al. 2013). Further, the self-guided and personalized process of e-learning helps in building individuals' confidence in a secure environment (Bird and Akerman 2005; Verpoorten et al. 2010; Barkan et al. 2011). Such provisions support them in acquiring or improving extant IT skills, overcoming basic literacy and numeracy deficits, acquiring employability skills, and receiving subject-specific knowledge, all of which are capabilities offenders tend to appreciate (Barkan et al. 2011). In turn, these can support those offenders who wish their successful resettlement upon their release in reaching the desired functionings—i.e., to avoid future exclusion and recidivism (Barkan et al. 2011).

4.5 Underprivileged Women

For years there has been an interest on technology's role in empowering women, helping them improve their life conditions and participating equally in their communities (Masika and Bailur 2015). Women in developing countries, and especially those residing in rural areas, are at greater risk of being socially excluded due to their gender role within the society, as evidenced through education, employment, health and poverty rates (Gonzalez Ramos and Arroyo Prieto 2014). However, even in urban contexts, gender-based inequalities still exists; recent findings show that poor women, controlling for education, and income, have 50% less chances than men to have an Internet connection (A4AI 2016).

Along these lines, ICTs can help financially disadvantaged women toward improving their life conditions in the long term; through the Internet, they can access information about employment opportunities, personally interact with local authorities and acquire IT and employability skills. For example, the mobile phone can help women in gaining their livelihood, as it can act as a communication channel with suppliers (in the case of peddlers), and a payment channel, by fulfilling micropayments and money transfers to other dependents (e.g., parents in remote areas). The mobile phone can also empower them, because through it women may organize gatherings and support groups among them, communicate with relatives and remain updated (e.g., health issues, lifestyle) (Masika and Bailur 2015).

The Internet can significantly support underprivileged women. Within strong patriarchal societies, where there are normative, family, and governmental constraints, women may prefer to build or maintain their friendships through the Internet because it provides them with a sense of freedom. In such contexts, they may prefer to use Web 2.0 tools (e.g., forums), where they can form and join groups which help them feel that they actively participate in social activities and can express themselves as citizens (Shahzeidi et al. 2013). Further, within this type of societies, women may use such Web 2.0 tools in order to access information, gain education, and attend online courses, socialize with others, and even run their business while remaining anonymous and refraining from meeting men, when their culture dictates so (Alhareth, Alhareth, and Dighrir 2015). A study on Singapore's Migrant Brides[2] has found that mobile phones can act as instruments for obtaining a sense of control over their lives and for interacting more effectively with the authorities in order to reduce their political exclusion from the Singaporean society (Ling et al. 2012).

Further, women may be marginalized due to their geographical location—i.e., when residing in remote and/or rural areas. The evaluation phase of an ICT for development (ICT4D) project in a remote rural area of Cambodia showed that ICTs assist women empower themselves. The particular project allowed women to attend typing and digital literacy courses, which are skills particularly valued; to the extent that participants mentioned that they made their lives easier and expressed an interest in teaching others (Grunfeld, Hak, and Pin 2011). Another study, examining ICTs in the rural areas of Galicia, Spain, showed that women were getting excluded both socially and professionally due to lack of public transportation and difficulties in accessing and using the Internet (Novo-Corti, Varela-Candamio, and García-Álvarez 2013). Following training courses on health issues, online banking, information seeking, e-learning, and online social networking, Galician women felt they could overcome these barriers, because the Internet in particular can help indirectly to raise their self-esteem and improve their social relationships, their self-esteem, and so forth.

In all the aforementioned examples, women are constrained by their society's hierarchy and have fewer opportunities due to poverty, geographical location, or any combination of these. As a result, they may lack the resources, such as a computer, an Internet connection, a mobile phone, or associated skills and competences, and are then prohibited from meeting their capabilities. However, shall these resources be available, underprivileged women can take advantage of them; based on their personal, social, and environmental conversion factors, they can build upon them in order to enrich their capability set and lead a life they value more. Namely, these

ICTs can help them conduct transactions, access various messaging applications, seek valuable information, interact with local authorities, receive education, and support their dependents. As a result, they can achieve their preferred functionings, such as an empowered life, where they can support each other, achieve gender balance in their communities, access the labor market, and participate in their communities.

4.6 The Homeless

Across cities and countries, homelessness is a major problem that is bound to get exacerbated as the financial crisis deepens. Homeless people are excluded from society and social activities due to their appearance, the lack of permanent residence and social interactions (Buré 2005). As a result, this group of individuals may be deprived of the necessary resources that would allow them to build upon their capabilities and achieve their social inclusion through ICT-enabled products and services.

From the few studies that exist today, the homeless are not necessarily deprived of basic ICTs. A U.S.-based study found that 44% to 62% of homeless people own a mobile phone, and that 47% to 55% of them have access to a computer, while 93% of young adult homeless use e-mail and online social networks on a weekly basis (McInnes, Li, and Hogan 2013). Similarly, the mobile phone is the most frequently used ICTs among the homeless in Madrid, Spain; namely, findings from a recent study show that 59% of the homeless use a mobile phone, 40.4% use a computer, 37.8% have Internet access in some way, 33.5% have an e-mail address and 17% are online social networks users (Vázquez et al. 2015).

In the past, public networks have worked toward helping the homeless with their job seeking activities, their access to public services and exercising their rights through the provision of e-mail addresses and electronic contact points (Graham 2002). These tools are generally envisaged to help the homeless in participating in a more integral fashion in the society and in improving their life conditions (Le Dantec 2010). McInnes et al. (2013), for example, developed a service based on text messaging for helping homeless veterans keep up with their outpatient appointments. Another project, focused on housing issues, aimed at offering homeless people ICTs tools toward accessing health care, among other things, and entailed setting up Internet cafes in hostels, where the homeless could set up e-mail accounts and register with local services (Gaved and Anderson 2006).

It thus seems that the homeless are to a certain extent digitally included; however, this doesn't necessarily mean that they can benefit of ICTs (Buré 2005). In contrast, previous findings show that the homeless, while not necessarily deprived from accessing ICT-enabled products and services, continue being socially excluded because they cannot consistently rely on ICTs for maintaining a form of permanent address in the digital world; while they own and use mobile phones, it is difficult to keep them from getting stolen, lost, or discharged (from battery or credit), and they frequently decide to trade or sell them. Similarly, because they have fewer opportunities to log into their online accounts, they often forget their usernames and passwords. They thus continue being excluded from institutions and deprived from the life skills that can help them be self-sufficient (Buré 2005).

Having access to ICTs and being able to use them would mean that homeless individuals are in a position to build upon the available resources and secure their capabilities. Namely, this would mean that they have the opportunity to access information through their local services, to communicate with those who matter through ICT-enabled communication and that they can maintain their digital presence. In turn, based on personal factors and choices, they can take advantage of these capabilities in order to achieve several important functionings, such as the development of a safety net, improvement of their well-being, overall empowerment, and control over their lives, all of which lead into a life they value.

In closing, it should be noted that there are great differences within this particular group. There are those cases of individuals who genuinely wish to change their life for the better, and for which cases ICTs can indeed become a lifeline. At the same time, there are those who appropriate ICTs in their own ways, and rather than profiting from them, they merely domesticate them, further reinforcing their everyday practices (Buré 2005).

5. Discussion

This chapter has highlighted how and why ICTs can help different groups of people in mobilizing their lives and becoming or remaining active within their communities. Namely, ICTs can enable communication with the "outside world," significantly support participation in everyday activities and help people conduct a life they value by attaining the necessary functionings. Even in cases of extreme social isolation, ICTs afford the aforementioned advantages because they are flexible, adaptable, and resilient, while offering anonymity and speed if necessary (Shahzeidi et al. 2013).

Emphasis should be placed on whose needs these ICTs are aimed to serve (e.g., Zheng 2009). At the level of resources, the findings exhibit that, across all groups, the most valued are the mobile phone, personal computer, and Internet access. It is clear that people do not necessarily require advanced ICTs, but simply those that help them meet their needs in an efficient and easy-to-use manner. This is best illustrated through the case of older adults. Similarly to the group of ethnic minorities, they, too, wish to use ICTs in order to participate in social activities, to maintain communication channels with loved ones and their support network, and to have access to valuable information. To do that, they need to have Internet access and mobile phones that serve their particular needs (e.g., affected by aging motor skills, interaction modalities that feel more natural or less awkward (Teixeira et al. 2012)). For ethnic minorities, similar capabilities are achieved through resources that are directed to their own needs, such as multilingual digital resources (Zinnbauer 2007). Depending on each group's particular conversion factors, these resources can be then used in order to meet their own set of capabilities, which leads to the desired functionings. Table 1.1 illustrates that these conversion factors comprise of a mixture of personal, social, and environmental issues, with the most pronounced being the personal ones, such as health and financial issues, lack of basic education, and employability skills.

With regard to the identified capabilities, in most cases these have to do with the means to interact with the outside world (communication or support) and to sustain themselves financially or otherwise, by maintaining a business or developing employability skills. The most prominent was the use of simple communication technologies such as SMS, instant messaging services, and the like, while access to information is also appreciated. These capabilities, based on particular conversion factors, are interpreted into relevant functionings, that is an overall improvement of life conditions, ranging from independent living (for older adults), to a more stable life in general (single parents) and legitimate career prospects (for offenders). In other words, people desire to use ICTs to feel and be socially included and to better their circumstances. There is however an exception to this; offenders appear to express no interest regarding the "C" in ICTs, possibly because following their incarceration, either their social ties were severed or other things became more important for them, as for example the decreasing of chances for recidivating.

Further, the findings showcase that experiencing deprivation of ICT-related capabilities entails other social deprivations as well (e.g., unemployment). This was not unexpected because ICTs are "so tightly woven into the fabric of society today" (Helsper 2008: 8), which makes ICT skills and opportunities necessary in our lives (Broadbent and Papadopoulos 2013). ICTs deprivation, in the form of lack of ICTs skills and the opportunities to attain them, unsuitable interfaces and modalities, lack of Internet connection and necessary accessories (e.g., device chargers), suggest that people are not able to satisfy their communication needs, access the labor market, or participate equally in their communities.

Also, the findings reveal that not all groups have received the same attention from researchers and practitioners. Older adults, for example, are one of the most investigated groups of individuals as far as ICT access is concerned. However, extant studies typically focus on more practical issues, such as usability dimensions, rather than issues of empowerment, possibly because older adults are considered marginalized due to access and use aspects (Hernández-Encuentra, Pousada, and Gómez-Zúñiga 2009). Yet, recent research suggests that older adults find more important to have access to interesting, rather than just easy-to-use activities, because they encourage them to use them more, which makes them feel empowered (Hur 2016).

Table 1.1 Conversion Factors, Capabilities, and Functionings per Group

Group	*Conversion Factors*	*Capabilities*	*Functionings*
Older adults			
Jaeger (2006), Righi et al. (2011), Teixeira et al. (2012), Roberts (2010), Waterworth et al. (2009), Choudrie, Grey, and Tsitsianis (2010), Pirkl (2009)	• Personal health issues, such as mental impairment, mobility, motor skills, etc. • Environmental issues, such as support structures	• Read the news online • Devices and applications they can use (for impaired vision, motor skills, etc.) • Make video calls with the family • Access to support network • ICT-enabled communication (family, friends, caregivers) e.g., send/receive text messages, instant messages	• Independent living and successful aging • Remain active and mentally stimulated • Continued participation in society
Single parents			
Doczi (2000), De Graaf-Zijl and Nolan (2011), Haddon (2000), Hardey (1989), Tripp (2011), Russo Lemor (2005), NTIA (2011), Haddon and Silverstone (2000), Schofield Clark (2009)	• Personal issues, such as financial problems • Social issues, such as single parent status	• Support dependents (online skills and self-efficacy) • ICT-enabled communication (family, friends) • Teleworking	• Stable life • Combat loneliness • Improvement of financial conditions
Ethnic minorities			
Lupianez, Codagnone, and Dalet (2015), Broadbent and Papadopoulos (2013), Ling et al. (2012), Codagnone and Kluzer (2011), Webb (2006), Zinnbauer (2007), Diaz Andrade and Doolin (2016), Khorshed and Imran (2015)	• Personal issues, such as basic literacy, numeracy, employability • Social issues, such as minority status	• Local information and services • Language skills development • Support dependents (online skills, language skills, self-efficacy) • Basic language skills • ICT-enabled communication (family, friends), e.g., online social networks	• Maintain connection with the homeland • Integration in the hosting country • Access to the labor market

(*Continued*)

Table 1.1 (Continued)

Group	*Conversion Factors*	*Capabilities*	*Functionings*
Offenders Fox (2014), Hurkmans et al. (2013), Bedford, Dearden, and Dorman (2005), Pike and Adams (2012), Eikeland, Manger, and Asbjornsen (2009), Barkan et al. (2011), Verpoorten et al. (2010), Bird and Akerman (2005), Monteiro, Barros, and Leite (2015)	• Personal issues, such as severe literacy and numeracy deficits • Social issues, such as offender status	• Basic employability training • Access to educational material (recorded material such as offline lectures and office-related applications) • Personalized process for e-learning • Administrative support	• A legitimate career and life after incarceration
Underprivileged women Masika and Bailur (2015), Gonzalez Ramos and Arroyo Prieto (2014), Shahzeidi et al. (2013), Alhareth et al. (2015), Ling et al. (2012), Grunfeld, Hak, and Pin (2011), Novo-Corti et al. (2013), A4AI (2016)	• Personal issues, such as employability skills, loneliness, financial problems • Social issues, such as the role of women in society • Environmental issues, such as geographical location and lack of relevant infrastructures	• Micropayments, mobile money • Messaging services (e.g., SMS, online forums, chat rooms) • Access to health care and lifestyle information • Communication with local authorities • ICT-focused education • Support dependents	• Do business and access to labor market • Gender balance • Empowerment and autonomy • Participation in society
Homeless Buré (2005), McInnes, Li, and Hogan (2013), Vazquez et al. (2015), Graham (2002), Le Dantec (2010), Gaved and Anderson (2006)	• Personal issues, such as employment, homeless, and financial problems • Environmental issues, such as lack of a physical, permanent address • Social issues, such as homelessness	• Access to information through local services • ICT-enabled communication (family, friends) e.g., e-mail accounts • Device chargers	• Access to labor market • Improve well-being • Empowerment, control over one's life

Further, as far as the other identified groups are concerned, extant studies tend to focus on the individuals' ICTs skills and competences, the delivery of online courses and so forth. Yet, quite often, it appears that individuals may choose to appropriate ICTs in ways that merely replicate previous practices (e.g., Buré 2005) or to refrain altogether from using ICTs because they lack interest. In such occasions, individuals may have available a series of capabilities and yet they purposefully decide not to capitalize on them and reject their conversion into functionings. While this population diminishes each year, this type of individuals will experience severe exclusion from the e-society in the future (Helsper and Reisdorf 2017). It follows that the importance of social desirability and individual choice in the process of converting capabilities to functionings is further underlined.

6. Conclusions

ICTs alone are not enough to combat social exclusion; however, being deprived of them in today's information society is among the reasons for being socially excluded (Broadbent and Papadopoulos 2013). Taking part in continuous education, seeking a job, participating in community activities and so forth are activities that take place largely online (Vázquez et al. 2015). As a result, being deprived of ICTs could suggest that one will be socially excluded as well. It is not enough to have access to ICT-related resources; people need to be able and allowed to use them. Specifically, ICT infrastructure in itself is simply a commodity (Hatakka and De' 2011), but being able and knowing how to use ICTs suggests that individuals can exercise their agency—i.e., capabilities (Zheng and Walsham 2008).

This chapter contributes to the literature of social inclusion through ICTs by identifying groups at the risk of social exclusion, namely older adults, single parents, ethnic minorities, offenders, underprivileged women, and the homeless. The findings showcase that the most cherished capability is that of maintaining communication with others, be it friends, family, or a support network. This is not unexpected, since the most frequently used ICT services include messaging applications, such as text messaging, instant messaging, and online social networks. In all occasions, people seek to use ICTs and enrich their capability sets in a way that will empower them and help them in securing a more stable life.

Deprived individuals are characterized by quite different needs and expectations. The various ICTs most likely will not fit all users. As a result, what may seem to help a group to overcome some impediments, it may significantly inhibit another group's access. Taking these two together in the future, and building upon this study's classification, can help designers and researchers toward theorizing and delving into the technological features of ICTs for social inclusion. The consolidation of existing findings and their interpretation within a single study can increase their pragmatic value for policy makers to design policy actions toward strengthening social inclusion.

6.1. Limitations

Some groups that are typically deprived of their capabilities are missing from this analysis, such as the disabled. Examining disabilities is a challenging task, as the term itself is an umbrella concept that refers to cognitive, sensory, physical restrictions, or any combination of these. Therefore, tracing the possible deprivations of a disabled person would be a non-realistic task within the context of a single study.

Notes

1 See at www.who.int/topics/disabilities/en/,
2 "Migrant Brides" are women from developing Asian countries, who migrate to Singapore to get marry. They typically lack education and financial means and due to Singaporean regulation, the Brides can essentially be "perpetually transient outsiders" (Chong 2014: 333).

References

A4AI. "Affordability Report 2015/16." Alliance for Affordable Internet, 2016. http://a4ai.org/wp-content/uploads/2016/04/A4AI-2015-16-Affordability-Report.pdf

Alhareth, Yahya Al, Yasra Al Alhareth, and Ibtisam Al Dighrir. "Review of Women and Society in Saudi Arabia." *American Journal of Educational Research* 3, no. 2 (2015): 121–5.

Barkan, M., E. Toprak, A. T. Kumtepe, E. Genc Kumtepe, M. Ataizi, H. Pilanci, M. E. Mutlu, I. Kayabas, and B. Kip Kayabas. "Eliminating Language Barriers Online at European Prisons (ELBEP): A Case-Study." *Educational Media International* 48, no. 3 (2011): 235–48.

Bedford, T., R. Dearden, and M. Dorman. "Offender Rehabilitation and Information Literacy: A Case for Providing Appropriate Prisoner Access to Contemporary ICT." In *Australasian Corrections Education Association (ACEA) Conference*, Darwin, NT, Australia, 2005. www.researchgate.net/publication/228662181_Offender_rehabilitation_and_information_literacy_A_case_for_providing_appropriate_prisoner_access_to_contemporary_ICT

Bird, Viv, and Rodie Akerman. "Every Which Way We Can: A Literacy and Social Inclusion Position Paper." National Literacy Trust, 2005.

Broadbent, Robyn, and Theo Papadopoulos. "Bridging the Digital Divide—an Australian Story." *Behaviour & Information Technology* 32, no. 1 (2013): 4–13.

Buré, C. "Digital Inclusion Without Social Inclusion: The Consumption of Information and Communication Technologies (ICTs) in Homeless Subculture in Central Scotland." *The Journal of Community Informatics* 1, no. 2 (2005): 116–33.

Chong, Amanda Wei-Zhen. "Migrant Brides in Singapore: Women Strategizing Within the Family, Market, and State." *Harvard Journal of Law & Gender* 37 (2014): 331–405.

Choudrie, J., S. Grey, and N. Tsitsianis. "Evaluating the Digital Divide: The Silver Surfer's Perspective." *Electronic Government, an International Journal* 7, no. 2 (2010): 148–67.

Codagnone, Cristiano, and Stefano Kluzer. "ICT for the Social and Economic Integration of Migrants into Europe." *JRC Technical Reports*, European Commission,

Joint Research Centre, Institute for Prospective Technological Studies, Luxembourg, 2011.

Diaz Andrade, Antonio, and Bill Doolin. "Information and Communication Technology and the Social Inclusion of Refugees." *MIS Quarterly* 40, no. 2 (2016): 405–16.

Eikeland, O-J., T. Manger, and A. Asbjornsen, eds. *Education in Nordic Prisons: Prisoners' Educational Background, Preferences and Motivation*. Copenhagen: Nordic Council of Ministers, 2009.

Eisma, R., A. Dickinson, J. Goodman, A. Syme, L. Tiwari, and A. F. Newell. "Early User Involvement in the Development of Information Technology-Related Products for Older People." *Universal Access in the Information Society* 3, no. 2 (2004): 131–40.

Fernback, Jan. "Information Technology, Networks and Community Voices." *Information, Communication & Society* 8, no. 4 (2005): 482–502.

Fox, Kathryn J. "Restoring the Social: Offender Reintegration in a Risky World." *International Journal of Comparative and Applied Criminal Justice* 38, no. 3 (2014): 235–56.

Gaved, M. B., and B. Anderson. "The Impact of Local ICT Initiatives on Social Capital and Quality of Life." *Chimera Working Paper 2006–6*, University of Essex, Colchester, 2006.

Gingrich, Luann Good, and Naomi Lightman. "The Empirical Measurement of a Theoretical Concept: Tracing Social Exclusion Among Racial Minority and Migrant Groups in Canada." *Social Inclusion* 3, no. 4 (2015): 98.

Gonzalez Ramos, Ana M., and Lidia Arroyo Prieto. "Digital Inclusion of Low-Income Women: Are Users of Internet Able to Improve Their Life Conditions?" ACM, 2662331, 2014, 1–5.

Graham, S. "Bridging Urban Digital Divides? Urban Polarisation and Information and Communications Technologies (ICTs)." *Urban Studies* 39, no. 1 (2002): 33–56.

Grunfeld, Helena, Sokleap Hak, and Tara Pin. "Understanding Benefits Realisation of iREACH from a Capability Approach Perspective." *Ethics and Information Technology* 13, no. 2 (2011): 151–72.

Guildford, Janet. "Making the Case for Social and Economic Inclusion." Health Canada, Population & Public Health Branch, Atlantic Regional Office, 2000.

Haddon, Leslie. "Social Exclusion and Information and Communication Technologies: Lessons from Studies of Single Parents and the Young Elderly." *New Media & Society* 2, no. 4 (2000): 387–406.

Haddon, Leslie, and R. Silverstone. "Information and Communication Technologies and Everyday Life: Individual and Social Dimensions." In *The Information Society in Europe: Work and Life in an Age of Globalization*, edited by Ken Ducatel, Juliet Webster, and Werner Herrmann. Lanham, MD: Rowman and Littlefield Inc., 2000.

Hardey, M. "Lone Parents and the Home." In *Home and Family: Creating the Domestic Sphere*, edited by G. Allen Graham and G. Crow, 122–40. Basingstoke: Macmillan, 1989.

Harris, Jennifer, and Claire Bamford. "The Uphill Struggle: Services for Deaf and Hard of Hearing People-Issues of Equality, Participation and Access." *Disability & Society* 16 no. 7 (2001): 969–79.

Hatakka, M., and R. De.' "Development, Capabilities and Technology: An Evaluative Framework." In *IFIP WG9. 4: 11th International Conference on Social Implications of Computers in Developing Countries*, 2011.

Helsper, Ellen. *Digital Inclusion: An Analysis of Social Disadvantage and the Information Society*. London: Department for Communities and Local Government, 2008.

Helsper, Ellen, and Bianca C. Reisdorf. "The Emergence of a "Digital Underclass" in Great Britain and Sweden: Changing Reasons for Digital Exclusion." *New Media & Society* 19, no. 8 (2017): 1253–70.

Hernández-Encuentra, Eulàlia, Modesta Pousada, and Beni Gómez-Zúñiga. "ICT and Older People: Beyond Usability." *Educational Gerontology* 35, no. 3 (2009): 226–45.

Hur, Mann Hyung. "Empowering the Elderly Population Through ICT-Based Activities: An Empirical Study of Older Adults in Korea." *Information Technology & People* 29, no. 2 (2016): 318–33.

Hurkmans, G., M. M. Bernal, A. C. Hoces, and M. Kert. "In & Out—Comparing Strategies of Prison Education in Belgium, Spain and Turkey." In *5th International Conference on Education and New Learning Technologies (EDULEARN13)*, 2013, 6259–66.

Jaeger, Paul T. "Telecommunications Policy and Individuals with Disabilities: Issues of Accessibility and Social Inclusion in the Policy and Research Agenda." *Telecommunications Policy* 30, no. 2 (2006): 112–24.

Jaeger, Paul T., and Bo Xie. "Developing Online Community Accessibility Guidelines for Persons with Disabilities and Older Adults." *Journal of Disability Policy Studies* 20 no. 1 (2009): 55–63.

Khorshed, Alam, and Sophia Imran. 2015. "The Digital Divide and Social Inclusion Among Refugee Migrants: A Case in Regional Australia." *Information Technology & People* 28, no. 2: 344–65.

Le Dantec, Christopher A. "Exploring Mobile Technologies for the Urban Homeless." In *CHI '10 Extended Abstracts on Human Factors in Computing Systems*, ACM, 1753876, 2010, 2883–6.

Ling, Abigail See Shyang, Rebecca Jia-Hui Cheang, Goh Siew Luan, Chloris Qiaolei Jiang, Ng Xiao Xuan, and Arul Indrasen Chib. "ICT Influence on Foreign Wives' Social Integration into Singaporean Society." In *14th Annual International Conference on Electronic Commerce,* ACM, 2346576, 2012, 209–10.

Lupiañez, Francisco, Cristiano Codagnone, and Rosa Dalet. "ICT for the Employability and Integration of Immigrants in the European Union. Results from a Survey in Three Member States." *JRC Technical Reports*, European Commission, Joint Research Centre, Institute for Prospective Technological Studies, Luxembourg, 2015.

Masika, Rachel, and Savita Bailur. "Negotiating Women's Agency Through ICTs: A Comparative Study of Uganda and India." *Gender, Technology and Development* 19, no. 1 (2015): 43–69.

McInnes, D. Keith, Alice E. Li, and Timothy P. Hogan. "Opportunities for Engaging Low-Income, Vulnerable Populations in Health Care: A Systematic Review of Homeless Persons' Access to and Use of Information Technologies." *American Journal of Public Health* 103, no. S2 (2013): e11–24.

Monteiro, Angélica, Rita Barros, and A. Leite. "Lifelong Learning Through E-Learning in European Prisons: Rethinking Digital and Social Inclusion." In *Proceedings of INTED2015 Conference*, 2015.

Notley, Tanya. "Young People, Online Networks, and Social Inclusion." *Journal of Computer-Mediated Communication* 14, no. 4 (2009): 1208–27.

Novo-Corti, Isabel, Laura Varela-Candamio, and María Teresa García-Álvarez. "Breaking the Walls of Social Exclusion of Women Rural by Means of ICTs: The Case of "Digital Divides" in Galician." *Computers in Human Behavior* 30 (2013): 497–507.

Pike, Anne, and Anne Adams. "Digital Exclusion or Learning Exclusion? An Ethnographic Study of Adult Male Distance Learners in English Prisons." *Research in Learning Technology* 20 (December), 2012. www.researchinlearningtechnology.net/index.php/rlt/article/view/18620

Pirkl, James Joseph. "The Demongraphics of Ageing." Transgenerational Design Matters, 2009. http://transgenerational.org/aging/demographics.htm.

Roberts, Simon. *The Fictions, Facts and Future of Older People and Technology*. London: The International Longevity Centre (ILC), 2010.

Robeyns, I. "The Capability Approach: A Theoretical Survey." *Journal of Human Development* 61, no. 3 (2005): 93–114.

Russo Lemor, Anna Maria. "Making a 'Home.' The Domestication of Information and Communication Technologies in Single Parents' Households." In *Domestication of Media and Technology*, edited by Thomas Berker, Maren Hartmann, and Yves Punie, 165–84. Berkshire: McGraw-Hill Education, 2005.

Schofield Clark, Lynn. "Digital Media and the Generation Gap." *Nformation, Communication & Society* 12, no. 3 (2009): 388–407.

Sen, A. *The Standard of Living*. Cambridge: Cambridge University Press, 1987.

Sen, A. *Social Exclusion: Concept, Application and Scrutiny*. Manila: Asian Development Bank, 2000.

Shahzeidi, Mehri, Elahe Shabani, Younghoon Chang, and Myeong-Cheol Park. "Can Women Benefit from Cyberspace as a Public Space?" In *CPRsouth8/ CPRafrica 2013: Innovation & Entrepreneurship in ICT: Changing Asia/Africa*, 2013.

Sourbati, M. "Disabling Communications? A Capabilities Perspective on Media Access, Social Inclusion and Communication Policy." *Media, Culture and Society* 34, no. 5 (2012): 571–87.

Stephens-Reicher, Justine, Atari Metcalf, Michelle Blanchard, Cheryl Mangan, and Jane Burns. "Reaching the Hard-to-Reach: How Information Communication Technologies Can Reach Young People at Greater Risk of Mental Health Difficulties." *Australasian Psychiatry* 19, sup1 (2011): S58–61.

Teixeira, V., C. Pires, F. Pinto, J. Freitas, M. Sales Dias, and E. Mendes Rodrigues. 'Towards Elderly Social Integration Using a Multimodal Human-Computer Interface.' In *2nd International Living Usability Lab Workshop on AAL Latest Solutions, Trends and Applications*, 2012.

Tripp, Lisa M. " 'The Computer Is Not for You to Be Looking Around, It Is for Schoolwork': Challenges for Digital Inclusion as Latino Immigrant Families Negotiate Children's Access to the Internet." *New Media & Society*, 13, no. 4 (2011): 552–67.

Vázquez, José Juan, Sonia Panadero, Rosa Martín, and María del Val Diaz-Pescador. "Access to New Information and Communication Technologies Among Homeless People in Madrid (Spain)." *Journal of Community Psychology* 43, no. 3 (2015): 338–47.

Verd, Joan Miquel, and Martí Lopez Andreu. "The Rewards of a Qualitative Approach to Life-Course Research: The Example of the Effects of Social Protection Policies on Career Paths." *Forum Qualitative Sozialforschung/Forum: Qualitative Social Research* 12, no. 3 (2011).

Verpoorten, Dominique, Jean-Marie Renson, Wim Westera, and Marcus Specht. "Personalised Learning: A Familiar Concept to Secondary Teachers? And Which Teachers?" In *Proceedings of the E-Learning@ Greenwich Conference (Making It Personal!)*, University of Greenwich, 2010, 61–7.

Waterworth, John A., Soledad Ballesteros, Christian Peter, Gerald Bieber, Andreas Kreiner, Andreas Wiratanaya, Lazaros Polymenakos, et al. "Ageing in a Networked Society: Social Inclusion and Mental Stimulation." ACM, 1579160, 2009, 1–5.

Webb, Sue. "Can ICT Reduce Social Exclusion? The Case of an Adults' English Language Learning Programme." *British Educational Research Journal* 32, no. 3 (2006): 481–507.

Zheng, Yingqin. "Different Spaces for e-Development: What Can We Learn from the Capability Approach?" *Information Technology for Development* 15, no. 2 (2009): 66–82.

Zheng, Yingqin, and G. Walsham. "Inequality of What? Social Exclusion in the e-Society as Capability Deprivation." *Information Technology & People* 21, no. 3 (2008): 222–43.

Zinnbauer, Dieter. "What Can Social Capital and ICT Do for Inclusion?" *JRC Technical Reports*, European Commission, Joint Research Centre, Institute for Prospective Technological Studies, Luxembourg, 2007.

2 Connectivity

A Socio-technical Construct to Examine ICT-Enabled Service

Christoph F. Breidbach

1. Introduction

Advances in ICT have changed, and continue to change, interactions between service providers and customers. Service industries like health care or consulting traditionally relied on interpersonal "high touch, low tech" (Bitner, Brown, and Meuter 2000: 138) exchanges. Today, however, service providers and customers increasingly interact through virtual, rather than physical interfaces (Breidbach, Kolb, and Srinivasan 2013a). But, service research to date has focused predominantly on face-to-face settings (e.g., Froehle and Roth 2004), while technology-enabled value co-creation processes remain largely unexplored and misunderstood (Breidbach and Maglio 2015). Consequently, the understanding of ICT-enabled service is incomplete, and exploring the broader role and implications of ICT in service represents a key research priority for service science (e.g., Srinivasan, Breidbach, and Kolb 2015) and IS scholars alike (Maglio and Breidbach 2014).

While focusing on the role of ICT in service is important, exploring the human side of ICT-enabled service is equally relevant (Edvardsson, Tronvoll, and Gruber 2011). Social and interpersonal relationships may even be more significant for successful value co-creation, than ICTs enabling these interactions (Ostrom et al. 2010). However, a significant empirical gap in knowledge exists, since existing research focuses predominantly on "face-to-face service encounters [. . .] but not on service encounters involving both technology and the human touches" (Makarem, Mudambi, and Podoshen 2009: 144).

This chapter introduces and explores how the "connectivity" construct can provide the much-needed socio-technical lens, terminology, and understanding of ICT-enabled service (Breidbach, Kolb, and Srinivasan 2013a). Initially introduced in the organizational behavior literature, connectivity aims to understand intra-and inter-organizational interactions, as well as performance (e.g., social inclusion) of a distributed workforce on social and technical levels (Kolb et al. 2008). Connectivity, therefore, helps to address calls by Makarem, Mudambi, and Podoshen (2009) and Edvardsson, Tronvoll, and Gruber (2011), and contributes to this volume on *Innovative*

ICT-Enabled Services and Social Inclusion by providing a novel analytical lens for scholars and practitioners interested in understanding, assessing, and ultimately reinforcing social inclusion of human actors in ICT-enabled service.

This chapter first explores how ICTs alter service interactions, and delineates the resulting challenges. It subsequently outlines how and why established schools of thought like communication and media studies, or research on virtual teams, are inherently limited when attempting to understand and assess the ICT-enabled service landscape of the 21st century. Finally, the chapter introduces the connectivity construct, including its attributes and dimensions, and highlights how various connective states can influence the performance of ICT-enabled service. The chapter closes with a summary and future research opportunity.

2. ICT-Enablement in Service

ICTs are known as having implications for service firms on the strategic, developmental and executional levels (Ostrom et al. 2010), increasing the profitability for service businesses (Rust and Miu 2006), being a source of innovation by itself (Sheehan 2006), or enabling new types of interactions between customers and service providers (Davis, Spohrer, and Maglio 2011). While ICTs traditionally played a minute role in service encounters, the advancement, ubiquity, and sophistication of modern ICTs have changed, and are likely to continue to change how service providers and customers interact. For example, service exchanges in networked societies are characterized by platforms, with ICTs ranging from e-mail to videoconferencing and social networking sites representing the dominant interface for economic exchange between physically dispersed service firms and customers (Breidbach, Brodie, and Hollebeek 2014).

While some firms use ICTs to "replace or substantially diminish personal interaction in service" (Walker and Johnson 2004: 564), resulting in self-service, ICTs also enable interpersonal communication and interaction that closely resembles face-to-face interactions, thus potentially increasing social inclusion (Makarem, Mudambi, and Podoshen 2009). For example, the consulting industry, which, like most knowledge-based services, typically requires close face-to-face interaction, is undergoing a technology-driven transformation. Specifically, the need for physical contact between a consultant and customer—as a mediator for customer input—has become less relevant due to technological advancements (Lee and Park 2009). While this industry is beginning to shift into virtual realms, the social implications for individuals working within this context remain unexplored.

Donofrio, Sanchez, and Spohrer (2010) identify three drivers that explain the shift from physical toward virtual service interfaces: greater opportunities for customer-provider collaboration through commonly available ICTs (network ubiquity), open standards ensuring accessibility and usability of

ICTs, and new business designs (Donofrio, Sanchez, and Spohrer 2010). New ICTs, most notably the Internet and associated communication applications, are ubiquitous and standardized, so their accessibility has become relatively simple. New business designs consequently emerged and influenced the ways organizations co-create value with customers (Blomberg 2010). Therefore, the physical location of human economic actors is irrelevant today because, "as the ability to communicate increases, the need for transport decreases" (Lusch, Vargo, and Tanniru 2010: 23).

While ICT-enabled service is an emerging reality, service scholars have, for the past decade, articulated the critique that academic research has not caught-up with this development (Lovelock and Gummesson 2004). Especially the dominant focus on face-to-face service interactions is now recognized as an issue resulting in a (technology-driven) gap in knowledge. While previous research contributed to our understanding of service interactions in face-to-face contexts, "considerably less work has been done to improve our understanding of [. . .] technology-mediated settings (e.g., via telephone, instant messaging (IM), or e-mail" (Froehle and Roth 2004: 1). This is an important gap in knowledge because the types of ICTs that service providers rely on to interact with their customers are not limited to standardized technologies, such as e-mail. While IM or videoconferencing is utilized as well, the extent to which these technologies impact the performance of ICT-enabled service remains unclear.

Before normative research can provide recommendations on how to manage ICT-enabled service effectively, it is more important to explore how service systems operate and interact by means of ICTs (Vargo, Maglio, and Akaka 2008). Understanding ICT-enabled service requires a new perception of the socio-technical context of value creation (Vargo, Maglio, and Akaka 2008). This is especially relevant when taking the significance of social environments of human communication into consideration, which are, especially in the case of ICT-enabled service, potentially more important for successful value creation than the technology enabling the interaction (Breidbach, Kolb, and Srinivasan 2013a). For example, differences in time-zones or privacy concerns are suspected to inhibit the ability of human actors to interact effectively (Ostrom et al. 2010). Ultimately, our understanding of how service systems can ideally interact via ICTs is still very limited (Ostrom et al. 2010). Edvardsson, Tronvoll, and Gruber (2011) therefore suggest that there should be a focus on the social reality in which a service is co-created, as well as the social implications for individuals involved in ICT-enabled service.

An evolution from interpersonal to virtual interactions has previously been documented in research on virtual teams (e.g., Martins, Gilson, and Maynard 2004). Consequently, the argument has been brought forward for service research to widen its boundaries, and to incorporate insights gained in this field (Froehle 2006). Specifically, ICT-enabled service should be perceived as a "type of virtual team," and hence "opportunities arise for employing some of the findings from the virtual team literature to technology-mediated [service]" (Froehle 2006: 12). However, while the established body of literature associated with virtual team research has some merit in regard to ICT-enabled service, several substantial shortcomings prevail, which are discussed in the subsequent section.

3. Established Perspectives on ICT-Enabled Interactions

3.1 Virtual Team Studies

The virtual team literature explores human interaction in ICT-enabled work environments. The defining features of location, temporal, and relational independence, as well as technology use, distinguish virtual teams from other marginally related organizational forms (Martins, Gilson, and Maynard 2004). Location independence explains that virtual teams are not constrained to one physical location, but can be located throughout the world (i.e., global virtual teams). Temporal independence describes that team members are located in different time-zones (Kayworth and Leidner 2002), while relational independence describes the often different organizational backgrounds and affiliations of members in virtual teams that are brought together in selected projects (Maznevski and Chudoba 2001). All of these characteristics are, in fact, applicable to ICT-enabled service contexts, such as consulting, which implies that interactions between consultants and their customers may be perceived as a type of virtual team, as suggested by Froehle (2006).

Research on virtual teams typically focuses on team design, socio-emotional processes, or the performance of virtual teams (Martins, Gilson, and Maynard 2004). Studies investigating the initial design of a virtual team typically attempt to identify means that can structure the interactions within the team, especially in the earlier phases of the team's formation (e.g., Powell, Piccoli, and Ives 2004). Socio-emotional process research mainly explores relationship-building (e.g., Warketin and Beranek 1999), or trust (e.g., Ariss, Nykodym, and Cole-Laramore 2002). Finally, research on the performance of virtual teams is typically related to the quality of decisions, or the effectiveness of the interaction (e.g., Cappel and Windsor 2000).

Empirical findings in virtual teams research have been extensively criticized for being theoretically limited (Hertel, Geister, and Konradt 2005; Martins, Gilson, and Maynard 2004; Powell, Piccoli, and Ives 2004). First, empirical research "relied on media richness and social presence theories" (Martins, Gilson, and Maynard 2004: 821), whose shortcomings are now evident. Second, several studies compared virtual to face-to-face teams, which limits the generalizability of findings, since teams relying on face-to-face interactions only, are rare (Griffith, Sawyer, and Neale 2003). Third, methodological shortcomings related to laboratory experiments limit the generalizability of most empirical virtual team studies, and challenges the findings of these studies in regard to their applicability in real-world contexts (Powell, Piccoli, and Ives 2004).

3.2 Communication Media and Technology Studies

Communication media and technology studies represent the second body of research that is relevant to distributed human economic exchange. These studies explore the characteristics of communication media and the means by which these are chosen and used. Early work explored communication media through media richness, with media being considered rich if they can "overcome different frames of reference [. . . and] change understanding" (Daft and Lengel 1986: 560). Whenever a medium requires "a long time to enable understanding" (Daft and Lengel 1986: 560), it is considered lower in richness, with face-to-face interactions being considered the richest, followed by telephone and, finally, impersonal documents.

Despite its popularity, media-richness research has not been without its critics. For example, Carlson and Zmud (1999) introduced the channel expansion theory and argued that a seemingly lean media can increase in richness, depending on its use by an actor. Furthermore, new communication media often provide distinct features that make them superior to seemingly richer media. E-mails can, for example, be edited and re-read if necessary, which is impossible for a telephone call. Yet, according to media-richness theory, the telephone would be considered superior (Froehle 2006). The key challenge lies in "comparing two different features or channels while holding all other things constant [. . . which] is nearly impossible to do outside of a laboratory experiment" (Froehle 2006: 12). Consequently, due to the inherent weaknesses of this approach, investigating ICT-enabled service through a media-richness lens is inappropriate, as it is conceptually limited.

Focusing on the means by which actors choose and use a particular communication medium or media choice represents the second school of thought in communication and media studies. Media choice represents "an individual's decision to use a medium" (Trevino, Webster, and Stein 2000: 163), with media use defined as "an individual's general pattern of use over time" (Trevino, Webster, and Stein 2000: 163). Studies following this approach attempt to support and operationalize media choices, typically assuming that actors choose communication media after a coherent evaluation of the medium and task at hand. They are typically related to media richness (Daft and Lengel 1986) and follow a "fit rationale" (Riemer and Filius 2009: 166), in regards to task-media fit.

Just like media-richness studies, media choice, and media use studies have been widely criticized. This body of research attempts to generalize communication behavior and perceives all interactions as analytically dissectible. However, human actors behave differently in their media choice, use communication media not in isolation, but in combination, which implies that investigating the choice and use of a single medium is neither feasible nor realistic in today's technology dominated environment. Ultimately, this stream of research fails to provide a holistic understanding of technology-enabled interactions, and it is considered "incomplete for it fails to examine [. . .] relationships between media and communication in organizations over time" (Yates and Orlikowski 1992: 310). Approaches originating purely out of communication media and technology studies are consequently not suitable when investigating ICT-enabled service.

Ultimately, communication media and technology, as well as virtual team studies, have significant empirical, conceptual, and methodological shortcomings. These research streams focus on the technical *or* human elements *only*, yet without providing a holistic understanding of the interrelationship of these two dimensions, as is deemed necessary by service scholars calling for a multidimensional socio-technical understanding of ICT-enabled service (e.g., Makarem, Mudambi, and Podoshen 2009). Neither communication media and technology studies, nor virtual team research is truly applicable to service research, as initially suggested by Froehle (2006). However, recent research related to the connectivity construct extends the discourse by stating, "We used to ask which media were best for certain tasks [. . .] we must now ask the question: 'how much' connectivity do we need?" (Kolb, Caza, and Collins 2012: 5). The following section discusses connectivity as a construct to investigate ICT-enabled service in general and social inclusion, in this context, in particular.

4. Connectivity as a Novel Construct to Examine ICT-Enabled Service

4.1 Defining Connectivity

Connectivity is a socio-technical lens to advance our "thinking about how we connect and disconnect in an increasingly interconnected world" (Kolb 2008: 141). Since a focus on technical connections only fails to provide sufficient insights into ICT-enabled service and distributed work contexts (e.g., Waverman, Dasgupta, and Brooks 2009), connectivity "can and should be applied to social interactions" (Kolb 2008: 140) as well. The socio-technical multidimensionality of connectivity represents its key advantage when compared to the limited approaches originating from the virtual team literature. Connectivity hence provides a novel construct and analytical lens to explore, assess, and ultimately understand ICT-enabled service interactions and social inclusion (Breidbach, Kolb, and Srinivasan 2013a).

Both academics and practitioners have used the term connectivity in a variety of contexts. For example, it has been used to describe non-technical characteristics of human interaction (Cartwright 2002; Kanter 1999), as well as in research about mergers and acquisitions (Schweiger and Goulet 2005), systems thinking and development (Mason 2005), or socio-ecological systems (Janssen et al. 2006). Most importantly, the connectivity construct is recognized as a suitable socio-technical lens when investigating intra- and inter-organizational interactions (Quan-Haase and Wellman 2005; Wajcman and Rose 2011). Connectivity, however, is not unanimously defined and is often used without further clarification as to what it entails. Table 2.1 provides a list of definitions.

Angwin and Vaara (2005) and Kolb (2008) define connectivity within an organizational context, while Janssen et al. (2006) perceive connectivity more abstract and in relation to systems thinking and network theory. Waverman, Dasgupta, and Brooks (2009), however, divert from the understanding of connectivity as a socio-technical construct and follow a third approach, which focuses on technical interactions on a nation-state level only, while delineating connectivity therein. Despite these seemingly unrelated contexts, several similarities are eminent in all definitions. According to Breidbach, Kolb, and Srinivasan (2013a), in the context of ICT-enabled service, connectivity is defined as: *1) a holistic perspective on interactions within a system that 2) explicitly recognizes entities at the core of these interactions and is 3) understood as a multidimensional socio-technical construct that can 4) vary in its levels of intensity between human actors, and therefore impact the performance of the entire system.* Each pillar is discussed in turn.

Table 2.1 Overview of Connectivity Definitions

Author	*Definition*
Angwin and Vaara (2005: 1445)	"Highlights the complexities, interconnected processes and synchronized activities in organisations and their contexts"
Janssen et al. (2006: 4)	"One characteristic represented by the level of connectivity is the density of the links within the network, i.e., the number of links divided by the maximum possible number of links. Another aspect of connectivity is reachability, or the extent to which all the nodes in the network are accessible to each other"
Kolb (2008: 128)	"The mechanisms, processes, systems and relationships that link individuals and collectives (e.g. groups, organizations, cultures, societies) by facilitating material, informational or social exchange. It includes geo-physical (e.g. space, time and location), technological (e.g. information technologies and their applications) as well as social interactions and artifacts, including shared histories, travel, trade, migration, culture, politics and other social activities"
Waverman, Dasgupta, and Brooks (2009: 6)	"The totality of interaction between a nation's telecommunications infrastructure, hardware, software, networks, and users of these networks, hardware and software"

Connectivity Provides a Holistic Perspective on Interactions Within a System

Connectivity provides a holistic perspective on interactions within a distinct system. Mason emphasizes this notion by arguing, "The idea of connectivity is central to systems thinking" (2005: 69). This system-notion is represented through a network (Janssen et al. 2006), an organization (Angwin and Vaara 2005), "groups, organizations, cultures, societies" (Kolb 2008: 128), "a nation's telecommunications infrastructure, hardware, software, networks, and users" (Waverman, Dasgupta, and Brooks 2009: 6) or, within service research, as a service system (Maglio and Spohrer 2008). While the purpose or resources exchanged during interactions are rarely defined, Kolb states, "Material, informational or social exchange" (2008: 128) is central to interactions in systems. This corresponds with the understanding of interactions in service systems, which are also based on the material, informational, or social exchange among human economic actors (Maglio and Spohrer 2008).

Connected Entities Are Central to a System-Perspective of Connectivity

The understanding of connectivity as a holistic perspective on interactions within a system implies that this system must consist of connected entities (e.g., human economic actors). Any definition of system "appears to include reference, directly or indirectly, to connected entities." These are "things in the world sustained by a network of connections" (Mason 2005: 70). Jordan adds, "The only things [. . .] common to all systems are identifiable entities and identifiable connections between them" (1969/2001, 64). All authors who define connectivity refer to connected entities, such as actors (Janssen et al. 2006), or distinguish between human and technical entities (Waverman, Dasgupta, and Brooks 2009). The same rationale applies for service systems, which consist of entities or "dynamic configurations of resources" (Spohrer and Maglio 2008: 147), embodied through people, technology, organizations, and shared information.

Connectivity Includes Multiple Socio-technical Dimensions

Waverman et al. argue that connectivity is often regarded as "the key enabler of the flow of information" (2009: 6). However, this flow of information depends on elements other than technology, and understanding connectivity in regard to technical links alone is not sufficient. ICTs represent, in the context of connectivity, "only part of the connective equation" (Kolb 2008: 140). Human skills relevant for the use of technologies, or interpersonal links among employees, customers and other stakeholders are also crucial and need to be equally taken into consideration (Waverman, Dasgupta, and Brooks 2009). This dyadic perspective on the social *and* technical dimension inherent in connectivity is central to the definitions of both Kolb (2008) and Waverman, Dasgupta, and Brooks (2009). It is this multidimensionality which makes connectivity particularly suitable as an analytical lens for investigating ICT-enabled service from the dyadic socio-technical perspective called for in the service science literature (e.g., Makarem, Mudambi, and Podoshen 2009; Edvardsson, Tronvoll, and Gruber 2011).

The Levels of Connectivity Within a System Influence Its Performance

Connectivity is not static, but rather a fluctuating construct that can be represented through various connective states. The fact that levels of socio-technical connectivity within a system can vary throughout connective states suggests a link between connectivity and the performance of a system (Janssen et al. 2006; Kolb, Caza, and Collins 2012; Breidbach, Kolb, and Srinivasan 2013a). For example, performance can be measured through the motivation or creativity between individuals (Kolb, Caza, and Collins 2012), work performance (Breidbach, Kolb, and Srinivasan 2013a), sustainability of the system itself (Janssen et al. 2006), or social inclusion, which is the focus of this volume. The link between connectivity and system performance is fundamental to the applicability of the construct of connectivity in ICT-enabled service contexts. The following sections explore this relationship in more depth.

4.2 Connective Attributes and Dimensions

While the four pillars of connectivity sufficiently describe and define the construct, it is also important to understand the evolution from the origins of connectivity as a description of technical connections only, to its application to social phenomena.

The term connectivity is based on the adjective connective, which Kolb relates to "connecting or serving to connect" (2008: 129). The related expression of connectedness however, implies joined together or fastened (past tense), suggesting an established connection or already completed action. Connectivity, in contrast, in present tense, implies future options and thus a latent potential to connect. Connectivity is consequently, as a condition, transferred metaphorically from the background to the foreground, depending on an actor's needs. Connectedness, however, refers to "states-of-being" (Kolb 2008: 129) and not to potential states.

Actor agency suggests that, despite the often-abundant prevalence of technical connections in a system, entities in that system might not be highly connected. Using or not using a technical link depends on human agency (Cousins and Robey 2005). This implies that human actors in any system can choose if, when and how to use technology. Consequently, a high level of connectivity might simply not be achieved because entities refuse to utilize a particular technological link (Kolb 2008).

Temporal intermittency implies that connections between human actors can always vanish through technical breakdowns or because different time zones keep us from calling each other in the middle of the night. Connectivity can, therefore, only to a certain extent be influenced and controlled by connected actors (Kolb 2008). Unknowable pervasiveness refers to the fact that, despite any given level of technical connectivity, and the attribute of latent potentiality, nobody can be aware of all of their real or potential connections. A world that is increasingly interconnected inherently bears the likelihood for connective uncertainties and accidents.

In conclusion, connectivity consists of social and technical connectivity. This chapter defines technical connectivity as *the degree to which ICTs are readily available for all entities in the system and adequate for the successful exchange of resources*, and social connectivity as the *strength of social ties between human actors that is necessary for the successful exchange of resources*. Both technical and social connectivity are input factors that influence the level of connectedness between human actors. Figure 2.1 summarizes the relationship between the connective dimensions of technical and social connectivity, and the various states of connectivity that are explained in the next section.

4.3 Connective States and Performance

Varying levels of connectivity, or connective states, affect the performance of a distributed work team—for example, physically distributed service providers and customers (Breidbach, Kolb, and Srinivasan 2013a). Kolb, Collins, and Lind's (2008) model suggests that the connective states of hypo (insufficient connectivity) and hyper-connectivity (excessive connectivity) negatively influence performance (e.g., information exchange, social inclusion). On the contrary, requisite connectivity implies a threshold condition of just enough connectivity for the given task or goal, while connective flow represents an optimum condition. These four connective states are understood as outcomes, or quantitative measures of connectivity, influenced by the previously discussed social and technical input factors. Figure 2.2 illustrates the model.

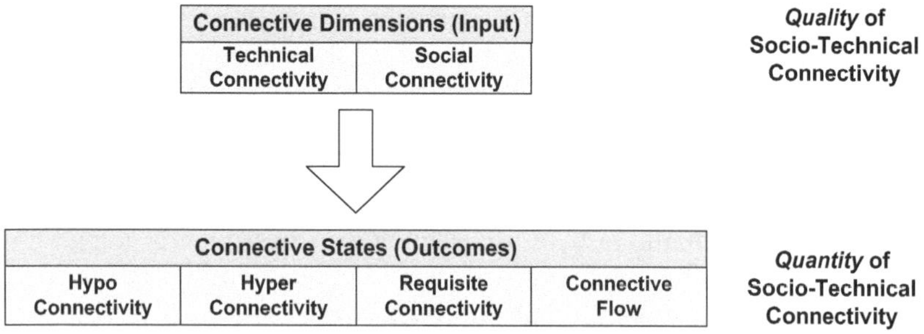

Figure 2.1 Connective Dimensions as Input Factors on Connective States.

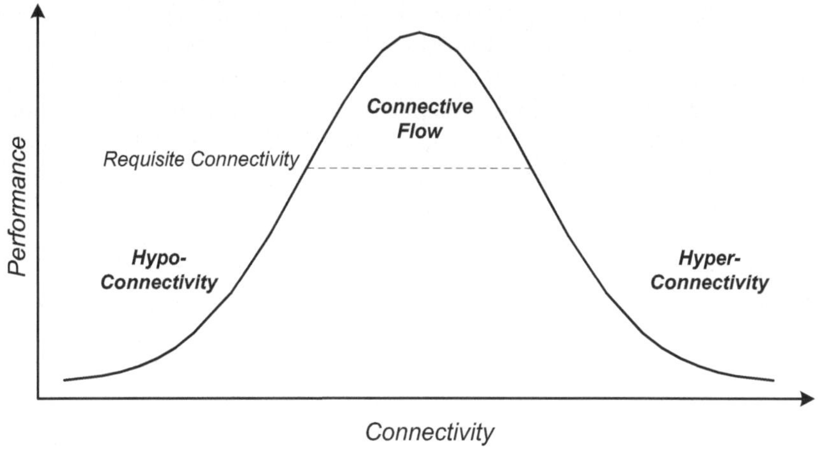

Figure 2.2 Connective States and Performance (Based on Kolb, Collins, and Lind 2008).

Hyper-connectivity represents a connective state where individuals experience too much social and/or technical connectivity, they "have too much of a good thing [and are] too connected" (Murphy 2007: 17). Hyper-connectivity is also referred to as super-connectivity (Redman and Kinzig 2003), and relates to an experiential state that involves "the instant availability of people for communication anywhere anytime" (Quan-Haase and Wellman 2005: 215). Quan-Haase and Wellman (2005), and Wajcman and Rose (2011) relate hyper-connectivity to the ubiquitous availability of ICTs in modern organizations and knowledge work, namely "the pervasive presence of information and communication technologies" (Wajcman and Rose 2011: 941). A more extensive description of hyper-connectivity, however, is provided by Kolb et al. who link this connective state to "information overload, attention-taxing workflow and interruptions in collocated spaces, [. . .] including pervasive and ubiquitous computing applications such as wireless e-mail and 24/7 telephone accessibility" (2008: 182).

Hyper-connectivity is well defined on a conceptual level, yet empirical evidence and further insights on the causes and consequences of this connective state are scarce. For example, Quan-Haase and Wellman (2005) explored hyper-connectivity empirically while investigating media use in software development teams, and found that high level of task complexity and task interdependence are positively correlated with an individual being hyper-connected. While their study focuses on the technical dimension of connectivity only, their findings confirm that technologically hyper-connected individuals struggle to complete tasks, which is typically related to constant interruptions and the increasing necessity for coordination required between team members (Quan-Haase and Wellman 2005).

Too little, or hypo-connectivity, on the contrary, is also suspected to have negative effects on the performance of systems (Kolb, Collins, and Lind 2008). Hypo-connectivity is defined as "not having sufficient connections for the task or job at hand" (Kolb, Collins, and Lind 2008, 181), and is conceptually linked to technical issues, such as weak Internet connections, insufficient mobile phone reception, limited travel options between subsidiaries or members of distributed teams, or a lack of cross-cultural understanding (Kolb, Collins, and Lind 2008).

Both hypo and hyper-connectedness can be viewed as connective gaps (Breidbach, Kolb, and Srinivasan 2013a), or temporal instances of disruption that inhibit interactions between individuals. As indicated by Quan-Haase and Wellman (2005), hyper-connectivity can cause interruptions, while hypo-connectivity is related to a lack of connectivity, respectively. The argument has been brought forward that connective gaps ultimately matter more than location, which implies that exploring the emergence and means to overcome connective gaps is crucial when attempting to optimize the performance of any system. One suggestion in order to overcome connective gaps is to control the degree of density. Janssen et al. (2006) define density as "the number of links [in a system] divided by the maximum possible number of links" (Janssen et al. 2006: 4), while Kolb, Collins, and Lind (2008: 184) define connective density as "the combined viable modes of social and technical connections between two or more persons or collectives," an absolute number, quantified by the "number of links between actors" (Kolb, Collins, and Lind 2008: 184).

Requisite connectivity is the connective state in which organizations experience a sufficient level of socio-technical connectivity. This threshold condition is a condition of "robust and reliable communication and/or transportation media/modes, with operable alternative work around options, so that contact may be initiated or maintained at the rate, richness and intensity required for a given task" (Kolb, Collins, and Lind 2008: 182). Requisite connectivity not only means not having too much or too little connectivity, but is, as a connective state, also suspected to be "contingent and relative to the situation, person and task" (Kolb, Collins, and Lind 2008: 184). Other factors that may have an impact on what constitutes requisite connectivity in a given scenario are the rank or seniority of an individual, group maturity, and profession (Kolb, Collins, and Lind 2008).

Connective flow is an ideal theorized condition where "communication is highly effective and highly efficient, and balanced in accordance with our needs and the demands of the task or situation at hand" (Kolb, Collins, and Lind 2008: 183). Rooted in Csikszentmihalyi's (1975) theory of flow, a state where individuals experience "the holistic sensations that people feel when they act with total involvement" (Csikszentmihalyi 1975: 36), connective flow remains empirically un-investigated. Kolb, Collins, and Lind (2008) argue that very little is known about which factors actually influence the emergence of requisite connectivity and connective flow. Individual rank and seniority or profession (Kolb, Collins, and Lind 2008), or the means by which individuals utilize ICTs (Froehle 2006) are suspected to have an impact. This argument corresponds with arguments stated in the virtual team literature, where a lack of research on diversity, such as the organizational tenure of team members or the "impact of dispersed organizational affiliations on team functioning" (Martins, Gilson, and Maynard 2004: 820), has been criticized.

The Information Systems (IS) literature investigated flow two decades ago when Hoffman and Novak (1996) applied it in the context of online shopping. Flow has since been explored in other ICT-enabled service contexts, such as online-learning (Shin 2006), online advertising (Sicilia and Ruiz 2007), and shopping (Pilke 2004). However, Hoffman and Novak (2009) argue that the current research on flow has not kept up with the dramatic technological changes. Web 2.0, social networks, or ICT-enabled work, are areas in which flow is likely to occur; yet, it remains un-investigated. Kolb et al. consequently argue that "more empirical work is now needed in order to answer the crucial question of how certain [connective] states influence or determine productivity" (2012: 5).

5. Conclusions, Future Research Opportunities, and Managerial Implications

This chapter introduced connectivity as a novel construct and analytical lens that can, and should, be used to investigate ICT-enabled service from a holistic socio-technical perspective. The connectivity construct provides service researchers with a terminology and understanding through which to view service systems that are transformed by ICTs. However, many aspects of connectivity are not fully understood and thus provide significant opportunities for future research. For example, future research in this area could provide recommendations on how to plan, build, and manage service systems that consistently deliver optimal levels of connectivity and social inclusion. This will require ab understanding of what the performance implications of each individual service system imply. While improved information exchange might be relevant in the context of a consulting engagement, outcomes such as social inclusion of individuals may be the desired outcome in other contexts.

Future scholars should utilize connectivity as a socio-technical lens in service research and empirically verify previously unexplored aspects of connectivity. For example, to date, the link between time and connectivity has not been discussed or further explored. There may be ways to reconceptualize connective states in relation to time, and thereby provide insights into the evolution of ICT-enabled service interactions and systems. Investigating the link between connectivity and time may therefore provide insights into the emergence of connective flow, a construct that is still not fully understood and empirically explored. Ultimately, future work should investigate if simply avoiding connective gaps is sufficient for connective flow to emerge, or if other alternative factors exist. It also remains unclear what constitutes connective flow in service.

Future methodological contributions may be longitudinal case studies, which could provide explicit insights into the communication behavior of all human actors within an ICT-enabled service system. Data could be collected by logging every instance of ICT usage and by comparing the resulting dataset to external critical incidents. This approach could be applied to service contexts like service supply chains (e.g., Breidbach, Reefke, and Wood 2015a), home entertainment (e.g., Breidbach, Chandler, and Maglio 2015b), or professional service firms (e.g., Breidbach, Chandler, and Maglio 2013b). Finally, insights from systems theory and systems dynamics can provide yet another additional methodological angle. For example, causal loop diagrams could be used to model the interactions within service systems, and subsequent studies could use casual loop diagrams to simulate interactions between human actors, which may provide us with a clearer picture about the parameters and conditions under which connective gaps occur. All of these approaches have the potential to provide a better understanding of varying levels of connectivity in service research and beyond.

Finally, practitioners can also benefit by adopting a connectivity lens through which to view, analyze, and ideally improve their technology-enabled value co-creation processes. The first step when attempting to achieve requisite levels of connectivity and connective flow is to critically review, and potentially alter, team and organizational processes. Small and stable teams that frequently interact with one another are particularly resistant to connective gaps. This is because these groups typically display a better understanding of roles. Project levels of connectivity may be improved by minimizing interpersonal relationship barriers, which tend to be a major driver of hypo and hyper-connectivity, and practitioners should ensure that all team members have access to a diverse technology-repertoire, which can help to overcome hypo-connectivity (Breidbach, Kolb, and Srinivasan 2013a).

References

Angwin, Duncan, and Eero Vaara. "Introduction to the Special Issue: 'Connectivity' in Merging Organizations: Beyond Traditional Cultural Perspectives." *Organization Studies* 26 (2005): 1445–53.

Ariss, Sonny, Nick Nykodym, and Aimee A. Cole-Laramore. "Trust and Technology in the Virtual Organization." *Advanced Management Journal* 67 (2002): 22–5.

Bitner, Mary Jo, Stephen W. Brown, and Matthew L. Meuter. "Technology Infusion in Service Encounters." *Journal of the Academy of Marketing Science* 28 (2000): 138–49.

Blomberg, Jane. "Work in the Service Economy." In *Introduction to Service Engineering*, edited by G. Salvendy and W. Karwowski, 48–71. Hoboken, NJ: Wiley, 2010.

Breidbach, Christoph F., Roderick J. Brodie, and Linda D. Hollebeek. "Beyond Virtuality: From Engagement Platforms to Engagement Ecosystems." *Managing Service Quality* 24: 592–611, 2014.

Breidbach, Christoph F., Jennifer Chandler, and Paul P. Maglio. "The Duality of Second Screens: A Phenomenological Study of Multi-Platform Engagement and Service Experiences." In *Proceedings of the 48th Hawaiian Conference on Systems Sciences (HICSS)*, Kauai, January 5–8, 2015b, 1432–41.

Breidbach, Christoph F., Darl G. Kolb, and Ananth Srinivasan. "Connectivity in Service Systems: Does Technology-Enablement Impact the Ability of a Service System to Co-Create Value?" *Journal of Service Research* 16 (2013a): 428–41.

Breidbach, Christoph F., and Paul P. Maglio. "A Service Science Perspective on the Role of ICT in Service Innovation." In *23rd European Conference on Information Systems* (ECIS), Research-in-Progress Papers, Paper 33, 2015.

Breidbach, Christoph F., and Paul P. Maglio. "Technology-Enabled Value Co-Creation: An Empirical Analysis of Actors, Resources, and Practices." *Industrial Marketing Management* 56 (2016): 73–85.

Breidbach, Christoph F., Hendrik Reefke, and Lincoln C. Wood. "Investigating the Formation of Service Supply Chains." *The Service Industries Journal* 35 (2015a): 5–23.

Breidbach, Christoph F., Peter Smith, and Lisa Callagher. "Advancing Innovation in Professional Service Firms: Insights from the Service-Dominant Logic." *Service Science* 5 (2013b): 263–75.

Cappel, James J., and John C. Windsor. "Ethical Decision Making: A Comparison of Computer-Supported and Face-to-Face Group." *Journal of Business Ethics* 28 (2000): 95–107.

Carlson, John R., and Robert W. Zmud. "Channel Expansion Theory and the Experiential Nature of Media Richness Perceptions." *Academy of Management Journal* 42 (1999): 153–70.

Cartwright, Phillip A. "Only Converge: Networks and Connectivity in the Information Economy." *Business Strategy Review* 13 (2002): 59–64.

Cousins, Karlene C., and Daniel Robey. "Human Agency in a Wireless World: Patterns of Technology Use in Nomadic Computing Environments." *Information and Organization* 15 (2005): 151–80.

Csikszentmihalyi, Mihaly. "Play and Intrinsic Rewards." *Journal of Humanistic Psychology* 15 (1975): 41–63.

Daft, Richard L., and Robert H. Lengel. "Organisational Information Requirements, Media Richness, and Structural Design." *I Science* 32 (1986): 554–71.

Davis, Mark M., Jim C. Spohrer, and Paul P. Maglio. "Guest Editorial: How Technology Is Changing the Design and Delivery of Services." *Operations Management Research* 4 (2011): 1–5.

Donofrio, Nicholas, Calline Sanchez, and Jim Spohrer. "Collaborative Innovation and Service Systems: Implications for Institutions and Disciplines." In *Holistic Engineering Education*, edited by D. Grasso and M. B. Burkhins, 243–69. New York: Springer, 2010.

Edvardsson, Bo, Bard Tronvoll, and Thorsten Gruber. "Expanding Understanding of Service Exchange and Value Co-Creation: A Social Construction Approach." *Journal of the Academy of Marketing Science* 39 (2011): 327–39.

Froehle, Craig M. "Service Personnel, Technology, and Their Interaction in Influencing Customer Satisfaction." *Decision Sciences* 37 (2006): 5–36.

Froehle, Craig M., and Aleda Roth V. "New Measurement Scales for Evaluating Perceptions of the Technology-Mediated Customer Service Experience." *Journal of Operations Management* 22 (2004): 1–21.
Griffith, Terri L., John E. Sawyer, and Neale, M. A. "Virtualness and Knowledge in Teams: Managing the Love Triangle of Organizations, Individuals, and Information Technology." *MIS Quarterly* 27 (2003): 265–87.
Hertel, G., S. Geister, and U. Konradt. "Managing Virtual Teams: A Review of Current Empirical Research." *Human Resource Management Review* 15 (2005): 69–95.
Hoffman, Donna L., and Thomas P. Novak. "Marketing in Hypermedia Computer-Mediated Environments: Conceptual Foundations." *Journal of Marketing* 60 (2009): 50–68.
Janssen, Marco A., Orjan Bodin, John M. Anderies, T Elmquvist, Henrik Ernstson, Ryan R. J. McAllister. "Toward a Network Perspective of the Study of Resilience in Socio-Ecological Systems." *Ecology and Society* 11 (2006): 1–20.
Kanter, Rosabeth M. "Change Is Everyone's Job: Managing the Extended Enterprise in a Globally Connected World." *Organizational Dynamics* 28 (1999): 7–23.
Kayworth, Timothy R., and Dorothy E. Leidner. "Leadership Effectiveness in Global Virtual Teams." *Journal of Management Information Systems* 18 (2002): 7–40.
Kolb, Darl G. "Exploring the Metaphor of Connectivity: Attributes, Dimensions and Duality." *Organization Studies* 29 (2008): 127–44.
Kolb, Darl G., Aarran Caza, and Paul D. Collins. "States of Connectivity: New Questions and New Directions." *Organization Studies* 33 (2012): 267–73.
Kolb, Darl G., Paul D. Collins, and E. Allan Lind. "Requisite Connectivity: Finding Flow in a Not-So-Flat World." *Organizational Dynamics* 37 (2008): 181–9.
Lee, Sungjoo, and Yongtae Park. "The Classification and Strategic Management of Services in e-Commerce: Development of Service Taxonomy Based on Customer Perception." *Expert Systems with Applications* 36 (2009): 9618–24.
Lovelock, Christopher H., and Evert Gummesson. "Whither Services Marketing? In Search of a New Paradigm and Fresh Perspectives." *Journal of Service Research* 7 (2004): 20–41.
Lusch, Robert F., Stephen L. Vargo, and M. Tanniru. "Service, Value Networks and Learning." *Journal of the Academy of Marketing Science* 38 (2010): 19–31.
Maglio, Paul P., and Christoph F. Breidbach. "Service Science: Toward Systematic Service System Innovation." In *Bridging Data and Decisions*, INFORMS Tutorials Series, edited by A. Newman, J. Leung, and J. C. Smith, 161–70. Catonsville, MD: INFORMS, 2014.
Maglio, Paul P., and Jim Spohrer. "Fundamentals of Service Science." *Journal of the Academy of Marketing Science* 36 (2008): 18–20.
Makarem, Suzanne C., Susan M. Mudambi, and Jeffrey S. Podoshen. "Satisfaction in Technology-Enabled Service Encounters." *Journal of Services Marketing* 23 (2009): 134–44.
Martins, Luis L., Lucy L. Gilson, and M. Travis Maynard. "Virtual Teams: What Do We Know and Where Do We Go From Here?" *Journal of Management* 30 (2004): 805–35.
Mason, Geoffrey L. "Connectivity as a Basis for a Systems Modelling Ontology." *Systems Research and Behavioral Science* 22 (2005): 69–80.
Maznevski, Martha L., and Katherine M. Chudoba. "Bridging Space Over Time: Global Virtual Team Dynamics and Effectiveness." *Organisation Science* 11 (2001): 473–92.

Murphy, P. "You Are Wasting My Time: Why Limits on Connectivity are Essential for Economies of Creativity." *University of Auckland Business Review* 9, no. 2, (2007): 17–26.

Ostrom, Amy L., Mary Jo Bitner, Stephen W. Brown, Kevin A. Burkhard, Michael Goul, Vicky Smith-Daniels, and Elliot Rabinovich. "Moving Forward and Making a Difference: Research Priorities for the Science of Service." *Journal of Service Research* 13 (2010): 4–36.

Pilke, E. M. "Flow Experiences in Information Technology Use." *International Journal of Human-Computer Studies* 61 (2004): 347–57.

Powell, Anne, Gabriele Piccoli, and Blake Ives. "Virtual Teams: A Review of Current Literature and Directions for Future Research." *Advances in Information Systems* 35 (2004): 6–36.

Quan-Haase, Anabel, and Barry Wellman. "Hyperconnected Net Work: Computer-Mediated Community in a High-Tech Organization." In *The Corporation as a Collaborative Community: Reconstruction Trust in the Knowledge Economy*, edited by C. Heckscher and P. S. Adler, 281–333. New York: Oxford University Press, 2005.

Redman, Charles L., and Ann P. Kinzig. "Resilience of Past Landscapes: Resilience Theory, Society, and the Longe Duree." *Conservation Ecology* 7 (2003): 14.

Riemer, Kai, and Stephanie Filius. "Contextualizing Media Choice Using Genre Analysis." *Business & Information Systems Engineering* 2 (2009): 1–13.

Rust, Roland T., and Carol Miu. "What Academic Research Tells Us About Service." *Communications of the ACM* 49 (2006): 49–54.

Schweiger, D. M., and P. K. Goulet. "Facilitating Acquisition Integration Through Deep-Level Cultural Learning Interventions: A Longitudinal Field Experiment." *Organization Studies* 26 (2005): 1477–99.

Sheehan, Jerry. "Understanding Service Sector Innovation." *Communications of the ACM* 49 (2006): 43–7.

Shin, Namin. "Online Learner's Flow Experience: An Empirical Study." *British Journal of Educational Technology* 37 (2006): 708–20.

Sicilia, Maria, and Salvador Ruiz. "The Role of Flow in Web Site Effectiveness." *Journal of Interactive Marketing* 8 (2007): 33–44.

Spohrer, Jim, and Paul P. Maglio. "The Emergence of Service Science: Toward Systematic Innovations to Accelerate Co-Creation of Value." *Production and Operations Management* 17 (2008): 1–9.

Srinivasan, Ananth, Christoph F. Breidbach, and Darl G. Kolb. "Service Science." In *Wiley Encyclopedia of Management*, 3rd ed., Vol. 7, 1–3. Chichester: Wiley, 2015.

Trevino, Linda K., Jane Webster, and Eric W. Stein. "Making Connections: Complementary Influences on Communication Media Choices, Attitudes, and Use." *Organization Science* 11 (2000): 163–182.

Vargo, Stephen L., Paul P. Maglio, and Melissa Akaka. "On Value and Value Co-Creation: A Service Systems and Service Logic Perspective." *European Management Journal* 26 (2008): 145–52.

Wajcman, Judy, and Emily Rose. "Constant Connectivity: Rethinking Interruptions at Work." *Organization Studies* 32 (2011): 941–61.

Walker, Rhett H., and Lester W. Johnson. "Managing Technology-Enabled Service Innovations." *International Journal of Entrepreneurship and Innovation Management* 4 (2004): 561–74.

Warketin, Merill E., and Peggy M. Beranek. "Training to Improve Virtual Team Communication." *Information Systems Journal* 9 (1999): 271–89.

Waverman, Leonard, Kalyan Dasgupta, and Nicholas Brooks. *Connectivity Scorecard 2009*. Calgary: Nokia Siemens Networks, 2009.

Yates, Joanne, and Wanda J. Orlikowski. "Genres of Organizational Communication: A Structurational Approach to Studying Communication and Media." *Academy of Management Review* 17 (1992): 299–326.

3 Re-conceptualizing Social Inclusion in the Context of 21st-Century Smart Cities

H. Patricia McKenna

1. Introduction

The rapid growth of cities worldwide is giving rise to the emergence of complex and wicked challenges for urban areas (Charoubi et al. 2012: 2289). Around the world, ICTs are being employed as a strategy by cities to innovate themselves (Nam and Pardo 2011: 185) and achieve a smarter urban agenda (Gil-Garcia, Pardo, and Nam 2016: 1). Gil-Garcia, Pardo, and Nam (2016: 5) look beyond "a dichotomy between 'being smart' or 'not being smart' " to "a continuum in which local government officials, citizens, and other stakeholders think about initiatives that make a city a better place to live." Scholl (2016: viii) describes smartness as it relates to a 21st-century urban agenda as including "sophisticated and effective infrastructures of all kinds", adding that smartness "is an inherent human capacity" and that ICTs and other technologies "need to be in the hands and under the purview of smart and savvy human actors to be deployed in the ways that they can enact their full potential and make a real difference" (2016: ix).

In order to make sense of ICT-enabled innovation in smart cities it is important to consider the concept of urbanizing. According to Sassen (2012), urbanizing refers to how technologies, spaces, and objects are being used and adapted in cities for contemporary needs, purposes, and contexts. The notion of urbanizing technologies provides opportunities for inclusion because, according to Sassen (2012), to "urbanize" "draws and needs all types of people—children, professionals, and tourists alike" as a kind of open source urbanizing with innovation potential. As such, this chapter highlights the importance of ICT appropriation in a smart city context and argues that the urbanizing of innovative ICT-enabled services for social inclusion is a key knowledge gap requiring exploration and understanding in contemporary urban environments.

Re-conceptualizing ICT-enabled social inclusion is now a necessity in that it responds to three contemporary phenomena. First, the increased blurring of boundaries across work, learning, and everyday spaces enabled by ICTs and other emerging technologies (COST 2015) gives rise to the need for more flexible and fluid understandings of social inclusion beyond traditional organizational settings and structures (Misuraca et al. 2015; Taket et al. 2014; Duxbury, Moniz, and Sgueo 2013). Second, inclusion in cities, now and in the future, is a concern highlighted by the Institute For The Future (IFTF 2011) and other researchers (Duxbury, Moniz, and Sgueo 2013; UrbanLab+ 2013; Albino, Berardi, and Dangelico 2015: 13). Third, researchers (UrbanLab+ 2013) note the complexity of issues relating to inclusion and cities, calling for an expansion of the discourse space for social inclusion to that of urban inclusion. Thus, the main aim of this chapter is to explore innovative ICT-enabled social inclusion in terms of i) aspects, ii) emerging forms and attributes, and iii) benefits, with a view to re-conceptualizing the concept for 21st-century urban environments. Using a case study approach, this work is guided by the following research questions.

Q1: Why is inclusion and social inclusion important in 21st-century cities in terms of benefits?

Q2: How does social inclusion manifest in innovative ICT-enabled contemporary urban environments, in terms of emerging forms and attributes?

In responding to these research questions, this chapter re-conceptualizes the social inclusion concept in the context of smart cities by developing and operationalizing an ambient inclusion framework for innovative ICT-enabled spaces, services, and designs.

The remainder of this chapter includes a section describing the theoretical perspective for this work through a review of the scholarly literature on inclusion and the digital divide, social inclusion, urban inclusion and smart cities, and ambient inclusion. A framework for ambient inclusion in smart cities is then described and operationalized for use in this work. The methodology section describes the research design, study recruitment, rationale for the combining of parallel research techniques, and a description of the cities and individuals involved. Findings from the research are presented in relation to two propositions under exploration and in response to the research questions. A discussion of findings in relation to the ambient inclusion framework for smart cities follows. The chapter concludes with the identification of key contributions and implications, future directions for researchers and practitioners, and limitations and mitigations.

2. Theorizing Ambient Inclusion in Smart Cities

A review of the digital divide, social inclusion, and urban inclusion literature is presented in relation to ICT-enabled services. Identification of key constructs and theories associated with the inclusion concept are discussed in developing a conceptual framework for ambient inclusion in smart cities for 21st-century urban environments.

2.1 Inclusion and the Digital Divide

Wolske et al. (2010: 312) define the digital divide as "the inability of underserved populations to access and use ICT, furthering social, economic, and educational inequities." Trauth and Howcroft (2006: 5) noted, "ICTs are portrayed as either exacerbating exclusion or are presented as the solution to greater inclusion" and that "globally, ICT penetration is highly uneven." To address this and other related societal disparities, Trauth and Howcroft (2006: 3) pointed to the importance of exploring ICTs "beyond the boundaries of the corporation." UNESCO describes inclusion through access to public space, defined as "an area or place that is open and accessible to all peoples, regardless of gender, race, ethnicity, age or socio-economic level" (2016).

Graham (2015: 160) addressed inclusion—referred to as "contours of unevenness" from a human geographies and information perspective—focusing on the Internet to explore geographies of access, participation, and representation that reveal "data shadows and digital divisions of labor" (2015: 164). Graham argues that the "removal of the tether between content and container" affects the "very geography that it represents" and "untethers it from the digital contexts of its genesis and production." Warschauer (2003: 6) identified "human and social systems that must also change for technology to make a difference" seeing "meaningful access to ICT" as "embedded in a complex array of factors encompassing physical, digital, human, and social resources and relationships."

2.2 Social Inclusion

Urquhart and Underhill-Sem (2009) draw on the work of Warschauer (2003) to describe social inclusion as "a more critical and less binary way of understanding digital divides." From a socio-economic perspective, social inclusion is defined as "the situation in which individuals and communities are fully involved in the society in which they reside/occur/exist, including the economic, social, cultural, and political dimensions of that society" (CMHC 2007: 2). Asgarkhani (2007: 144) discussed social inclusion in 21st-century cities in relation to digital government, suggesting, "Access to ICTs for enabling social inclusion is essential but insufficient" on its own. Asgarkhani (130) added that ICT-enabled solutions are hampered by issues of access, awareness, and training, as well as by public sector processes (144).

UrbanLab+ (2013) describes social inclusion as a complex and contested concept that is "fundamentally linked to the degree of political, social, and economic participation within society" where interpretations of participation differ widely. Focusing on how ICTs contribute to processes of social inclusion, Urquhart and Underhill-Sem (2009) describe the interdisciplinary perspectives of IS, geography, media studies, management, development studies, and sociology. Wolske et al. (2010: 314) argue that implementation and design of community spaces are critical to effective ICT use for social inclusion. From a public space perspective, contemporary manifestations of social inclusion may be found in the placemaking movement (www.pps.org/reference/what_is_placemaking/) with inclusion and social as key defining elements, and the potential for ICT-enabled interventions. UNESCO (2016) links the social with ICT-enabled inclusion, claiming, "In the 21st century, some even consider the virtual spaces available through the Internet as a new type of public space that develops interaction and social mixing." From a human geography perspective, Beaude (2015) argues that the Internet is a spatial innovation and offers a reminder "that society is made of social ties, and of contact and interaction."

From an education perspective, Selwyn (2008) acknowledged the rapid pace of change evident with emerging technologies and contemporary society, and, in light of the uncertainty, fluidity, and complexity of issues, he called for a stance that is more attentive to "the open possibilities of new technologies and social inclusion." In the context of higher education, James (2014: 567) identifies the lack of appropriation as the main hindrance to the participatory potential for ICTs and pointed to the need for further study.

2.3 Urban Inclusion and Smart Cities

Urban inclusion is described by Duxbury, Moniz, and Sgueo (2013) as "a meeting space between local governance efforts and bottom-up initiatives" contributing to the potential for "new forms of collective action and community experimentation in reshaping cities in different contexts at both urban and extra-urban scales." The work of the European Commission (UrbanLab+ 2013) provides an overview of the inclusion and social inclusion concepts in preparation for "making sense of "urban inclusion," a background on the origin of the social inclusion concept, a glimpse of the complexities involved in defining and conceptualizing social inclusion, and initiates preliminary discussion of the urban realm where a rights-based approach, viewed "as a critical approach for achieving 'a more inclusive city' " is advanced, for study in the next project phase.

Muggah (2016) discusses inclusive public spaces as a way of mitigating the fragility of cities in relation to risks, such as "the pace of urbanization, income and social inequality, poor access to key services." Regarding the use of technology in cities, an exploration of civic apps by Lee, Almirall, and Wareham (2016: 82–9) reveals an evolving range of issues and complexities associated with the development, adoption, and value of software applications for use in urban environments. Whereas the first generation of these apps led to many failures, the second generation is resulting in improvements based on "stronger management and consideration of the motivations of external collaborators" (82) along with the inclusion of more actors.

It is worth noting that the Canada Mortgage and Housing Corporation (CMHC 2007: 2) defines and explores social inclusion in combination with *urban form*, enabling a focusing on cities that "highlights awareness of the importance of place and interest in the role of neighborhoods." CMHC draws on the definition of *settlement form* by Lynch (1989) as "the spatial arrangement of persons doing things" and "the physical features which modify space in some way significant to those actions, including enclosures, surfaces, channels, ambiences, and objects" (CMHC 2207, 2). Lynch (1996: 476) described ambiences as "the set of encompassing conditions" in the city. CMHC (2007: 2) identified "aspects of urban form that received the most attention" related to social inclusion as "land-use mix, public transportation, public space, and social mix" (CMHC 2007: 2). Habitat III, the United Nations Conference on Housing and Sustainable Urban Development, generated a series of issue papers (http://unhabitat.org/issue-papers-and-policy-units/), several of which include the concept of inclusion (e.g., smart cities, inclusive cities, urban and spatial planning and design, and public spaces). Other concepts figuring strongly in these papers (e.g., access, awareness, compactness, connection/connectivity, content, complexity, mixed-use, openness, and social mix) are important because they appear in various combinations in the literature for smart cities and social inclusion and contribute to theorizing ambient inclusion.

2.4 *Inclusive Innovation and Smart Cities*

Foster and Heeks (2013: 338) describe the evolving research domain of inclusive innovation as consisting of the "five core structure and process components" of "innovation, actors, learning, relations, and institutions." Heeks et al. (2013: 21) claim, "The new models of inclusive innovation that are emerging all involve information and communication technologies in some way." By drawing upon inclusive innovation in urban contexts, Bencardino and Greco (2014) address social inclusion in smart cities as social innovation. Specifically, Bencardino and Greco (2014: 39) define social innovation as a way to "innovate practice with the aim of creating a positive impact for society that is as wide as possible," contributing to the concept of social cities. Bencardino and Greco (2014: 44) point to the importance of public space "interpreted as an experience created by the interaction between people." Craglia and Granell (2014: 41) discussed inclusion-related and participation concepts based on Haklay's Levels of Citizen Science where level 4 (Extreme) participation features public engagement as "collaborative science" involving "problem definition, data collection and analysis."

2.5 *An Ambient Inclusion Framework for Smart Cities*

The dynamic and emergent nature of neighborhood in cities gives way to the notion of the ambient, which is defined by McCullough (2013: 18) as "that which surrounds but does not distract" and as "an awareness of continuum and a continuum of awareness." Contributing to an understanding of the ambient in relation to culture and citing the work of Marzano (2006), Rauterberg (2007: 37–8) stated, "Ambient culture focuses nowadays on the development of open systems that understand and support the rituals of our living and adapt themselves to people through time and space." Innovative ICT-enabled pervasive and aware technologies in the form or smartphones, social media spaces, and sensors give rise to the need for a closer probing of the ambient culture concept. The nature of ambient culture in 21st-century cities is found (McKenna 2015) to be more dynamic and active when compared to more traditional taxonomic views of culture as static and passive. As such, ambient culture complements and extends current understandings of culture, enabling an emergent, continuous unfolding "in real-time, anywhere, anytime through the interactivity of people, technologies, and cities" (McKenna 2015: 1507).

The dynamic, fluid, and evolving context articulated by Selwyn (2008) (see Section 2.2), opens the way for an ambient inclusion discourse in the context of smart cities. However, Sassen (2012), in referring to open source urbanism, points to "the extent that these technologies" for smart cities "have not been sufficiently 'urbanized'" and "have not been made to work within a particular urban context." As such, the usability of ICT-enabled products and services in an urban context can be considered to be a kind of urbanizing.

Graham, De Sabbata, and Zook (2015: 88–9) note that "information has always had geography"; yet in a digital age, "ICTs have facilitated an evolution of information beyond immutable mobiles," such as maps and tourist guides, by creating "(im)mutable augmentations characterized by the layering of dynamic information across and over geographic space." Graham et al. note, "(Im)mutable augmentations of information about places matter because they shape how we are able to find and understand different parts of the world" (89), such that "places invisible or discounted in representations are equally invisible in practice to many people" (89). As such, "how places are presented within informational augmentations fundamentally affects how they are used or brought into being" (89). In an age of pervasive technologies, "the rise of information geographies" (89) as an emerging area of inquiry for "informational augmentations to places" (89) gives rise to the need for development of an inclusion framework for smart cities.

In response, this chapter presents an ambient inclusion framework for smart cities, depicted in Figure 3.1, accommodating ICT-enabled spaces, designs, and services that involve an interactive dynamic of people, technologies, and cities.

Key elements of ambient inclusion for smart cities are presented in terms of aspects, mechanisms, emerging forms and attributes, and benefits, as follows:

- *Aspects*: Adaptive, dynamic, collaborative, complex, and open.
- *Mechanism(s):* A city-focused social media space (app/platform) serves as an engagement and participation mechanism, as well as workshops (events) and discussions.

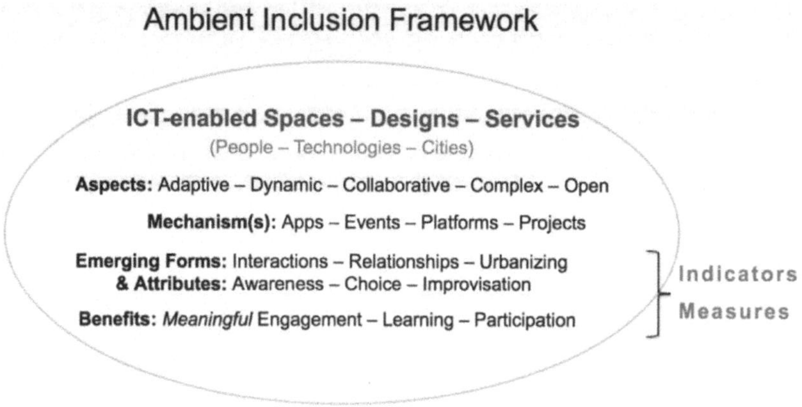

Figure 3.1 Ambient Inclusion Framework for Innovative ICT-Enabled Smart Cities.

- *Indicators* of ambient inclusion include *emerging forms* of interactions, relationships, and urbanizing along with the *emerging attributes* of awareness, choice, and improvisation.
- *Measures* of ambient inclusion are identified in terms of *benefits* that constitute one or more forms of meaningful engagement, learning, and participation.

What people notice (awareness), interactions in public spaces, and sharing serve as proxies for indicators, mechanisms, and aspects. This provides the basis for operationalizing the ambient inclusion framework for smart cities, thus addressing the two research questions, which are articulated here as propositions for exploration.

P1: Inclusion and social inclusion are important aspects of the smart city, serving to enhance urban experiences, spaces, designs, and services.

P2: Emerging forms and attributes of social inclusion in innovative ICT-enabled contemporary urban environments, as in smart cities, are multiple and diverse.

3. Methodology

The research design for this work emphasized an emergent, exploratory approach involving two avenues for data collection. Implemented in two phases, as depicted in Figure 3.2, phase I consisted of the organizing of group and individual discussions resulting in the systematic gathering of anecdotal evidence over a 1.25-year period during February 2015 to June 2016.

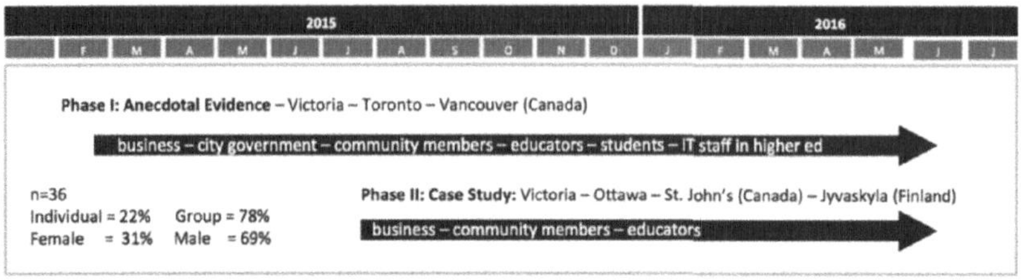

Figure 3.2 Study Time Lines, Participants, and Locations.

In parallel and beginning six months later, Phase II involved an exploratory case study approach, conducted over a one-year period during July 2015 to June 2016.

In support of the rationale for collecting the first type of data, Cubitt (2013) makes a "claim for the anecdote as a viable and vital form of evidence," arguing that "the power of anecdote is to bring us to the absolute specificity of experience," forcing "us to confront the materiality of people, things, and events" in order to understand that "the human cannot be separated from the technical, physical, or organic environments." For the second phase of data collection, an exploratory case study approach was used because it is particularly suited to the study of contemporary and emergent phenomena in real-world contexts (Yin 2014: 16), such as, innovative ICT-enabled services for social inclusion in smart cities.

Recruitment for the anecdotal evidence relied upon serendipitous invitations to engage in discussions about the city and response to workshop initiatives focusing on smart cities and learning cities (e.g., Greater Victoria, Toronto, Vancouver). Recruitment for the case study targeted a diverse range of individuals, 18 years of age and older across the city, including city officials (e.g., mayors, city councilors), business, community members, and educators who were invited by e-mail to participate in the study. Cities were decided upon based initially on convenience to the researcher who is based in the City of Victoria, while remaining open to opportunities to recruit participation from individuals in other small, medium, and large cities across Canada (e.g., Ottawa, St. John's) and even internationally (Finland). As such, formalized access to the city was intentionally not sought for this research so as to keep up with the emergent and exploratory nature of the study.

In phase one, group and individual interviews focused the discussion on noticing the city, thoughts about smarter cities, and opportunities for learning, sharing, and action. In phase two the exploratory case study used a website equipped with a registration page for study signup where basic demographic data were gathered (e.g., city, age range, self-categorization). Access to a minimally viable, city-focused web space was then granted and used as an early stage type of social media platform. Specifically, participants were invited to contribute content about their city through an *ideas* webpage and a *noticing* webpage. Website contributions were immediately viewable to all participants enabling the sharing of ideas across multiple cities. Thus, a city-focused social media web space was generated during the research study as a simple mechanism to prompt people to think about, focus on, and engage with their city and with each other through a sharing experience. Web space use was followed up with an interview focused on the use experience and on noticing the city; thoughts about smarter cities; and opportunities for learning, sharing, and action. The interview questions were formed based on the research questions, as well as the theoretical and conceptual framework for the study, and an interview protocol guided the discussions.

As qualitative data were generated in phase I through individual and interactive group discussion, transcription, and analysis activities began immediately and iteratively. Similarly, as qualitative data were generated through content contributed to the web space and follow-up individual interviews from case study participants in phase II, transcription began immediately. Iterative data analysis was conducted via content analysis. Deductive analysis was conducted based on terms from the research literature and inductive analysis was based on emergent concepts and terms from the collected data. The two streams of data from the parallel research techniques enabled simultaneous analysis, comparison, and triangulation of data.

Overall, data were analyzed for an n = 36 spanning age ranges from people in their 20s to 70s with 31% females and 69% males. Individual (22%) and group (78%) participation consisted of business actors, government officials, educators, students, IT staff in higher education, and community members.

4. Findings

Findings are presented in relation to the two propositions and in response to the two research questions.

4.1 Social Inclusion Enhances Urban Experiences, Spaces, Designs, and Services in Smart Cities

In exploring proposition 1—*inclusion and social inclusion are important aspects of the smart city, serving to enhance urban experiences, spaces, designs, and services*—the findings focus on several types of public space.

First, in Greater Victoria, an educator highlighted and described a fountain "as a touchstone" that "brought people out" and "made an awareness of something in the community." It was noted that the fountain "made people talk" to each other and "slowed people down" for "that second of connectivity." As such, the fountain was seen as a space for social inclusion to occur, functioning as "a reason to be there" and to gather, interact, and enjoy the city. The fountain was described as an important example of a public space, an object, and a service with social inclusion elements in that it is accessible to all, usable as in viewable with seating and technology components where people can gather together.

Second, in Toronto, sports events give rise to the setting up of display screens in public spaces. A community member discussed the phenomenon in terms of fostering connectivity, sharing, interactivity, and social inclusion in urban spaces. Through city efforts, the placing of jumbotron screens outside of major sports events was described as an experimental and highly effective way to bring people together in the city.

Third, in St. John's, the loop trail in Bannerman Park, referred to as the infinity loop, was highlighted as a socially inclusive urban space. Described as "unique to the city," the loop illustrates an urban element contributing to identity, a dimension of social inclusion. It is worth noting that the use of drone technology enabled the capturing and sharing of multiple dimensions of the loop on New Year's Eve, such as the infinity figure eight shape along with lighting, the activity of skaters of all ages, and the sense of vibrancy of the space (http://bit.ly/29cIK4Q).

Fourth, in Jyvaskyla, Finland, a range and variety of festivals as inclusive events and spaces for residents and visitors alike were described, occurring throughout the summer months and attracting people from across the city and beyond. An open-air gym was also described with enthusiasm by an educator in Jyvaskyla, highlighting social inclusiveness in terms of being free and accessible to all, while encouraging and supporting healthy practices and well-being. Of note is the innovation of "durable equipment" available in open-air gyms, supporting outdoor weather conditions and use along with the soon to be available online-posting of "video instructions and a sample exercise programme" (http://bit.ly/2904khn).

4.2 Emerging Forms and Attributes of Social Inclusion as Ambient in Smart Cities

In exploring proposition 2—*emerging forms and attributes of social inclusion in innovative ICT-enabled contemporary urban environments, as in, smart cities, are multiple and diverse*—the findings focus on the interplay of people, technologies, and urban spaces. The design of urban space is highlighted as multi-modal, multi-purpose, and mixed-use, where usability and appropriation are associated with identity and urbanizing.

Multi-modality

Multi-modality emerged in Victoria in terms of how a meeting was organized for social inclusion by the city hall as an eTownHall event. A city councilor stated,

> We held our first interactive eTownHall this year and we were able to get feedback from people watching the livestream so we have a packed house in person and also an overflow room with hundreds tweeting, sending direct messages that we could respond to.

The use of physical space, in combination with online spaces generated "an interactive experience that makes the city more real," as observed by the councilor. Multi-modality in urban areas was also described in terms of the range and variety of transportation options. An educator in Jyvaskyla, Finland, commented, "One thing I really like about the city is that it's compact." For the participant, that was important because "in terms of logistics" it enables one to "move very quickly" and the inclusive design allows for cycle lanes and pedestrian spaces. In Toronto, a community member commented, "You're not just moving one way, you're walking; a lot of times I'll ride my bike somewhere and I'll park my bike and jump on a streetcar and the subway."

The example of using social media at a conference where a speaker also includes a screen with Twitter feeds emerged as a multi-modal form of education and sharing. In the course of the discussion about this event, the educator realized the social inclusion potential whereby the social media space of Twitter extended the conference beyond itself. The educator observed, "You've got 500 people at a conference but you've actually got 5000 that are participating in that conference through the Twitter feed." The educator added, "It's a fabulous idea and I think it adds great richness particularly when that conversation" is moderated and compiled to record, so "what really took place" was "not just fun, but I think very meaningful."

Multi-purpose

In terms of urban form, multi-purpose aspects of the city were identified by a community member using the example of the Path in Toronto, an underground pedestrian space including 30 kilometers of tunnels, walkways, shopping, and a range of other service and entertainment spaces. This type of socially inclusive design was described in terms of how "the city allows you to make choices" so as to "interact with it and what it allows." IT professionals in the postsecondary sector in Vancouver identified concerns with inclusion from a data sharing perspective in terms of "how we exchange the data," while an educator at the city level articulated the need for inclusion in relation to the solving of complex problems related to an uncertain future that "needs all of us working together."

Mixed-Use

Mixed-use spaces in urban areas were discussed as mechanisms to engage and include people in the city. An educator in Greater Victoria made reference to the "conscious things" that governments, organizations, and communities do, including the housing, the coffee shops, the restaurants, the things that encourage people to come out and "because we've blended it all together it's exciting." From a city councilor perspective, the use of ICTs is viewed as valuable for meaningful urban inclusion where "we're working on a youth engagement strategy and we will be looking to use online tools in a new way that will be designed for youth to create" and "to bring them into some sort of online public and engage them on the issues of the city."

Identity

Social inclusion as identity, cadence, sense of place, and how people connect with and relate to the city was explored in terms of feeling the pulse of the city. A postsecondary educator in Finland spoke of the rhythm of the city connected with the academic cycles. The city filling with students and how "the nightlife is bustling with energy" were mentioned along with "walking on the streets" as inclusive spaces. In Victoria, an educator equated the pulse of the city with "the people, first and foremost the people, the way that people walk and talk and engage each other, the coffee shops, the businesses, all those things" encompassing "people and daily goings on." Explored in another way, associated with the pulse of the city, is the feeling of vibrancy of the city. A city councilor in Victoria stated, "Vibrancy is created by people and connections between people," adding, "The way that comes to life" is through "arts and culture, sitting and having a coffee discussing the city and things you're passionate about." Critical to the pulse and vibrancy of the city, added the councilor, "is people interacting with each other and with their surroundings because its creating activity." Asked about social media, the councilor responded, "It enhances the personality of a city on the Internet" and noted, "The city has a personality on the Internet and whether that's Facebook or 10s of 1000s of Twitter followers, it makes it all more accessible."

Urbanizing

A community member in Victoria spoke of "ways to animate a space," providing the example of a city parkade embedded with sensor technology that "plays different sounds as you go up based on where you are in the stairwell" and the "lighting changes." Referred to as the "musical railing," this city-supported, cross-sector urban collaboration (e.g., local technology, arts, culture) characteristic of smart cities is described as "art in public places." A city councilor said that the musical railing is "designed to ensure that civic parkades are safe and welcoming" pointing to safety, engagement, cultural, and other benefits (http://bit.ly/290HwIg). A local community group in Greater Victoria highlighted one of their most active and interactive social media blog posts about "library boxes," describing the free sharing and exchange of books. The post triggered multiple forms of spontaneously generated and unexpected interest, along with interactivities described in terms of Twitter posts, Google mapping, new connections, engagement and participation, and video sharing about the library boxes.

5. Discussion

A discussion of these findings is presented through a visual that illustrates how the ambient inclusion framework for smart cities was operationalized in this study. As depicted in Figure 3.3, examples of social inclusion are described in terms of the *aspects* of—access, identity, implementation, and usability. In a smart cities context, the aspects of adaptive, dynamic, collaborative, complex, and open are taken into consideration, affecting the nature of social inclusion. Urban experiences described in the data include sports events, a fountain, a loop trail, a city council eTownHall, an academic conference, and two forms of emergent ambient culture (musical railings and library boxes). How each form operates in relation to technology is described in the ambient inclusion column on the right. *Mechanisms* for exploring social inclusion in the city are identified as the web space/platform (described in Section 3), interviews, and workshop discussions. *Emerging forms and attributes* of social inclusion are identified as appropriation in terms of interactions, relationships, and urbanizing, based on awareness, choice, and improvisation, contributing to the development of indicators for ambient inclusion. *Benefits* are identified and shown to be emerging from all elements of the framework contributing to the development of potential measures for ambient inclusion in terms of meaningful engagement, learning, and participation.

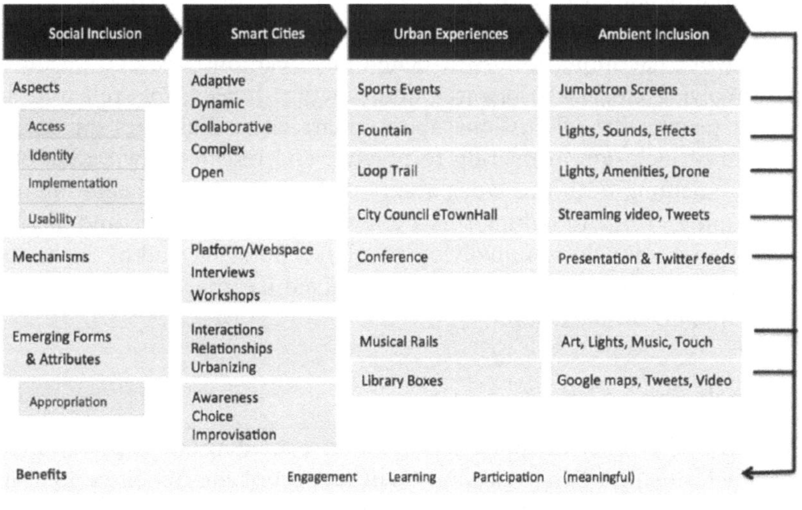

Figure 3.3 Ambient Inclusion Framework in Action for Smart Cities.

Urban spaces were found to become multi-purpose beyond traditional, physical notions, incorporating the innovations in space afforded by the Internet, social media, and other aware technologies. This intermingling offers opportunities for the evolving and enriching of relationships and partnerships across the city where barriers and silos may previously have existed. Using social media and other innovative ICT-enabled services, the spaces for action, experience, and choice broadens, enabling new forms of inclusion, such as ambient inclusion, and the potential for increased and meaningful engagement, learning, and participation.

The benefits and value of inclusion in innovative ICT-enabled smart urban environments should not be underestimated, given the expanding and ever-evolving nature of emerging technologies in relation to education and to smart cities. Learning cities as inclusive ICT-enabled smart spaces afford associated literacies for 21st-century urban environments. Combining more aware people with aware technologies contributes further to the ambient inclusion concept as a smarter adaptive mechanism where learning can occur more readily and meaningfully in the city. As such, the learning city can become an enabler of ambient inclusion.

The dynamic nature of cities, enabling malleable and porous spaces, contributes to the potential for non-fixed notions of inclusion. Indeed, social and urban inclusion can be dynamic, with ambient inclusion opening the potential for emerging understandings, challenges, and opportunities. The ambient inclusion framework for smart cities calls for a re-conceptualizing of inclusion-related and participation concepts. The originality of the ambient inclusion conceptualization is that it provides a real-world context for what Haklay's level 4 participation (Extreme) (Craglia and Granell 2014: 41) may involve. As such, the contribution of this chapter lies in the way it is able to articulate scenarios where all types of actors/stakeholders across the city become involved.

This work has implications for definitions, relationships, and meaningfully involving citizens in research and practice. Indeed, this research, by inviting people into discussions about smart cities illustrates an emerging form of inclusion in relation to practice and research. Two additional insights emerging from this work are: a) the importance of the notion of a continuum as a thread linking smart cities, ambient inclusion, and innovative ICT-enabled inclusion involving aware technologies; and b) the importance of multi-purpose, multi-modal, and mixed-use spaces as enablers and drivers of ambient inclusion.

6. Conclusion

This chapter is interdisciplinary in nature, providing an exploration of social inclusion in urban environments in support of smarter cities. In terms of contributions, a conceptual framework is advanced for ambient inclusion in the smart city as a way of re-conceptualizing current understandings of inclusion, social inclusion, and the emerging area of urban inclusion. The ambient inclusion framework that is theorized and operationalized in this chapter is intended to support the potential for solutions to emerge in addressing the complex and wicked challenges posed by the rapid and unprecedented growth of 21st-century cities. Through the interactive dynamic of *people—technologies—cities*, key aspects, attributes, and benefits of social inclusion in urban environments are identified in this study using the mechanisms of a city-focused social media space, workshops, and interviews. Sassen's (2012) notion of the necessity of urbanizing technology is extended in this work as a way of appropriating innovative ICT-enabled social inclusion to support the opening of a space for ambient inclusion. The ambient inclusion framework provides opportunities for further exploration and debate while inviting further testing, development, and validating by researchers and practitioners. Implications of this work for academia, industry, and policy makers pertain to definitions, relationships, and the nature of meaningfully involving citizens in research and practice.

The key take away from this chapter is the emergence of evolving perspectives on social inclusion in the context of smart cities—accommodating technology-rich, mobile, wearable, and other emerging technologies that underlie urban designs, interactions, spaces, and services. This chapter will be of interest to practitioners and researchers in the areas of smart cities, learning cities, urban design, planning and development, Internet of Things (IoT), and anyone concerned with inclusion in innovative ICT-enabled urban spaces. Limitations of this exploratory study relating to sample size, number and size of cities, and geographic location are mitigated by the depth and richness of the data and the potential to extend this type of study and include other geographic locations, cities of greater scale (e.g., megacities and regions with populations exceeding ten million people), and urban contexts.

References

Albino, V., U. Berardi, and R. M. Dangelico. "Smart Cities: Definitions, Dimensions, Performance, and Initiatives." *Journal of Urban Technology* 22, no. 1 (2015): 3–21. DOI: 10.1080/10630732.2014.942092.

Asgarkhani, M. "The Reality of Social Inclusion Through Digital Government." *Journal of Technology in Human Services* 25, nos. 1–2 (2007): 127–46. DOI: 10.1300/J017v25n01_09.

Beaude, B. "Internet: A Unique Space of Coexistence." In *EPFLx: SpaceX Exploring Humans' Space: An Introduction to Geographicity*, edited by J. Lévy et al. Massive Open Online Course (MOOC), edX, Fall, 2015.

Bencardino, M., and I. Greco. "Smart Communities: Social Innovation at the Service of Smart Cities." TeMA INPUT 2014, *Journal of Land Use, Mobility and Environment* 2014: 39–51. Special Issue, Eighth International Conference INPUT, Smart City—Planning for Energy, Transportation and Sustainability of the Urban System.

Charoubi, H., T. Nam, S. Walker, J. R. Gil-Garcia, S. Mellouli, K. Nahon, T. A. Pardo, and H. J. Scholl. "Understanding Smart Cities: An Integrative Framework." In *Proceedings of the 45th HICSS*, 2012, 2289–97.

CMHC. *Urban Form and Social Inclusion: Research Highlights*. Socio-Economic Series 07–007. Ottawa, Canada: Canada Mortgage and Housing Corporation, 2007.

COST. "Dynamics of Virtual Work." European Cooperation in Science and Technology (COST), 2015. Accessed March 24, 2015. http://dynamicsofvirtualwork.com/aboutus/

Craglia, M., and C. Granell, eds. *Citizen Science and Smart Cities*. European Commission, *JRC Technical Reports*, Report of Summit, Ispra, Italy, February 5–7, 2014.

Cubitt, S. "Anecdotal Evidence." *NECSUS: European Journal of Media Studies*, 2013. Accessed March 21, 2016. www.necsus-ejms.org/anecdotal-evidence/

Duxbury, N., G. C. Moniz, and G. Sgueo. "Rethinking Urban Inclusion: Spaces, Mobilizations, Interventions." *Cescontexto, Debates* 10, no. 2 (2013): 832pp.

Foster, C., and R. Heeks. "Conceptualising Inclusive Innovation: Modifying Systems of Innovation Frameworks to Understand Diffusion of New Technology to Low-Income Consumers." *European Journal of Development Research* 25, no. 3 (2013): 333–55.

Gil-Garcia, J. R., T. A. Pardo, and T. Nam, eds. *Smarter as the New Urban Agenda: A Comprehensive View of the 21st Century City*. Public Administration and Information Technology Series, Vol. 11. New York: Springer, 2016.

Graham, M. "Information Geographies and Geographies of Information." *New Geographies* 7 (2015): 159–65.

Graham, M., S. De Sabbata, and M. A. Zook. "Towards a Study of Information Geographies: (Im)Mutable Augmentations and a Mapping of the Geographies of Information." *Geo: Geography and Environment* 2 (2015): 88–105.

Heeks, R., M. Amalia, R. Kintu, and N. Shah. "Inclusive Innovation: Definition, Conceptualization and Future Research Priorities." *Working Paper Series*, Paper No. 53, Development Informatics. Manchester, UK: University of Manchester, Manchester Centre for Development Informatics by the Institute for Development Policy and Management, SEED, 2013.

IFTF. *A Planet of Civic Laboratories: The Future of Cities, Information, and Inclusion*. Palo Alto, CA: Institute for the Future, 2011. Accessed June 20, 2015. www.iftf.org/our-work/global-landscape/human-settlement/the-future-of-cities-information-and-inclusion/ (overview); http://iftf.me/public/SR-1352_Rockefeller_Map_reader.pdf (document).

James, R. "ICT's Participatory Potential in Higher Education Collaborations: Reality or Just Talk." *British Journal of Educational Technology* 45, no. 4 (2014): 557–70. DOI: 10.1111/bjet.12060.

Lee, M., E. Almirall, and J. Wareham. "Open Data and Civic Apps: First-generation Failures, and Second-Generation Improvements." *Communications of the ACM* 59, no. 1 (2016): 82–9.

Lynch, K. *Good City Form*. Cambridge, MA: MIT Press, 1989.

Lynch, K. *City Sense and City Design: Writings and Projects of Kevin Lynch*. Edited by Tridib Banerjee and Michael Southworth. Cambridge, MA: MIT Press, 1996.

Marzano, S. "Ambient Culture." In *True Visions: The Emergence of Ambient Intelligence*, edited by E. Aarts and J. Encarnação, 35–52. Berlin, Heidelberg: Springer-Verlag, 2006.

McCullough, M. *Ambient Commons: Attention in the Age of Embodied Information*. Cambridge, MA: The MIT Press, 2013.

McKenna, H. P. "Ambient Culture in 21st Century Urban Educational Environments: An Exploration of Awareness, Learning, Openness, and Engagement." In *Proceedings of the 9th International Technology, Education and Development Conference (INTED2015)*, 2015, 1502–12.

Misuraca, G., C. Colombo, R. Radescu, and M. Bacigalupo. "Mapping and Analysis of ICT-Enabled Social Innovation Initiatives Promoting Social Investment." In *Integrated Approaches to the Provision of Social Services. JRC Technical Reports Series*, European Commission's Joint Research Centre, Institute for Prospective Technological Studies, 2015.

Muggah, R. *How Fragile Are Our Cities?* New York: World Economic Forum, 2016. Accessed February 9, 2016. www.weforum.org/agenda/2016/02/how-fragile-are-our-cities

Nam, T., and T. A. Pardo. "Smart City as Urban Innovation: Focusing on Management, Policy, and Context." In *ICEGOV2011*, 2011, 185–94.

Rauterberg, M. "Ambient Culture: A Possible Future for Entertainment Computing." In *Interactive TV: A Shared Experience—Adjunct Proceedings of EuroITV 2007*, edited by A. Lugmayr and P. Golebiowski, 37–9. Tampere, Finland: TICSP series#35, 2007.

Sassen, S. *Global Networks, Linked Cities*. London: Routledge, 2002.

Sassen, S. "Urbanizing Technology." In *Urban Age Electric City Conference*, LSE Cities, London, 2012. Accessed May 26, 2015. https://lsecities.net/media/objects/rbanizingrbanising-technology/en-gb/

Scholl, H. J. "Foreword." In *Smarter as the New Urban Agenda: A Comprehensive View of the 21st Century City*, edited by J. R. Gil-Garcia, T. A. Pardo, and T. Nam, vii–x. Switzerland: Springer, 2016.

Selwyn, N. "The Use of ICTs in Education and the Promotion of Social Inclusion: A Critical Perspective from the UK." *Education and Society* 29, no. 104 (2008). Accessed June 15, 2016. www.scielo.br/scielo.php?pid=S0101-73302008000300009&script=sci_arttext&tlng=en

Taket, A., B. R. Crisp, M. Graham, L. Hanna, S. Goldingay, and L. Wilson, eds. *Practicing Social Inclusion*. New York: Routledge, 2014.

Trauth, E. M., and D. Howcroft. "Social Inclusion and the Information Systems Field: Why Now?" In *Social Inclusion: Societal and Organizational Implications for Information Systems*, edited by E. Trauth, D. Howcroft, T. Butler, B. Fitzgerald, and J. DeGross, 3–12. Boston, MA: Springer, 2006.

UNESCO. *Inclusion Through Access to Public Space*. Paris, France: United Nations Educational, Scientific, and Cultural Organization, 2016. Accessed February 24, 2016. www.unesco.org/new/en/social-and-human-sciences/themes/urban-development/migrants-inclusion-in-cities/good-practices/inclusion-through-access-to-public-space

UrbanLab+. *Making Sense of 'Urban Inclusion.'* Berlin, Germany: International Network of Urban Laboratories, European Commission, Erasmus Mundus, 2013. Accessed January 11, 2016. www.urbanlabplus.eu/project/inclusion

Urquhart, C., and Y. Underhill-Sem. "Special Issue on 'ICTs and Social Inclusion.'" *Journal of Information, Communication and Ethics in Society* 7, no. 2/3 (2009): 96–210.

Warschauer, M. *Technology and Social Inclusion: Rethinking the Digital Divide.* Cambridge, MA: The MIT Press, 2003.

Wolske, M., N. S. Williams, S. U. Noble, E. O. Johnson, and R. Y. Duple. "Effective ICT Use for Social Inclusion." In *Proceedings of the 2010 iConference*, 2010, 312–16.

Yin, R. K. *Case Study Research: Design and Methods*, 5th ed. Los Angeles, CA: Sage, 2014.

4 Enhancing Social Inclusion Through Optimal Community Participation Levels in ICT4D Projects

Arthur Glenn Maail, Sherah Kurnia, and Shanton Chang

1. Introduction

Information and Communication Technology for Development (ICT4D) initiatives reinforce social inclusion through impacting the delivery of health care, education, or community related services to individuals and communities that previously might not have taken advantage of such services. This may have been due to demographic, personal, or broad social constraints. However, ICT4D initiatives are not just about providing access to the services (Roode et al. 2004; Warschauer 2004). To achieve social inclusion objectives, ICT4D projects must establish inclusive, consultative, and, preferably, participative processes for citizens' involvement, both in the identification of issues and the development of goals, structures and decision-making processes (Avgerou and Madon 2005; Madon 2004). If the inequality between those who benefit and those who do not benefit from the presence of ICT4D projects and related services is too high, then it could lead to project failure (Kumar and Best 2006b).

There is a general consensus among researchers that ICT4D initiatives require participation from all groups within a community so that these initiatives can provide technological innovations that can benefit all groups within that community (Bailey 2009; Bhatnagar 2003; Harris et al. 2003; Oestmann and Dymond 2001; Roman and Colle 2003). However, maximizing community participation does not always lead to successful ICT4D projects (Heeks 1999). A particular development approach may require a higher level of community participation than another approach. Therefore, achieving community participation level that is consistent with the development approach is important for achieving success rather than merely maximizing community participation in all projects regardless of the development approach. Such a balance condition is known as "community participation equilibrium" (Doll and Torkzadeh 1989).

The difficulty in achieving participation equilibrium lies in the current limited understanding about two competing elements, which are the desired user participation inscribed by a particular development approach and the actual participation level that is affected by a number of conditional factors. With the existence of the various approaches in the development of

ICT4D projects that require different levels of user participation, further understanding of relevant conditional factors affecting user participation is critical to obtain the optimal level. Currently, no studies have examined factors affecting user participation for different approaches to ICT4D projects. Existing studies (e.g., Puri et al. (2004)) focus on identifying factors that affect user participation levels in ICT4D projects without differentiating the development approaches involved. The lack of understanding in this area contributes to inconsistencies in several studies regarding the impacts of those conditional factors and user participation levels on project success (Puri et al. 2004; Puri and Sahay 2007).

Therefore, this chapter aims to develop a conceptual framework that identifies the optimal user participation levels for different ICT4D approaches and the relevant conditional factors. ICT4D is regarded as one type of human actions that deals with "the purposeful crafting and construction of artifacts, which can be in the forms of physical hardware configurations, design and analysis documents, codes, user documentations, organization structure and procedures, etc." (Hirschheim et al. 1996: 7). Hence, understanding the nature of human beings and inherent human qualities has been suggested as the way to understand "users" in developing human-centered IS and the associated services (Hirschheim et al. 1995; Isomäki and Pekkola 2011). Habermas's Theory of Communicative Action (TCA) can help interpret the actions performed by the human actors associated with the development of such systems and services (Hirschheim et al. 1996; Klein and Huynh 2004; Lyytinen 1992; Mingers 1992). In this chapter, we use Habermas's typology of human action to identify different types of human actions involved in different ICT4D approaches. Then we apply the concept of "action constitutive resources" to systematically identify relevant conditional factors for each type of human action. Based on our systematic reasoning, we develop a framework to illustrate types of human actions involved in each ICT4 project development approach and the associated conditional factors.

The remainder of the chapter is organized as follows. The next section briefly describes the concept of social inclusion in relation to digital divide and socio-techno divide. It then continues with an explanation of the importance of user participation, the concept of user participation equilibrium, and elements affecting participation equilibrium in the development of ICT4D projects. Then the following section outlines the TCA, including the typology of social actions and the action constitutive resources. Next, we show how the TCA is applied to identify the relevant human actions for each development approach and relevant conditional factors affecting user participation. Finally, we discuss how the proposed framework can be useful to inform future research related to community participation in ICT4D projects.

2. Social Inclusion, Digital Divide, and Socio-techno Divide

The effort to ensure social inclusion using ICTs is underpinned by two important phases. The early ICT4D projects mainly focus on digital divide, which means addressing the issue of ICT infrastructure for those who have no access to ICTs. Governments, donors, and development agencies have focused on providing technical, financial, and political support, including ICT infrastructures (e.g., hardware and relevant software and content) and policy (DiMaggio and Hargittai 2001; Forestier et al. 2002; Kenny 2002; van Dijk and Hacker 2003). However, technological infrastructure alone is not enough to achieve a successful development of an ICT4D project given that it is not just a matter of ICT provision but also whether or not the relevant communities can utilize the ICTs and related services provided (Tsatsou 2011). Avgerou and Madon (2005), for example, argue that digital divide problem in developing countries can indeed be created by the provision of ICTs or ICT-enabled services that are in conflict with communities' aspirations and the way they live their lives. Likewise, Madon (2004) and Tipson and Fritteli (2003) warn about the tendency of defining the "digital divide" merely as a problem of scarcity of supply of ICT infrastructure. Thus, focusing on technology provision through the concept of digital divide does not necessarily warrant social inclusion since those technologies may not be utilized by the communities.

Therefore, since the second World Summit on the Information Society in Tunis in 2005, the issue in the development of ICT4D projects has now shifted toward addressing the other divide, that is "socio-techno divide" which concerns with ICT uptake and its impact on the social and economic development (Harris 2015; Heeks 2008, 2010; Unwin 2009; Warschauer 2004). Thus, the focus has shifted from simply access to the technology to actual usage that can contribute toward broader development goals. Roode et al. (2004) suggest that development efforts and ICT4D projects should aim to bridge the "socio-techno" divide, which focuses on people and their development needs. After putting all efforts toward addressing the availability of technological infrastructure, it is now necessary to focus on developing services and applications that meet these needs (Heeks 2010; Toyama 2010). Clearly, the development of technological innovations in ICT4D projects needs to move beyond a techno-centric approach toward a socio-centric one. Addressing both digital divide and the socio-techno divide is important in the ICT4D context since they affect the success and effectiveness of various development programs which in turn improves social inclusion in various economic and social activities, particularly among disadvantaged communities (Harris 2015; Walsham and Sahay 2006).

3. User Participation Equilibrium

There is a general consensus among researchers that ICT initiatives for communities require participation from all groups in a particular community so that they can provide technological innovations and services that can benefit all groups within the community (Bailey 2009; Bhatnagar 2003; Harris et al. 2003; Oestmann and Dymond 2001; Roman and Colle 2003). Colle (2005) argues that participation is important because "it conveys the sense of community ownership . . . it helps reflect community values and will help us identify information needs." It is recognized that direct engagement with potential users and stakeholders provides project staff with a sound and extensive knowledge base about their needs and characteristics (Rao 2008; Roman and Colle 2002; Roman and Colle 2003). Heeks (2002) suggests that user participation bridges the design-actuality gap, which is caused by disparate assumptions inscribed by designers during system development, and the prevailing way and state of organizational life in the local context in which technologies are implemented (Akinnuwesi et al. 2013). Thus, user participation is important for the ICT4D projects to achieve their social inclusion objectives and ensure equity and improvement in quality of life of the communities involved. It facilitates learning, knowledge sharing and innovation within the communities (Unwin 2009). Hence, most practitioners assume that a higher degree of participation is desirable for the success of the projects, since it enhances psychological buy-in to the system and services (Mumford 1983; Schuler and Namioka 1993). However, a number of studies have indicated that higher participation may have a detrimental effect (Heeks 1999). For example, Karlsson et al. (2012) argue that unclear user target segments can impede the fulfillment of usability and relevance goals of the project, while lack of adequate user skills can reduce efficiency of service delivery.

In this chapter, an important concept from IS research introduced by Doll and Torkzadeh (1989) more than 20 years ago is adopted to explain these conflicting findings. Doll and Torkzadeh (1989) highlight that user participation has a significant impact on the success of IS development if the desired/optimal degree is equal to the actual degree of participation. Such a condition is referred to as "participation equilibrium" or "moderate deprivation." Doll and Torkzadeh began their investigation by showing that there have been some inconsistencies in previous findings regarding the impact of user participation on the success of IS development (He and King 2008; Lin and Shao 2000). Although, in general, user participation has a strong impact on the success of IS development, in some studies it sometimes has only a moderate impact on the overall success (Bano and Zowghi 2015; Iivari et al. 2010). Doll and Torkzadeh's (1989) hypothesis was confirmed by King and Lee (1991), who concluded that a higher degree of participation in IS development may even impact on user satisfaction negatively. Thus, it is important that participation equilibrium be achieved in managing user participation in any ICT4D projects. Harris and Weistroffer (2009) reviewed studies from 1996 to 2009 that investigated the relationship between user participation and system success, concluded that the participation level that passes the optimal level of participation does not add any value, but can rather be perceived as a waste of resources.

There are two important elements affecting the achievement of participation equilibrium: 1) the desired degree of user participation, which is prescribed in the approach employed for the development of the project and, 2) the actual degree of user participation, which is affected by the conditional factors surrounding the context of ICT4D projects. Therefore, the understanding of the development approaches and conditional factors affecting user participation for different ICT4D project development approaches are critical to bring positive impact of user participation on the success of ICT4D projects. These two important elements are explained next.

3.1 ICT4D Project Development Approaches

Both IS (Damodaran 1996; Hirschheim and Klein 1989; Mattia and Weistroffer 2008) and ICT4D researchers (OECD 2001; Puri and Sahay 2003a, 2007) define three approaches in the development of ICT4D projects based on three criteria: (1) degree of user participation, (2) key actors involved during the development of an ICT4D project, and (3) social relationships between key actors and during the development of an ICT4D project. These approaches, as summarized in Table 4.1, are associated with "developer-enacted" approaches (Mattia and Weistroffer 2008; Puri and Sahay 2003a, 2007) because in assessing the degrees of user participation, the developer's point of view toward the user is considered.

For the agency-driven approach (AD), users are not typically consulted about the development of ICT services, but they are only informed about the development process (Puri and Sahay 2007). The project staff may give specific directives to the participant to do something in relation to service development without providing the reasoning behind the action. The key actors in this approach are the donor agency(ies) and the project staff. The donor agency(ies) are responsible for providing the system objectives. At the same time, the staff takes the objectives and turns them into a constructed system. The guiding principle behind the AD approach is profit

Table 4.1 The Approaches Used in the Development of an ICT4D Project

Approach	*Criteria*		
	Degree of User Participation	*Key Actors*	*Social Relationship Among Key Actors*
Agency-Driven	*Informative, users provide and/or receive information requested by the donor agency/telecenter staff.*	*Donor agency and telecenter staff*	*Donor agency-led*
Shared-Driven	*Consultative, users comment on a predefined program set by the telecenter staff.*	*Users and telecenter staff*	*Telecenter staff-led*
User-Driven	*Participative, users influence decisions relating to the whole project.*	*Users and telecenter staff*	*Joint system development*

maximization, that is, to support rational organizational operation and effective and efficient project management.

For the shared-driven (SD) approach, users have the role to interpret and make sense of the system and their surroundings. However, it is the staff who acts as facilitator, interacting with users, working within their perspective and helping them find their preferred views of the system (Hirschheim and Klein 1989). The staff relies on their experience and insights to ease any conflicting views among users (Hirschheim and Klein 1989; Mattia and Weistroffer 2008). This is a two-way relationship in which users or the community provide feedback to the staff. Intended beneficiaries (users) are consulted in meetings organized by officials/donor agency experts about pre-identified initiatives (Damodaran 1996; Puri and Sahay 2007), but the development agenda is still externally driven. The consultation is mainly aimed at obtaining community feedback about project formulation (Puri and Sahay 2007)

For the user-driven (UD) approach, staff and users act as partners (Mattia and Weistroffer 2008). Users are encouraged to define and convey their aspirations and needs. The development agenda is set jointly and users' views and knowledge are deliberately sought and respected. The relevance of local practices and knowledge is more explicitly recognized. The development process is conducted based on joint decisions between staff and users (Hirschheim and Klein 1989; Mattia and Weistroffer 2008). This is a relation based on partnership with the staff, in which users are actively engaged in project development.

3.2 Factors Affecting User Participation

In the development of ICT4D projects and IS projects in general, the actual degree of user participation is affected by the conditional factors surrounding the development of such projects. Prior studies have extensively studied the conditional factors affecting user participation in the development of ICT4D projects (see Table 4.2).

The conditional factors are categorized into domains based on the theoretical model developed by Ives and Olson (1984). The model is based on the contingency theory of Participative Decision Making by Vroom and Yetton (1973), which posits that the way human actors behave in carrying out decision-making processes is a function of both the properties of the situation with which they are confronted and the relatively stable properties of the person including personal characteristics and attitudes. Ives and Olson's model describes the relational impact of user participation on system success, directly and indirectly, by improving users' cognitive and motivational factors toward the system, which then affects the overall system success. This theoretical model also recognizes that there are contextual factors surrounding systems development that may enable or prohibit user participation. These are called conditional factors.

The conditional factors are divided into three domains: organizational, technical, and user. The organizational domain covers work arrangements,

Table 4.2 Summary of Conditional Factors Affecting User Participation in the Development of ICT4D Projects

Domain	*Conditional factors*	*Possible value*	*References*
Organization	Constraint on the project duration approved by the donor agency (O1)	Present/ Absent	(Byrne and Sahay 2007; Carroll and Rosson 2007; Ramirez et al. 2005; Weiner et al. 2002)
	Constraint on the amount of funding given by the donor agency (O2)	Present/ Absent	(Benston 1990; Byrne and Sahay 2007; Carroll and Rosson 2007; Ramirez et al. 2005; Weiner et al. 2002)
	Direct involvement of government/donor officer (O3)	Present/ Absent	(Byrne and Sahay 2007; Carroll and Rosson 2007; Heeks 1999; Puri et al. 2004)
	Familiarity between telecenter staff and users (O4)	Yes/No	(Hearn et al. 2005; Kanungo 2004; Puri and Sahay 2003b; Van Belle and Trusler 2005)
	Support from community champion (O5)	Present/ Absent	(Hearn et al. 2005; Kanungo 2004; Kumar and Best 2006a; Puri and Sahay 2003b; Ramirez et al. 2005)
	Incentives given to the users (O6)	Present/ Absent	(Bailur 2007a, 2007b; Madon 2005)
Technical	Degree of task complexity given to the users (T1)	High/low	(Bailey and Ngwenyama 2011; Bailur 2007a)
	Availability of easy-to-use technology (T2)	Present/ Absent	(James 2004; Merkel et al. 2007; Ramirez et al. 2005; Roman and Colle 2003)
User	User experience with the ICT4D project (U1)	Present/ Absent	(Bailur 2007a; Byrne and Sahay 2007; Carroll and Rosson 2007; Puri and Sahay 2007; Van Belle and Trusler 2005),
	User-shared attitude about the ICT4D project (U2)	Present/ Absent	(Bailur 2007a; Heeks 1999)

roles, positions, power, values, norms, and cultures. The technical system domain covers the physical means and technical know-how by which various tasks are accomplished. The user domain covers user characteristics and their attitudes toward the system. These conditional factors have been identified to affect user participation in previous studies. However, none of the previous studies has investigated what specific factors are relevant for a particular ICT4D approach in order to obtain the optimal level of user participation. This chapter is specifically designed to address this knowledge gap by investigating appropriate level of user participation for each development approach and from the perspective of human action.

4. Theory of Communicative Action

The human action perspective in analyzing information systems development (ISD) was popularized in the early 1990s (Hirschheim et al. 1991). In a wider view, all design activities are social actions performed in a specific socio-historical context and therefore must be influenced by a set of practices that are shared by the local community (Avgerou 2001; Avgerou and Walsham 2000). Because design activity is a form of social action, therefore ISD can be more effectively understood by applying basic blocks of social action theories. The social action theories offer theoretical lenses to understand various elements of socially organized human behaviors. These theories can be distinguished from other social theories due to their focus on understanding human actors and their dynamic behaviors from the viewpoint of the involved actors (Hirschheim et al. 1996). The Habermas's TCA is one of those theories.

Two important concepts within Habermas's TCA are used in the development of the framework: (1) the typology of human action and (2) the action constitutive resources. They are important to ensure plausibility and cogency of logical reasoning for the proposed framework. The typology of human action categorizes complex human actions into several types based on the orientation of the action. We analyze the type(s) of human action associated with user participation for different ICT4D project development approaches, which then helps to identify a set of particular conditional factors that affect the ability of the human actors involved in the development of ICT4D projects to perform the action(s) associated with each approach. Both concepts in Habermas's TCA are described next.

4.1 Typology of Human Action

As a social theorist, Habermas is interested in how individual actions can be coordinated into pattern of interactions. In TCA, Habermas derives a typology of human action based on the observation of two human tendencies or orientations (Habermas 1984). The other distinction among different types of action tracks Max Weber's notions of "social" and "non-social" action—where "social action" means action in which the actor takes account of the behavior of others and orients his/her action accordingly (see Table 4.3).

The first type of action is based on the tendencies toward achieving own success, which is called the "purposive-rational action." The purposive-rational action is the action directed toward the achievement of given objectives. Success is measured by how nearly one achieves an objective. If the action is an intervention in the physical world and is achieved by following technical rules (or the non-social domain), the action is called "instrumental action." If the purposive-rational action is taken by considering the impact of the action on social situations or other actors who may engage in counteraction, the action is called "strategic action."

The second type of human action is based on human orientation to achieve mutual agreement. Here, the success orientation is replaced by a desire to

Table 4.3 Habermas's Basic Typology of Human Action (Habermas 1984)

Domains of action	*Type of action*		*Type of interaction*
	Purposive-rational or teleological One or more actors are oriented toward their own success	*Communicative Actors (at least two) are oriented toward mutual agreement*	
• Non-social	• Instrumental action	• N/A	• Technical Action
• Social	• Strategic action	• Communicative action • Discursive action	• Social Interaction

understand a communicating partner. There are two types of action in this orientation: communicative and discursive action. In communicative action, people reach agreement by having a common background of values, shared norms, conventions, habits, and assumptions about the world. The discursive action is oriented toward achieving clarification and justification of claims by providing reasons and evidence. It places an emphasis on the concept of argumentation where various assumptions must be carefully examined, clarified, and tested.

Habermas supports the view that all actions might be strategic in some sense because all actors pursue particular aims of their own (Outhwaite 2009). However, the mechanisms of coordination are different (Baxter 2011). The purposive-rational action (i.e., instrumental and strategic action) specifies coordination based on calculation of cost-benefits and consideration of the actor's self-interests. On the other hand, communicative and discursive action emphasizes cooperative processes as the mechanism for coordination. Habermas argues that the cooperative process of interpretation requires complex coordination of individual acts as well as their self-interests. Therefore, the concept of coordination separates the purposive-rational action from the communicative and discursive action.

4.2 Action Constitutive Resources

Each type of action maintains a specific orientation, which assumes a set of resources and understanding of the rules governing social interactions and procedures for enacting the action within specific institutional practices. The basic resources that an actor needs to skillfully perform the action within specific institutional contexts are called "action constitutive resources," which are discussed next.

Instrumental and strategic action is success-oriented. However, instrumental action focuses on control, manipulation, and transformation of the physical object. Hence, for the enactment of this type of action, the actor will depend upon the technical knowledge of input-output relationships and the tools needed to achieve the given ends. On the other hand, strategic action focuses on transforming the behavior of the other human actor. The actors who engage in strategic action also recognize that their opponent may engage in intelligent counteraction. Hence, they need to have an understanding of the feasible outcomes. In addition, the actors need to recognize their opponent's goals, positions, and potential for counteraction (Ngwenyama and Lee 1997). The transformation of the behavior of other human actors consequently needs domination of one actor over the others. This power can be attained through possession of social resources, such as social status, authority/power, and items of exchange value (e.g., time, expertise) (Habermas 1984).

Communicative and discursive action aims to achieve agreement among the participating actors. Both actions presuppose the existence of shared media for communication. Communicative action is generally oriented toward maintaining mutual understanding among participating actors engaged in coordinated action. It is enacted via language and other forms of symbolic interaction to seek possible bases for agreements and compromises, as well as for interpretations of shared norms, values, and the meanings of situated action. When the validity of the agreement is challenged, the actors enter the discursive action, which is oriented toward restoring agreement and condition for coordinated action. Discursive activity manifests through critical debate and argumentation. These two activities form the basis for joint decision making and agreement. Habermas argues that the effectiveness of discursive activity depends on the existence of rules of discourse and critical debate, as well as on the tools of analysis and evaluation of alternative arguments (Habermas 1984).

5. Conceptual Framework Development

In this section, we analyze the type(s) of human action associated with user participation for different ICT4D project development approaches, which helps identify a set of particular conditional factors that affect the ability of human actors involved in the development of ICT4D projects to perform the action(s) associated with each approach.

5.1 Human Action(s) Involved in Different ICT4D Project Development Approaches

Depending on the chosen approach in the development of ICT4D projects, certain human actions would be more prominent than others, as explained next. Table 4.4 summarizes the propositions (P1-P3) that suggest the human actions associated with each approach.

With the AD approach, the project management is responsible for developing ICT4D services without any inputs from the community (OECD 2001; Puri and Sahay 2003a, 2007). From the perspective of TCA, the action taken by the involved management follows the purposive-rational

Table 4.4 Human Action/Interaction Involved in User Participation for Each ICT4D Project Development Approach Based on Habermas's TCA

ICT4D Project Development Approach	*Type of Human Action/Type of Interaction*				*Proposition*
	Purposive-rational or teleological One or more actors are oriented toward their own success		*Communicative Actors (at least two) are oriented toward mutual agreement*		
	Instrumental/ (Technical Action)	*Strategic/ (Social Interaction)*	*Communicative/ (Social Interaction)*	*Discursive / (Social Interaction)*	
AD	✓	✓	✓	✓	P1
SD	✓	✓	✓	✓	P2
UD	✓	✓	✓	✓	P3

orientation. Two types of human action are involved here. Instrumental action is needed to create IC4TD or to perform manipulation on the hardware/software. Besides instrumental action, management needs to interact with the community to promote the ready-made ICT4D services. This is a form of strategic action, which is performed with the aim to influence and transform the behavior of the community to conform to the management's desires or goals (OECD 2001). Therefore, the following proposition is formulated:

P1: The AD approach to ICT4D projects involves user participation in terms of instrumental/technical action and strategic social interaction.

With the SD approach, an ICT4D project is run in consultation between the project management and the community. This indicates that there is an action associated with communicative orientation (i.e., communicative action). In SD projects, the consultation normally aims only to gather community inputs about the development process (Puri and Sahay 2003a). The management still maintains the authority to make the final decision. This is what Habermas refers to as a breakdown in communicative action, which occurs due to misunderstanding or possible differing opinions. In this situation, the actors can either pursue their own ends by choosing strategic action or enter discursive action (Habermas 1984). For the SD approach, obviously, the management choses strategic action, and, therefore, abandons discursive action. Instrumental action is still required for manipulating hardware/software related to IC4TD services. Therefore, the following proposition is formulated:

P2: The SD approach to ICT4D projects involves user participation in terms of instrumental/technical action as well as strategic and communicative action.

When the UD approach is implemented, the development of ICT4D projects is based on the agreement between project management and users (OECD 2001). The discussion (i.e., communicative action) is chosen as a way to reach an agreement. Any disagreement is resolved by continuing debate or even voting in order to reach the best possible solution (Puri and Sahay 2003b). From the perspective of TCA, both project management and users take discursive action instead of strategic action that can only benefit certain actors. Like the two previous two approaches, instrumental action is still required during the development of ICT4D services. Therefore, the following proposition is formulated:

P3: The UD approach to ICT4D projects involves user participation in terms of instrumental/technical action as well as communicative and discursive action.

5.2 Conditional Factors Affecting User Participation for Each ICT4D Project Development Approach

Action constitutive resources described in Habermas's TCA can be related to conditional factors affecting user participation, since conditional factors provide different types of resources for target users in ICT4D projects either to enact technical action (i.e., instrumental action) or to facilitate social interaction (i.e., strategic, communicative, and discursive action) between users and the project staff. This interpretation is accomplished by examining the list of conditional factors that have been reported in the literature (Table 4.2). The examination of this list of conditional factors is intended to reveal a particular type of action that each conditional factor enables.

First, the ability to perform an instrumental action is affected by technical factors. As listed in Table 4.5, these factors include the degree of task complexity given to users and the availability of easy-to-use technology. Here, task complexity to perform an instrumental action is related to technical knowledge required to produce a product/object in ICT4D projects. This product may take the form of a shared-database, web-blog, agricultural information web portal, or e-literacy training module. The degree of task complexity is important, since participation requires knowledge of procedures, processes, and the context in which the system is developed (McKeen et al. 1994). In addition, the availability of easy-to-use technology is essential to provide a tool to target users in ICT4D projects, who generally have little experience with ICTs to perform instrumental action. This tool helps users to participate in the project (Butler and Fitzgerald 1997; Cavaye 1995; Doll 1987). For example, a special module may be created for users with disabilities, or an operating system may be created using the local language (Forestier et al. 2002; Kenny 2002). Therefore, the following proposition is suggested:

P4: In technical action, Degree of task complexity given to users (T1) and Availability of easy-to-use technology (T2) affect user participation in ICT4D projects.

Second, looking through the lens of Habermas's typology of human action, to perform a strategic action toward users, project staff rely on the resources at their own disposal to try to persuade a user to participate in an ICT4D project. Such resources are what Habermas describes "social and material resources that are involved in the generation of power and dominion of some actors over others" (Ngwenyama and Lyytinen 1997: 76). Habermas contends that such resources may include charisma, social status, authority, time, and financial resources or items of exchange value (Habermas 1984). Hence, the time availability of both project staff and users, and the availability of pre-established funding for participation activities are important factors for strategic action to be enacted. Staff also relies on the direct involvement of government/donor officers to provide political support and legitimacy and to encourage users to participate. A government/donor officer is a principal stakeholder in an ICT4D project. Symbolic actions of support by governments/donor agencies contribute to the legitimization of project development and may persuade the community to expend the effort required during project development (Madon et al. 2007).

The incentive given to users is another important conditional factor that presupposes strategic action. The literature on user participation has argued that the perceived system impact is an important factor since any goal conflicts are posited to be an impediment for participation activities. The expected system impact is defined as the expected organizational changes or impact brought by the new system (Cavaye 1995; Damodaran 1996; Lu and Wang 1997; Mao and Pan 2009; Nasirin et al. 2005). Based on organizational change theory, the literature posits that people are generally used to the status quo and any disruptions may evoke human resistance (Tait and Vessey 1988). However, when the perceived impacts are in line with their goals, users are more willing to participate (Lin and Shao 2000). This is in line with what Habermas puts forward as "the actor's knowledge of what is feasible to achieve, and opponents' goals" (Ngwenyama and Lyytinen 1997: 76). Based on the argument made earlier, the following proposition is proposed:

P5: In strategic action, Constraints on the project duration approved by the donor agency (O1), Constraints on the amount of project funding approved by the donor agency (O2), Direct involvement of a government/donor officer (O3), and Incentives given to users (O6) affect user participation in ICT4D projects.

Third, to perform communicative action, the familiarity between project staff and users provides what Habermas posits as knowledge of language, shared norms, and action situation (Habermas 1984) to perform communicative action. For staff, such capability can only be obtained if they have developed familiarity with the target users or a high degree of people-oriented skills. Therefore, the familiarity between project staff and users allows staff to perform the strategic action described earlier, as well as communicative action with users. The project staff must have "people-oriented skills," defined as the extent to which managerial behavior results in mutual trust, friendship, respect, and warmth (Lin and Shao 2000). The people-oriented skills consider interpersonal relations and accept individual differences.

Furthermore, to perform communicative action, users need to have experiences with and positive attitudes about the ICT4D project. User experience with the ICT4D project is not about technical skills, but knowledge of the system to be developed (i.e., the ICT4D project), the task involved, and the environment within which the ICT4D project operates. Users with these experiences can overcome the knowledge gap for effective communicative

Table 4.5 Human Action and Associated Conditional Factors

Type of human action	*Action orientation*	*Action constitutive resources*	*Conditional factors affecting user participation*	*Proposition*
Technical Action / Instrumental action	Purposive-Rational Transformation, Manipulation, and Control of Objects	Technical knowledge Tools	a. Degree of task complexity given to the users (T1) b. Availability of easy-to-use technology (T2)	P4a-b
Social Interaction/ Strategic action	Purposive-Rational Influencing and Transforming the behavior of others	Authority/power Items of exchange value Knowledge of the rules of process and opponent	a. Constraint on the project duration approved by donor agency (O1) b. Constraint on the amount of project funding approved by the donor agency (O2) c. Direct involvement of government/donor officer (O3) d. Incentives given to the users (O6)	P5a-d
Social Interaction/ Communicative action	Communicative Maintaining understanding and coordinating action	Shared norms Knowledge of organizational context	a. User experience with the ICT4D project (U1) b. User-shared attitude about the ICT4D project (U2) c. Familiarity between telecenter staff and users (O4) d. Support from community champion (O5)	P6a-d
Social Interaction/ Discursive action	Communicative Restoring agreement and conditions for coordinated action	Knowledge of rules of discourse & critical debate, evaluation protocols	a. Support from community champion (O5)	P7

action with the staff, which enables meaningful user participation (Cavaye 1995). Hence, actors' knowledge of organizational context determines how well the actors can perform the communicative action.

Understanding of the system under development affects users' ability to participate in IS development (Mattia and Weistroffer 2008). Likewise, users' favorable attitude toward system development is positively related to their participation in IS development (Cavaye 1995; Damodaran 1996; Hartwick and Barki 1994; Hunton and Price 1997; Iivari and Igbaria 1997; Nasirin et al. 2005). The term "user attitude" concerns user perception of how good or bad the system is perceived to be (Damodaran 1996; Hartwick and Barki 1994; Nasirin et al. 2005; Saleem 1994; Tait and Vessey 1988). Barki and Hartwick (1994) demonstrate the importance of user attitude toward shared norms. Shared norms are important to achieve consensus, since an individual who holds certain beliefs thinks that others hold the same beliefs. In communicative action, people reach understanding through having a common background of assumptions about the world.

In addition, support from community champions is important for users engaging in communicative action. The literature argues that user champions have a vital part to play in helping to drive user participation and become lively promoters of the system among their peers (Nasirin et al. 2005). Prior work has also recognized champions as allies who use their power and influence to help ventures navigate the complex socio-political maze within their organizations (Vadapalli and Mone 2000). In the context of ICT4D projects, these individuals are what Colle (2005) calls "individuals who can translate and demonstrate the relevance and application of information and communication technologies to the realities of the community" (p. 11). The earlier discussion leads to the following proposition:

P6: In communicative action, Familiarity between project staff and users (O4), Support from the community champion (O5), User experience with the ICT4D project (U1), and User-shared attitude about the ICT4D project (U2) affect user participation in ICT4D projects.

Finally, in order for discursive action to take place between staff and users, the literature has identified that the ICT4D project needs to provide a platform for open and free dialogue between target users, staff, government/donor agencies, and other parties (Puri et al. 2004). This platform should provide users and staff with opportunities to express doubts and argue, as well as to query about each other's opinion. In the context of ICT4D projects, such platforms are provided by the village council, which is "a body constituted of all adult members of the village" (Puri and Sahay 2003b: 188), also known as the community council. Such a platform is generally provided if the development of the ICT4D project has the full support from community champions, including the village head, religious leaders, and other important stakeholders. Here, the community council maintains rules of discourse and critical debate as well as evaluation protocols needed to perform discursive action. The following proposition is hence formulated:

P7: In terms of discursive action, Support from the community champion (O5) affects user participation in ICT4D projects.

6. Conclusion

This chapter discusses the importance of ICT4D initiatives to enhance social inclusion of disadvantaged communities. It highlights the need for achieving an optimal level of user participation based on the chosen project development approach for maximizing social inclusion. The appropriate user participation level for each development approach is important to invoke the required stakeholder engagement. Excessive and unnecessary involvement of stakeholders may lead to frustration among the community and may have a negative impact on project success. Likewise, inadequate participation may lead to the lack of interest among the target users, which will impede the achievement of project success and social inclusion objectives. Therefore, managing the fit between the desired user participation inscribed by each development approach and the actual level of participation affected by a number of conditional factors is critical for project success rather than simply maximizing the actual level of participation alone (King and Lee 1991). User participation equilibrium is achieved when the optimal user participation level for a given ICT4D approach is obtained.

The central argument of this chapter is that issues related to user participation in the development of ICT4D projects are caused by difficulties in achieving the equilibrium condition. The development of ICT4D projects may employ different approaches, each of which implies a different degree of user participation. The existing frameworks, although useful to identify conditional factors affecting user participation level, lack explanatory power when different ICT4D project development approaches are taken into consideration. This chapter thus offers a contribution by enhancing the existing knowledge about ways to achieve participation equilibrium for different approaches to ICT4D projects. In particular, it innovatively applies the human action perspective to understand the two competing influencers that affect achievement of participation equilibrium in the development of ICT4D projects. The proposed framework will help researcher to further investigate conditional factors for different approaches to ICT4D project development.

The knowledge gained through this research is also useful for practice. The identification of a specific set of conditional factors helps increase ICT4D project success. With such knowledge, the Non-Government Organizations (NGOs) staff, ICT4D practitioners, and project managers can focus on the important "set" of conditional factors to achieve an equilibrium condition for each desired ICT4D project development approach.

However, the framework only outlines a set of conjectures on the possible conditional factors affecting user participation for different ICT4D project development approaches. Hence, it needs to be further refined through empirical investigations. Multiple case studies involving ICT4D projects with different project development approaches can be adopted. It will also be useful to conduct similar studies in different countries to improve generalizability.

References

Akinnuwesi, Boluwaji A., et al. "An Empirical Analysis of End-User Participation in Software Development Projects in a Developing Country Context." *The Electronic Journal of Information Systems in Developing Countries* 58, no. 6 (2013): 1–25.

Avgerou, Chrisanthi. "The Significance of Context in Information Systems and Organizational Change." *Information Systems Journal* 11 (2001): 43–63.

Avgerou, Chrisanthi, and Shirin Madon. "Information Society and the Digital Divide Problem in Developing Countries." In *Perspectives and Policies on ICT in Society*, edited by Jacques Berleur and Chrisanthi Avgerou, 205–18. New York: Springer, 2005.

Avgerou, Chrisanthi, and Geoff Walsham, eds. *Information Technology in Context: Studies from the Perspective of Developing Countries*.. London: Ashgate.

Bailey, Arlene. "Issues Affecting the Social Sustainability of Telecentres in Developing Contexts: A Field Study of Sixteen Telecentres in Jamaica." *The Electronic Journal on Information Systems in Developing Countries* 36, no. 4 (2009): 1–18.

Bailey, Arlene, and Ojelanki Ngwenyama. "The Challenge of E-Participation in the Digital City: Exploring Generational Influences among Community Telecentre Users." *Telematics and Informatics* 28 (2011): 204–14.

Bailur, Savita. "The Complexities of Community Participation in Rural Information Systems Projects: The Case of "Our Voices"." Paper presented at the 9th International Conference on Social Implications of Computers in Developing Countries, Sao Paulo, Brazil, 2007.

Bailur, Savita. "Using Stakeholder Theory to Analyze Telecenter Projects." *Information Technologies and International Development* 3, no. 3 (2007): 61-80.

Bano, Muneera, and Didar Zowghi. "A Systematic Review on the Relationship Between User Involvement and System Success." *Information and Software Technology* 58 (2015): 148–69.

Baxter, Hugh. *Habermas: The Discourse Theory of Law and Democracy*. Stanford, CA: Stanford University Press, 2011.

Benston, Margaret. "Participatory Design by Non-Profit Groups." In *Pdc '90 Conference on Participatory Design*, edited by Aki Namioka and Douglas Schuler. 107-13. Palo Alto, CA: Computer Proffesionals for Social Responsibility, 1990.

Bhatnagar, Subhash. "Development and Telecommunnications Access: Cases From South Asia." In *Information Systems and the Economics of Innovation*, edited by Chrisanthi Avgerou and Renata Lebre La Rovere, 33–52. Cheltenham: Edward Elgar Publishing Limited, 2003.

Butler, Tom, and Brian Fitzgerald. "A Case Study of User Participation in the Information Systems Development Process." *Proceedings of the Eighteenth International Conference on Information Systems*, Atlanta, 1997.

Byrne, Elaine, and Sundeep Sahay. "Participatory Design for Social Development: A South African Case Study on Community-Based Health Information Systems." *Information Technology for Development* 13, no. 1 (2007): 71–94.

Carroll, John M, and Mary Beth Rosson. "Participatory Design in Community Informatics." *Design Studies* 28 (2007): 243–61.

Cavaye, Angele L. M. "User Participation in System Development Revisited." *Information & Management* 28 (1995): 311–23.

Colle, Royal D. "Memo to Telecenter Planners." *The Electronic Journal on Information Systems in Developing Countries* 21, no. 1 (2005): 1–13.

Damodaran, Leela. "User Involvement in the Systems Design Process—a Practical Guide for Users." *Behaviour & Information Technology* 1996, no. 15 (1996): 6.

DiMaggio, Paul, and Eszter Hargittai. "From the 'Digital Divide'to 'Digital Inequality': Studying Internet Use as Penetration Increases." *Princeton: Center for Arts and Cultural Policy Studies, Woodrow Wilson School, Princeton University* 4, no. 1 (2001): 4-2.

Doll, William J. "Encouraging User Management Participation in Systems Design." *Information & Management* 13, no. 1 (1987): 25–32.

Doll, William J., and Gholamreza Torkzadeh. "A Discrepancy Model of End-User Computing Environment." *Management Science* 35, no. 10 (1989): 1151–71.

Forestier, Emmanuel, et al. "Can Information and Communication Technologies Be Pro-Poor?" *Telecommunications Policy* 26, no. 11 (2002): 623–46.

Habermas, Jurgen. *The Theory of Communicative Action: Reason and the Rationalization of Society*, 2 vols., Vol. 1. Boston, MA: Beacon Press.

Harris, Mark A., and Heinz Roland Weistroffer. "A New Look at the Relationship Between User Involvement in System Development and System Success." *Communication of the Association for Information Systems* 24, no. 42 (2009): 739–56.

Harris, Roger W. "How ICT4D Research Fails the Poor." *Information Technology for Development* 22, no. 1 (2015): 177–92.

Harris, Roger W., et al. "Sustainable Telecentres? Two Cases from India." In *The Digital Challenge: Information Technology in the Development Context*, edited by S. Krishna and S. Madon, 103–23. Aldershot: Ashgate Publishing, 2003.

Hartwick, John, and Henri Barki. "Explaining the Role of User Participation in Information System Use." *Management Science* 40, no. 4 (1994): 440–65.

He, Jun, and William H. King. "The Role of User Participation in Information Systems Development: Implications from a Meta-Analysis." *Journal of Management Information Systems* 25, no. 1 (2008): 301–31.

Hearn, Greg, Megan Kimber, June Lennie, and Lyn Simpson. "A Way Forward: Sustainable Icts and Regional Sustainability." *The Journal of Community Informatics* 1, no. 2 (2005): 18-31.

Heeks, Richard. "The Tyranny of Participation in Information Systems: Learning from Development Projects." In *Working Paper Serics, Paper No.4*. Manchester: Institute for Development Policy and Management, University of Manchester, 1999.

Heeks, Richard. "Information Systems and Developing Countries: Failure, Success, and Local Improvisations." *The Information Society* 18 (2002): 101–12.

Heeks, Richard. "ICT4D: The Next Phase of Applying ICT for International Development." *Computer* 41, no. 6 (2008): 26–33.

Heeks, Richard. "Do Information and Communication Technologies (ICTs) Contribute to Development." *Journal of International Development* 22 (2010): 625–40.

Hirschheim, Rudy, et al. "Information Systems Development as Social Action: Theoretical Perspective and Practice." *Omega* 19, no. 6 (1991): 587–608.

Hirschheim, Rudy, et al. *Information Systems Development: Conceptual and Philosophical Foundations*. Cambridge: Cambridge University Press, 1995.

Hirschheim, Rudy, et al. "Exploring the Intellectual Structures of Information Systems Development: A Social Action Theoretic Analysis." *Accounting, Management and Information Technologies* 6, no. ½ (1996): 1–64.

Hirschheim, Rudy, and Heinz K. Klein. "Four Paradigms of Information Systems Development." *Communication of ACM* 32, no. 10 (1989): 1199–216.

Hunton, James E., and Kenneth H. Price. "Effects of the User Participation Process and Task Meanigfulness on Key Information System Outcomes." *Management Science* 43, no. 6 (1997): 797–812.

Iivari, Juhani, et al. "The User—the Great Unknown of Systems Development: Reasons, Forms, Challenges, Experiences and Intellectual Contributions of User Involvement." *Information Systems Journal* 20, no. 2 (2010): 109–17.

Iivari, Juhani, and Magid Igbaria. "Determinants of User Participation: A Finnish Survey." *Behaviour & Information Technology* 16, no. 2: 111–21.

Isomäki, Hannakaisa, and Samuli Pekkola. "Introduction: Reframing Humans and Information Systems." In *Reframing Humans in Information Systems Development*, edited by Hannakaisa Isomäki and Samuli Pekkola, 1–14. London: Springer, 2011.

Ives, Blake, and Margrethe H. Olson. "User Involvement and MIS Success: A Review of Research." *Management Science* 30, no. 5 (1984): 586–603.

James, Jeffrey. *Information Technology and Development: A New Paradigm for Delivering the Internet to Rural Areas in Developing Countries*. Abingdon, Oxfordshire: Routledge, 2004.

Kanungo, Shivraj. "On the Emancipatory Role of Rural Information Systems." *Information Technology & People* 17, no. 4 (2004): 407–22.

Karlsson, Fredrik, et al. "Exploring User Participation Approaches in Public e-Service Development." *Government Information Quarterly* 29, no. 2 (2012): 158–68.

Kenny, Charles. "Information and Communication Technologies for Direct Poverty Alleviation: Costs and Benefits." *Development Policy Review* 20, no. 2 (2002): 141–57.

King, William R., and Tsang-Hsiu Lee. "The Effects of User Participation on System Success: Toward a Contingency Theory of User Satisfaction." In *Proceeding ICIS'91 Proceedings of the Twelfth International Conference on Information Systems*, 1991.

Klein, Heinz K., and Minh Q. Huynh. "The Critical Social Theory of Jurgen Habermas and Its Implications for IS Research." In *Social Theory and Philosophy for Information Systems*, edited by John Mingers and Leslie Willcocks, 157–237. West Sussex: John Wiley & Sons, 2004.

Kumar, Rajendra, and Michael Best. "Impact and Sustainability of E-Government Services in Developing Countries: Lesson Learned from Tamil Nadu, India." *The Information Society* 22 (2006a): 1-12.

Kumar, Rajendra, and Michael Best. "Social Impact and Diffusion of Telecenter Use: A Study from Sustainable Access in Rural India Project." *The Journal of Community Informatics* 2, no. 3 (2006b): 1–21.

Lin, Winston T., and Benjamin B. M. Shao. "The Relationship Between User Participation and System Success: A Simultaneous Contingency Approach." *Information & Management* 37 (2000): 283–95.

Lu, Hsei-Peng, and Jyun-Yu Wang. "The Relationships Between Management Styles, User Participation, and System Success over MIS Growth Stages." *Information & Management* 32 (1997): 203–13.

Lyytinen, Kalle. "Information Systems and Critical Theory." In *Critical Management Studies*, edited by Mats Alvesson and Hugh Willmott, 159–180. London, Newbury Park, PA and New Delhi: Sage Publications, 1992.

Madon, Shirin. "Evaluating the Developmental Impact of e-Governance Initiatives: An Exploratory Framework." *The Electronic Journal on Information Systems in Developing Countries* 20, no. 5 (2004): 1–13.

Madon, Shirin. "Governance Lessons from the Experience of Telecentres in Kerala." *European Journal of Information Systems* 14 (2005): 401-16.

Madon, Shirin, et al. "Digital Inclusion Projects in Developing Countries Process of Institutionalization." In *Preceedings of the 9th International Conference on Social Implications of Computers in Developing Countries*, 2007.

Mao, Ji-Ye, and Mianzhen Pan. "Enabling Effective User Participation in ERP Implementation: A Case Study on the Role of Brainstorming Sessions." In *Pasific Asia Conference on Information Systems*, 2009.

Mattia, Angela, and Heinz Roland Weistroffer. "Information Systems Development: A categorical analysis of user participation approaches." In *Proceedings of the 41st Hawaii International Conference on System Sciences*, 2008.

McKeen, James D., et al. "The Relationship Between User Participation and User Satisfaction: An Investigation of Four Contingency Factors." *MIS Quarterly* 18, no. 4 (1339): 427–51.

Merkel, Cecelia B, Umer Farooq, Lu Xiao, Craig H Ganoe, Mary Beth Rosson, and John M Carroll. "Managing Technology Use and Learning in Nonprofit Community Organisations: Methodological Challenges and Opportunities." In *CHI-MIT'07*. Cambridge, MA, U.S.A, 2007.

Mingers, John. "Technical, Practical and Critical OR Past, Present, and Future?" In *Critical Management Studies*, edited by Mats Alvesson and Hugh Willmott, 90–112.. London, Newbury Park, PA and New Delhi: Sage Publications, 1992.

Mumford, Enid. "Participative Systems Design: Practice and Theory." *Journal of Occupational Behaviour* 4, no. 1 (1983): 47–57.

Nasirin, Syed, et al. "Factors Influencing User Involvement in DSS Projcet Implenmentation: Some Lessons from the UK Health Sector." In *European Conference on Information Systems (ECIS) 2005 Proceedings*, 2005.

Ngwenyama, Ojelanki K., and Allen S. Lee. "Communication Richness in Electronic Mail: Critical Social Theory and the Contextuality of Meaning." *MIS Quarterly* 21, no. 2 (1997): 145–67.

Ngwenyama, Ojelanki K., and Kalle Lyytinen. "Groupware Enviroments as Action Constitute Resources: A Social Action Framework for Analysing Groupware Technologies." *The Journal of Collaborative Computing* 6 (1997): 71–93.

OECD. *Citizens as Partners: Information, Consultation, and Public Participation in Policy-Making*. PUMA, OECD, 2001.

Oestmann, Sonja, and Andrew C. Dymond. "Telecentres—Experiences, Lessons and Trends." In *Telecentres: Case Studies and Key Issues*, edited by Colin Latchem and David Walker, 1–16. Vancouver: The Commonwealth of Learning, 2001.

Outhwaite, William. *Habermas: A Critical Introduction*. Cambridge: Polity, 2009.

Puri, Satish K., et al. "Contextuality of Participation in IS Design: A Developing Country Perspective." In *Participatory Design Conference Toronto*, Canada, 2004.

Puri, Satish K., and Sundeep Sahay. "Institutional Structures and Participation: Comparative Case Studies from India." In *Organization Information Systems in the Context of Globalization: IFIP TC8/WG8.2 & WG9.4 Working Conference on Information Systems Perspectives and Challenges in the Context of Globalization*, Athens, Greece, edited by Mikko Korpela, Ramiro Montealegre and Angeliki Poulymenakou, June 15–17, 2003a.

Puri, Satish K., and Sundeep Sahay. "Participation Through Communicative Action: A Case Study of GIS for Addressing Land/Water Development in India." *Information Technology for Development* 10 (2003b): 179–99.

Puri, Satish K., and Sundeep Sahay. "Role of ICTs in Participatory Development: An Indian Experience." *Information Technology for Development* 13, no. 2 (2007): 133–60.

Ramirez, Ricardo, Helen Aitkin, Galin Kora, and Donald Richardson. "Community Engagement, Performance Measurement, and Suistainability." *Canadian Journal of Communication* 30 (2005): 259-79.

Rao, Siriginidi Subba. "Social Development in Indian Rural Communities: Adoption of Telecentres." *International Journal of Information Management* 28 (2008): 474–82.

Roman, Raul, and Royal D. Colle. "Creating Participatory Telecenter Enterprise." In the *Participatory Communication Research Section in the annual meeting of International Association for Media and Communication Research*, Barcelona, 2002.

Roman, Raul, and Royal D. Colle. "Content Creation for ICT Development Projects: Integrating Normative Approaches and Community Demand." *Information Technology for Development* 10, no. 2 (2003): 85–94.

Roode, Dewald, et al. "It's Not the Digital Divide—It's the Socio-Techno Divide." In *Proceedings of the 12th European Conference on Information Systems*, 2004.

Saleem, Naveed. "Alternative Perspectives of User Participation: Practical Implications." *ACM SIGCPR Computer Personnel* 15, no. 2 (1994): 25–31.

Schuler, Douglas, and Aki Namioka, eds. *Participatory Design: Principles and Practices*.Hillsdale, NJ: L. Erlbaum Associates, 1993.

Tait, Peter, and Iris Vessey. "The Effect of User Involvement on System Success: A Contingency Approach." *MIS Quarterly* 12, no. 1 (1988): 91–108.

Tipson, Frederick S., and Claudia Fritteli. *Global Digital Opportunities: National Strategies of ICT for Development*, 2003.

Toyama, Kentaro. "Human-Computer Interaction and Global Development." *Foundation and Trends in Human Computer Interaction* 4, no. 1 (2010): 1–79.

Tsatsou, Panayiota. "Digital Divides Revisited: What Is New About Divides and Their Research?" *Media Culture and Society* 33, no. 2 (2011): 317.

Unwin, Tim. *Information and Communication Technology for Development*. Cambridge: Cambridge University Press, 2009.

Vadapalli, Anand, and Mark A. Mone. "Information Technology Project Outcomes: User Participation Structures and the Impact of Organisation Behavior and Human Resource Management Issues." *Journal Engineering and Technology Management* 17 (2000): 127–51.

Van Belle, Jean-Paul, and Jonathan Trusler. "An Interpretivist Case Study of a South African Rural Multi-Purpose Community Centre." *The Journal of Community Informatics* 1, no. 2 (2005): 140–57.

Van Dijk, Jan, and Kenneth Hacker. "The Digital Divide as a Complex and Dynamic Phenomenon." *The Information Society* 19, no. 4 (2003): 315–26.

Vroom, Victor H., and Philip W. Yetton. *Leadership and Decision-Making*. Pittsburgh: University of Pittsburgh Press, 1973.

Walsham, Geoff, and Sundeep Sahay. 2006. "Research on Information Systems in Developing Countries: Current Landscape and Future Prospects." *Information Technology for Development* 12, no. 1 (2006): 7–24.

Warschauer, Mark. *Technology and Social Inclusion: Rethinking the Digital Divide*, 2004.Weiner, Daniel, Trevor M Harris, and William J Craig. "Community Participation and Geographic Information Systems." In *Community Participation and Geograhic Information Systems*, edited by William J Craig, Trevor M Harris and Daniel Weiner. London and New York: Taylor & Francis, 2002.

Part II

ICT-Enabled Services of Value to Society and Organizations

5 Understanding the Impact of Politication Structure, Governance, and Public Policy on E-Government

David J. Yates, Girish J. "Jeff" Gulati, and Christine B. Williams

1. Introduction

E-government offers citizens the potential for greater access to their representatives and offers policy makers the ability to make citizen-to-government contact more inclusive.[1] In the developed world government portals and websites include more advanced technical features, and a large percentage of national governments regularly update their e-government offerings (Dwivedi, Weerakkody, and Janssen 2011; United Nations 2014). While we see the delivery of online public services becoming an established practice (Osborne, Radnor, and Nasi 2013), countries are less successful in implementing e-participation. Indeed, less than half of UN member countries provided a facility on their national portals for citizen feedback concerning the improvement of their online services in 2013 (United Nations 2014: 68–9). Of the countries that provided such a facility, only a limited number offered services that enabled citizens to be active participants in government decision-making and the policy-making process.

Even though the number of countries that are expanding initiatives to further public sector efficiencies and transparency with new ICTs and applications is increasing (Krishnan, Teo, and Lim 2013; United Nations 2014), a vast body of scholarly research has documented the significant global digital divide between high-income and low-income countries in diffusion of ICTs (Norris 2001; Pick and Sarkar 2015). This study seeks to analyze the impact of potential explanatory factors on recent United Nations' global measurement of (1) online government services and (2) electronic participation capabilities (United Nations 2014). The remainder of this section develops our hypotheses on the impact that political structure, public sector performance, and specific public policy initiatives have on a nation's e-government capability in the form of government services offered and of opportunities for citizens to participate in government decision making online. We then present our measures and analyses that estimate the effects of structure, performance, and policy variables on two measures of e-government at the national level. Section 3 describes the factors that affect the UN online service index (OSI) and e-participation index (EPI), which reveal both similarities and differences in the two sets of findings. Then we discuss our conclusions and their implications for e-government and e-participation.

Government agencies and political bodies use ICTs to improve the availability of information and delivery of services to citizens and to facilitate their communication with business and industry. These developments in turn improve efficiency and transparency in government (Weerakkody et al. 2015). ICTs also offer citizens the potential for greater access to government officials and offer policy makers the ability to make citizen-to-government contact more inclusive. For the less fortunate and more isolated members of society, advances in ICTs can help overcome the geographical, institutional, and social barriers to information and give marginalized groups a voice in the political sphere.

While a few scholars find the potential of e-government to have been elusive (Hindman 2008; Sunstein 2007), the greater concern is that a technological divide is emerging both between nations and within nations, creating groups of "information-rich" and "information-poor" societies. Even more disconcerting is that this divide seems to overlap considerably with the economic divide that already separates developed and developing countries and, thus, is reinforcing or even widening existing economic, political and social inequalities between the haves and have-nots (Norris 2001; Pick and Sarkar 2015).

E-government surveys provide data that show that the concerns of the pessimists have not been unwarranted (United Nations 2014; West 2008). Among the 25 global leaders in the implementation of e-government in 2013 (United Nations 2014), 20 are from either North America, Europe, or Oceana and five are from high-income countries in East Asia. Most low-income nations rank within the bottom tier. African nations, moreover, have showed slow and uneven improvement over the past few years. Sixteen of the 54 African countries are at the bottom 10% of the world E-Government Development Index (EGDI) ranking. This digital divide also appears along the lines of income and region for offerings of online government services (Chatfield and Alhujran 2009; Gascó 2005; Helbig, Gil-García and Ferro 2009) as well as in the opportunities for citizen e-participation (Boudjelida, Mellouli, and Lee 2016; United Nations 2014).

Democratic governments should be inclined to be more inclusive in an attempt to widen their electoral appeal. Moreover, societies that have a culture of encouraging political expression and facilitating communication between government and citizens are more likely to demand that their governments use technology to provide more transparency and avenues for participation (Lijphart 1999; Weerakkody et al. 2015). Recent studies suggest that e-government has developed more rapidly in countries with stronger democratic institutions (Azad et al. 2010; Gulati and Yates 2011; Stier 2015). Each of these studies used a different indicator for measuring political structure, however, demonstrating that a more careful measurement strategy is needed before making any firm conclusions about the relationship between e-government and democratic institutions.

The changing nature of the relationship between democratic political structure and e-government development (Bertot, Estevez, and Janowski 2016; Stier 2015) is illustrated by the cluster of countries that the UN has highlighted for their efforts in developing e-government resources. Countries exhibiting various levels of democratic freedoms are cited for their highly integrated national governmental portals that allow citizens quick and easy access to government information and services. Uruguay and Morocco were cited for developing leadership within their ministries and agencies that have produced significantly improved online services for their citizens since 2011. France ranked first in online service delivery in 2013 because of its excellence in all areas and stages of online service delivery, including expanded and improved mobile applications and integrated e-services. South Korea and Singapore showed similar advances. The UN also heaped praise on democracies such as Australia, Japan, and Spain for their e-government efforts, but also had praise for Bahrain, Oman, and Tunisia (United Nations 2014).

There is similar inconsistency in the relationship between democracy and more democratic participatory e-government features (Gainous, Wagner, and Abbott 2015), which is also reflected in the cluster of countries offering greater e-participation opportunities (United Nations 2014). For example, Slovenia has an "I suggest to the government" online tool to solicit, debate, and accept initiatives or proposals from citizens regarding service delivery. Citizens in Uganda can interact via U-Report, a free SMS-based social networking tool for community engagement that anonymously manages opinion poll questions, shares results, and reports responses and feedback. Although evidence about the connection between democratic institutions and processes and e-government in previous studies is mixed, a positive relationship is assumed in our first hypothesis:

H1: Countries that have a more democratic political structure will have more advanced e-government capabilities than countries that have a less democratic political structure.

If innovative e-government initiatives have been implemented successfully in a number of non-democratic countries, then factors other than resources and democratic institutions contribute to the availably of government services and participatory opportunities online (Gulati, Williams, and Yates 2014; Rose 2005; Shirazi 2013; Stier 2015). Since e-government largely is an extension of existing government institutions and administered by public sector personnel (Brown 2005; Layne and Lee 2001), it is likely that a more professional and effective public sector will offer more information and services online than a public sector that is poorly developed and inefficient (Krishnan, Teo, and Lim 2013; Moon and Norris 2005). Fountain (2001) argues, for example, that public sector organizations that are structured as Weberian bureaucracies are best able to develop technological innovations, which includes e-government applications (Osborne, Radnor, and Nasi 2013). But while these innovations are meant to enhance efficiency and functionality by facilitating improved collaboration, information sharing, and communication among organizational members, new technologies can be employed coercively to reinforce existing norms and relationships.

Evidence from U.S. municipalities supports Fountain's theoretical framework by showing that technological innovations are more likely to emerge in governments in which the managerial culture is more professional in terms of administrative personnel and procedures (Moon and Norris 2005; Norris and Moon 2005). At a cross-national level, Gulati and Yates (2011) and Stier (2015) found that government effectiveness was important in determining e-government performance. Other studies have found that poor governance, measured as the extent of corruption, reduced the quality of web-based government applications and services (Azad et al. 2010; Rose 2005). Lee, Chang, and Berry (2011) not only found similar results for corruption but also found that greater government corruption reduced the extent of e-participation development. The strong theoretical link between effective governance and e-government development coupled with the empirical evidence from these cross-national studies are the basis for our second hypothesis:

H2: Countries that have more professional public administration practices will have more extensive development of e-government than countries that have less professional administrative practices.

The recent e-government survey conducted by the United Nations highlights bridging the global digital divide as a significant challenge to successful e-government (United Nations 2014: ch. 6). In sum,

> While initially the digital divide was considered primarily an issue of access to relevant information technology infrastructure, it is increasingly about capability and ability to access and use ICT . . . the digital divide in one form or another affects people both in developed and developing countries.
>
> (United Nations 2014: 9)

King et al. (1994) emphasize the importance of supply-side and demand-side policies and regulations in bridging national and regional digital divides. Both of these sets of policy initiatives also help develop and deploy ICTs within a country (Guillén and Suárez 2005; Pick and Sarkar 2015), which in turn should increase demand for e-government capabilities (Dwivedi, Weerakkody, and Janssen 2011; Helbig, Gil-García and Ferro 2009).

On the supply-side, direct financial investment in the telecommunications sector in the form of financing relevant infrastructure projects or indirectly in the form of tax incentives to the private-sector firms that invest in research and development can be essential to bridging the various digital divides (Norris 2001; Pick and Sarkar 2015). The difficulty of parsing out how much of a nation's investment and spending is the result of public sector contributions has made it difficult to test empirically its impact on the development of e-government. Attempts to estimate the impact of investment indirectly with relatively crude indicators have found a positive relationship between Internet penetration and the amount of spending on research and development in a country (Norris 2001) and a positive relationship between e-government development and the number of scientists residing in a country (West 2005).

Privatization, competition, and deregulation of the telecommunications industry are seen as critical for the bridging of digital divides. Studies show that effective demand-side policies can promote healthy competition within and among ICT industries by increasing innovation in ICTs and their availability while lowering the cost of production and prices to consumers. Nations with greater ICT development are those that have more competition to provide basic telecommunication services (Guillén and Suárez 2005). Other cross-national studies have found that there is a relationship between competition in the telecommunications industry and e-government development (Gulati and Yates 2011; Gulati, Williams, and Yates 2014).

The theoretical link between specific policy initiatives and development of ICTs generally, together with growing empirical evidence connecting e-government with financial investment in ICTs and competition in the ICT sector, underlie our next two hypotheses:

H3: Countries that have invested more in ICTs will have more extensive development of e-government than countries that have invested less in ICTs.

H4: Countries that have a more competitive telecommunications sector will have more extensive development of e-government than countries that have less competitive or state-owned telecommunication enterprises.

Like this study, recent work assessing the impact of targeted policies—for example, encouraging competition in the ICT sector or financial investment in ICTs—has controlled for the relative impact of public sector performance, which can be essential for the effective implementation of such policies.

2. Data and Methods

This study relied primarily on secondary data acquired from the United Nations, including the International Telecommunication Union (ITU), which provide the most detailed and comprehensive information regarding national e-government capabilities, telecommunications infrastructure and human development (ITU 2011, 2015; United Nations 2010, 2014). Cross-national government (Pemstein, Meserve, and Melton 2010), governance

(Kaufmann, Kraay, and Mastruzzi 2009; ITU 2011), policy (Gulati and Yates 2011), as well as demographic and geographic data (CIA 2011) supplement our UN and ITU sources.

2.1 Dependent Variables

Our first indicator measuring the extent of e-government and the dependent variable in our first model is the United Nations' OSI, a subset of the UN's broader EGDI. Constructed for 193 countries, the OSI measures the extent of a nation's performance in online service maturity. More specifically, the index captures the extent of (1) emerging information services, (2) enhanced information services, (3) transactional services, and (4) connected services, with greater weight given to the higher stage, more sophisticated tools, and applications. Values of this index range from 0 to 1, with France, Singapore, and South Korea exhibiting the highest scores (all greater than 0.97), and Eritrea and Guinea exhibiting the lowest (0.0) (United Nations 2014). We did not use the UN's broader EGDI because many of its individual indicators measure telecommunications infrastructure and educational levels, which are theoretically linked to causal explanations of e-government capability and are included in our two models as independent variables.

Our second dependent variable is the United Nations' EPI. Constructed for the same countries as the OSI, the EPI measures the extent of a nation's performance in promoting citizen engagement and facilitating communication among government, citizens, businesses, and society. More specifically, the index captures the extent to which governments use the Internet for disseminating information about its proposals and activities, consulting with citizens on matters of public policy, and allowing for direct citizen participation in decision making. Values of this index can range from 0 to 1, with South Korea (1.0), the Netherlands (1.0), and Uruguay (0.98) exhibiting the highest scores, and eight countries exhibiting scores less than 0.02. While there is criticism of the EPI because its components only superficially seem to capture citizen involvement in government decision making (Macintosh and Whyte 2008; Medaglia 2012), the individual items that comprise the EPI not only are distinct from the OSI but also go beyond information dissemination and service provision. In a sense, the EPI represents the minimum that would be needed on government websites for any citizen participation to be possible.

We test our four hypotheses by estimating two ordinary least squares (OLS) multiple regression models. This modeling technique is the most direct and one of the simplest ways to determine the relative effect of the independent variables described next on our OSI and EPI dependent variables.

While there is a high correlation between a country's score on the OSI and EPI ($r = 0.95$), there are a number of countries that are ranked much lower on the e-participation index than on the online service index and a few countries that are ranked much higher on e-participation than online services. This indicates that participatory features and services may not necessarily require a sophisticated e-government infrastructure for a country to use the Internet to engage and include citizens more fully in government decision making and governing.

The large number of countries that have very few of their government services online or do not use the web for enabling citizen participation yields a long tail of small values in the distribution of both the OSI and the EPI. We therefore use the natural logarithm of both indices as the dependent variable in our OLS regression analysis.

We selected all independent variables used in the multiple regression analysis described next from prior years (i.e., 2007–2010); thereby taking into account the time lag in realizing their effects.

2.2 *Government and Governance*

To account for the impact of political structure and a culture of democratic politics (i.e., government-related indicators), we included the Unified Democracy Scores (UDS) for 2008. The UDS is derived through a Bayesian latent variable approach and draws from ten frequently used indicators of democracy (e.g., Polity IV and Freedom House) to produce a single composite scale (Pemstein, Meserve, and Melton 2010). Values of the UDS ranged from –1.478 to 2.02 in 2008. Saudi Arabia, North Korea, Myanmar, and Qatar had the four lowest scores, while Switzerland, Denmark, the Netherlands, and Norway had the four highest scores.

We measured the impact of governance with two independent variables. First, we used the six indicators from the Worldwide Governance Indicators (WGI) project (Kaufmann, Kraay, and Mastruzzi 2009) to assess the administrative professionalism and governmental performance in each country: (1) government effectiveness, (2) regulatory quality, (3) rule of law, (4) political stability and absence of violence, (5) control of corruption, and, (6) voice and accountability. Initially, we considered including these indicators separately in the model, but observed that many of the inter-variable correlations were above 0.70. We instead created a government performance index by computing the average among the standardized scores for each of the six indicators. Denmark (2.12), Finland (2.11) and Sweden (2.03) scored highest on this measure, while Somalia (–2.53), Myanmar (–1.98), and Afghanistan (–1.92) scored the lowest.

A more specific indicator of governance in the ICT policy sphere is the presence or absence of an independent national regulatory authority for telecommunications. A review of a number of case studies indicates that nations that have been most successful at utilizing ICTs for online government are those that have established an independent executive-level department or national-level agency responsible for promoting and managing the expansion of telecommunication products and services (Azad et al. 2010; ITU 2015). Thus, countries with some sort of independent regulatory authority should provide better e-government for their citizens. To perform our analysis, we encoded the presence of a national independent regulatory authority for each country, as published in (ITU 2011), as follows:

"0" if there is no independent regulatory authority; and
"1" if there is an independent regulatory authority.

In 2010, over 63% of the countries for which the ITU publishes these data had an independent national telecommunications regulatory authority.

Most of the cross-national studies reviewed earlier did not distinguish between e-government services and the more democratic features of e-government (Gulati and Yates 2011; Stier 2015). While Lee, Chang, and Berry (2011) did make this distinction, they found that both services and participation were positively related with better governance. Thus, we do not expect that the effects of governance should be different in the OSI and EPI models.

2.3 *Financial Investment*

We used Gulati and Yates' (2011) *financial investment index* to measure a nation's financial investment in ICTs. To determine this index we obtained seven indicators obtained from the World Bank's World Development Indicators database of variables that pertain to economic activity related to technological development or to either public or private investment in ICTs. The seven indicators included in the index are telecommunications revenue (as a percentage of GDP), ICT expenditures (as a percentage of GDP), telecommunications investment (as a percentage of revenue), research and development spending (as a percentage of GDP), natural log of international Internet bandwidth (bits per second per person), high-technology exports (as a percentage of manufacturing exports), and computer, communications and other services (as a percentage of service exports). Because most of these indicators represent a proportion of total spending rather than an absolute amount per capita, this index also indirectly gauges the priority that a nation places on developing ICTs and support for the technology sector. The values of this index ranged from—1.808 (Liberia) to 1.178 (Philippines).

2.4 *Competition*

To measure the extent of competition in the telecommunications sector, we used Gulati and Yates's (2011) *telecommunications competition index*. We derived this index from six indicators obtained from the ITU's ICT Eye database, which take into account the level of competition in the following telecommunications industries: basic telephone service, mobile services, narrowband Internet service, DSL-based Internet service, cable-based Internet service, and cross-platform competition. Values of this index range from—1.951 to 1.184, with 50 countries sharing the highest score and three countries sharing the lowest score (East Timor, Palau, & Saint Kitts and Nevis). Most of the 50 top-tier countries are very highly developed (United Nations 2010), however, the UN also classifies three of these countries as having a low level of human development—namely, Mauritania, Pakistan, and Uganda. However, telecommunications competition has a lower correlation with the level of democratization than expected (r = 0.42) when democratization is measured as described in Section 2.2.

2.5 Control Variables

We include four non-political variables in our regression models that have a theoretical or empirical link to e-government, as shown in previous research. Previous cross-national studies of e-government have assumed that countries with more wealth and an affluent population will be in a stronger position to spend more on e-government development (e.g., Rose 2005). In addition, people who have a higher level of education are more likely to demand that more services be made available over the Internet (Pick and Sarkar 2015). The empirical evidence linking resources and education to e-government has not shown a consistent pattern, however. We use the United Nations' Income Index and Education Index for 2010 (United Nations 2010) to capture and distinguish the impacts of a nation's economic resources and education on the dependent variables in our two models.

Furthermore, government and private industry are more likely to be successful in delivering e-government applications and services in urban areas, where the population is more concentrated and infrastructure for Internet connections and mobile devices is already in place. On the other hand, the need for e-government services and applications is greater in areas that are geographically larger, where personal contact between citizens and members of the national government can be difficult or inconvenient. We therefore measure urbanization as the percentage of residents living in urban areas. We account for distance by using a country's total size in square kilometers. We obtained the data for urbanization and land area from the CIA website (CIA 2011).

3. Data Analysis

3.1 Online Service Index

The regression analysis of the log-transformed value for the OSI in 175 countries based on the nine independent variables revealed that higher government performance, an independent telecommunications regulatory authority and greater competition in the telecom sector, all increased the extent to which countries have government information and services online. The relationships between the OSI and education and land area also were positive. The relationship was negative, however, between the OSI and the level of democratization. There was no relationship between the OSI and either higher financial investment in telecommunications or greater economic wealth. In this analysis, the five political and four control variables together explained 49% of the variance in the dependent variable.

Because the Income Index control variable was not statistically significant but was highly correlated with government performance (r = 0.73), the Education Index (r = 0.83), and urbanization (r = 0.73), we re-estimated the model without the income variable and report these results in Table 5.1. The coefficients for the democracy scores in the first row are statistically significant at the 0.01 level and indicate that there is a strong negative connection between the presence of democratic political institutions and processes and the extent of online services. The relationship is not in the direction that we expected in hypothesis H_1, however. Rather, e-government services are more extensive in more autocratic countries and less extensive in more democratic countries. When holding all other variables constant, a one-unit increase in a country's UDS score decreases a country's score on the OSI by about 37%. To further illustrate, a country that has the mean value (0.432) on the UDS would have a score that was about 70% lower on the OSI than a country that has the minimum score (–1.48). And a country that has the maximum value (2.02) would have an OSI score that was 58% lower than a country with the mean value. Our result goes beyond Stier's (2015) conclusion for the same United Nations (2014) data that "increasing political online activism in authoritarian contexts that is—among other instruments—countered by an expanded regime presence on the Web" (Stier 2015: 276). Specifically, we find evidence that some authoritarian regimes are developing more extensive online services than their democratic counterparts.

The coefficients for the government performance index in the second row are statistically significant at the 0.01 level and support the hypothesis that effective public sector governance and administration increases the level of information and services a nation provides online (H_2). When holding all other variables constant, a 0.10-unit increase in a country's WGI score increases a country's score on the OSI by 4.0%. A country that has the mean value (0) on the WGI index would have a score that was more than 100% higher on the OSI than a country that has the minimum WGI score (–2.53). And a country that has the maximum value (2.12) would have a score on the OSI that was 84% higher than a country with the mean value.

The coefficients in the next row indicate that countries that have an independent national regulatory authority for telecommunications have more of an online presence than countries that do not have such a national agency. When holding all other variables constant, countries with an independent national regulatory authority have a score on the OSI that is 20% higher on the unit scale than countries without a national regulatory authority or countries that have one that is not independent. The coefficients are statistically significant at the 0.05 level and further support our hypothesis that

Table 5.1 Multiple Regression Analysis Explaining Online Government Services

	B	*Std. Err.*	*Std. B*	*Sig.*
Democratic political structure [UDS] (H_1)	−0.365	0.103	−0.302	0.000
Government performance [WGI] (H_2)	0.396	0.104	0.387	0.000
Telecom regulatory authority [ITU] (H_2)	0.200	0.084	0.154	0.019
Financial investment index (H_3)	0.033	0.126	0.017	0.791
Telecom competition index (H_4)	0.194	0.073	0.177	0.009
Education Index [United Nations]	2.282	0.541	0.393	0.000
Urbanization [CIA]	0.001	0.003	0.028	0.700
Land area [CIA]	0.053	0.028	0.107	0.054
(Constant)	−2.705	0.318		0.000

Dependent variable: Natural log of OSI.
N = 175; Adjusted R Squared = 0.492; Std. Error of the Estimate = 0.717.

those countries with effective governance, particularly in the area of strategic support for the advancement of ICTs, are more likely to provide their citizens with services online (H_2).

The presence of a national-level bureaucratic organization dedicated to regulating communications suggests a national commitment to guiding the development and diffusion of technological innovations in society. The public sector is a beneficiary of these innovations and, therefore, is in a position to assist legislative and governing institutions to develop innovative ways to communicate with citizens, organizations and each other (Fountain 2001; Moon and Norris 2005). But for public sector organizations to leverage technology to deliver services online, there already must exist a professional administrative culture and practice of effective governance (Brown 2005; Layne and Lee 2001). The results presented in rows two and three of Table 5.1 indicate that countries that are having the most success in delivering online services are those that have had the most success administering public services in general.

On the other hand, the coefficients for the financial investment index are not statistically significant. This suggests that countries that devote more financial resources to develop and promote information technology, telecommunications, and related industries are no more likely to develop a greater online presence.

The fifth row of data reports the coefficients for the telecommunications competition index. These coefficients indicate that countries that have more open competition in their telecommunication industries also have a greater online presence than countries that have a more heavily regulated sector. A 0.10-unit increase in the telecommunications competition index increases a country's score on the OSI by 1.9% when controlling for all other variables. The coefficients are statistically significant at the 0.01 level and thus provide support for our hypothesis that countries that implement policies to promote ICTs have a better online presence than countries that are less supportive (H_4). Privatization and competition in the computing and communication industries can create a highly favorable environment for government to develop online communications with (and service delivery to) its citizens. In such an environment, service providers deploy more efficient telecommunications infrastructure to connect citizens and their government; governments can usually purchase superior products and services to implement e-government; and citizens have more choices in service providers.

The coefficients for the control variables are shown beginning in the sixth row of data in Table 5.1. An increased level of education has a substantial effect on a national government's online presence. The coefficients for the Education Index are statistically significant at the 0.01 level. A 0.10-unit increase on the Education Index increases the level of online services by a substantial amount, more than 22%. The coefficients for land mass are statistically significant at the 0.1 level and are positively related to the level of online services.

3.2 E-Participation Index

The results of the regression analysis of the log-transformed value for the EPI in 175 countries using the five political variables and three control variables are reported in Table 5.2.[2] The same eight independent variables used to estimate the model of online services together explain about 43% of the variance in the extent to which countries provide participatory opportunities online. The same four political variables were statistically significant in the model of e-participation capabilities as in the model of online services.

The coefficients for the democracy scores in the first row show a negative relationship between the presence of democratic political institutions and processes and government implementation of e-participation. As for online services, the presence of a statistically significant relationship between the democracy scores and e-participation scores (at the .1 level for e-participation) indicates that a more democratic political structure has a negative effect on the extent of a country's participatory e-government. This result, which provides additional evidence that contradicts hypothesis H_1, is discussed further in Section 4.

As the coefficients in the second row show, there is a strong and statistically significant relationship between the government performance index and the EPI. When holding all other variables constant, a 0.10-unit increase in a country's WGI score increases a country's score on the EPI by 2.1%. A country that has the mean value (0) on the WGI index would have a score that was about 53% higher on the participation index than a country that has the minimum WGI score (–2.53). A country that has the maximum value (2.12) would have a score on the participation index that was 44% higher than a country with the mean value. Figure 5.1 plots per-country OSI and EPI scores versus WGI governance in solid red circles and blue diamonds, respectively. This figure provides interesting visual information

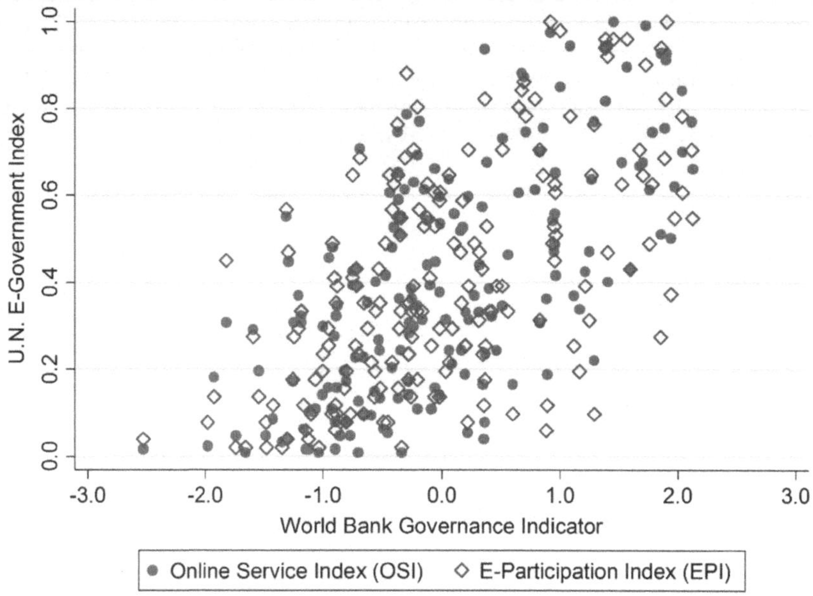

Figure 5.1 United Nations E-Government Indices Versus WGI Governance.

for three reasons. First, of the five independent variables, the governance variable on the *x*-axis has the greatest effect on both dependent variables, as indicated by the standardized coefficients in the second row of Table 5.1 (Std. B = 0.387) and Table 5.2 (Std. B = 0.235). Second, it is clear that the OSI and EPI show different scatter with respect to their *x*- and *y*-values. Third, with respect to this governance variable, the scatter of the OSI and EPI reflect the fact that the correlation coefficient of the OSI (0.66) is greater than that of the EPI (0.57).

Moving on to the role of the other independent variables, the coefficients in the next row of Table 5.2 indicate that there is a positive relationship between the presence of an independent national regulatory authority for telecommunications and e-participation capabilities (B = 0.145). Having an independent agency with responsibility of guiding the strategic development of communication technologies seems to be important for developing the infrastructure for e-government and for developing the corresponding higher level democratic means of citizen-to-government communication. The findings in rows two and three provide further support for the hypothesis that more professional public administration practices increase e-government (H_2). In addition, the results show that an efficient and effective public sector is necessary not only for electronic delivery of services but also for facilitating electronic avenues of participation.

Table 5.2 Multiple Regression Analysis Explaining E-Participation Capabilities

	B	*Std. Err.*	*Std. B*	*Sig.*
Democratic political structure (H_1)	−0.185	0.095	−0.176	0.054
Government performance (H_2)	0.208	0.096	0.235	0.032
Telecom regulatory authority (H_2)	0.145	0.078	0.128	0.064
Financial investment index (H_3)	0.070	0.119	0.041	0.558
Telecom competition index (H_4)	0.122	0.067	0.127	0.072
Education index	2.149	0.500	0.427	0.000
Urbanization	0.002	0.003	0.041	0.598
Land area	0.052	0.025	0.121	0.041
(Constant)	−2.617	0.294		0.000

Dependent variable: Natural log of EPI.
N = 175; Adjusted R Squared = 0.427; Std. Error of the Estimate = 0.663.

As for online services, the coefficients for the financial investment index are not statistically significant. This suggests that countries that devote more financial resources to invest in ICTs are no more likely to develop greater e-participation capabilities. Thus, our findings show that government performance and governance matter, providing strong support for hypothesis H_2, but do not show the same for financial investment (see H_3). This suggests that, with respect to e-government capability, how resources and processes are managed and targeted at the national level matters more than the overall amount invested in ICTs.

The fifth row of data reports the coefficients for the telecommunications competition index. These coefficients indicate that countries that have more open competition in their telecommunication and related industries also have greater e-participation opportunities than countries that have a more heavily regulated sector. A 0.10-unit increase in the telecom competition index increases a country's score on the EPI by more than 1.2% when controlling for all other variables. The coefficients are statistically significant at the .1 level and provide support for our hypothesis that countries that develop policies to promote the diffusion of ICTs have greater opportunities for citizen participation online than countries that are less supportive (H_4). In substantive terms, a country with the highest value on our index of competition (1.18) would score approximately 38% higher on the EPI than a country with the lowest score (−1.95), which corresponds to countries with the least competitive telecommunication markets.

The coefficients for the control variables are again shown beginning in the sixth row of data. An increased level of education has a positive effect on a nation's e-participation capabilities. The coefficients for land mass also are statistically significant and increase the EPI. Figure 5.2 plots each nation's OSI and EPI score versus their Education Index, again in solid red circles and blue diamonds. Of the three control variables, the UN Education Index on

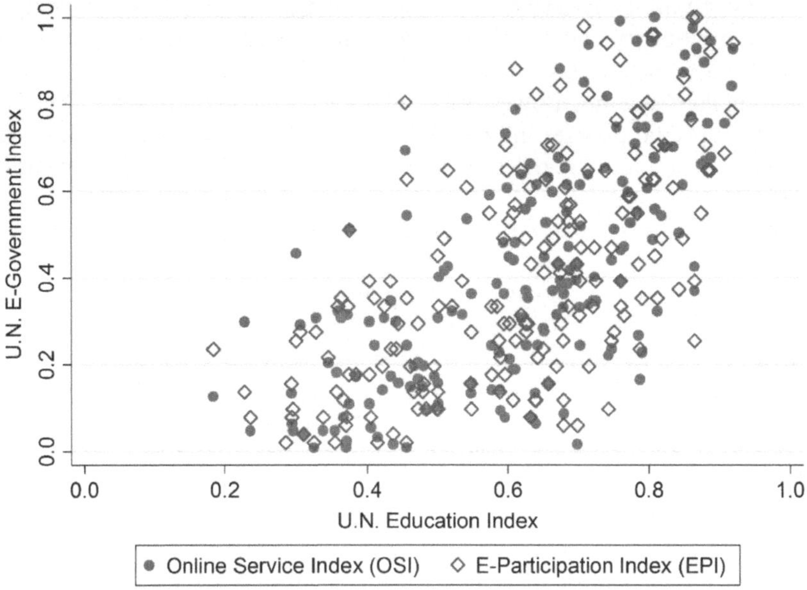

Figure 5.2 United Nations E-Government Indices Versus Education Index.

the *x*-axis has the greatest effect on the e-government dependent variables, as indicated by the standardized coefficients in the sixth row of Table 5.1 (0.393) and Table 5.2 (0.427). Finally, with respect to the level of education, the similar scatter of the OSI and EPI in Figure 5.2 reflects the fact that the correlation coefficient of the OSI (0.68) is similar to that of the EPI (0.63).

4. Discussion and Implications

This study of e-government implementation assesses the impact of effective governance using a broad range of governance indicators. Previous studies have relied on single indicators (Gulati and Yates 2011; Lee, Chang, and Berry 2011; Stier 2015) and, thus, omitted measurement of important aspects of public sector performance. Table 5.3 summarizes our findings for each of the four hypotheses. It is not surprising that we found that public sector performance has a strong positive impact on the extent of a country's e-government implementation since there were strong theoretical reasons for expecting this to be the case—for example, see Brown (2005) and Fountain (2001). It was not anticipated, however, that governance would have the largest positive impact of all variables in our models. These findings indicate that the culture and practices within the public sector are what will drive further implementation of e-government more so than external demands and needs. It is therefore unlikely that the promise of e-government, whether in terms of online services or citizen participation, can become a reality in laggard nations without transformations first occurring in their public sectors (Azad et al. 2010; Dwivedi, Weerakkody, and Janssen 2011). Elected policy makers are not powerless to make a difference, however. We also find that specific policies matter. While the policy initiatives we evaluated are intended to support the ICT sector as a whole and telecommunication industries more specifically, these initiatives also enhance e-government. Presumably, this is because the more available and the more affordable ICT products and services are, the more governments are able to operate and maintain an effective e-government capability.

Our most surprising finding is the negative relationship between the level of democracy and both online government services and e-participation capabilities. While researchers need to differentiate the drivers and outcomes of each, policy makers should be cognizant of the difference in developing strategies to advance them both. Our results also sound a cautionary note. Given that the path to e-government leverages different strategies depending on a nation's political structure, it is possible that those countries in which there is less democracy are utilizing e-government so as to maintain the status quo. Case studies of China, Costa Rica and Cuba (Hoffmann 2004; Kalathil and Boas 2001; Ma, Chung, and Thorson 2005); for example, demonstrate that these authoritarian governments have pursued successful reactive strategies to control Internet content and access. At the same time, authoritarian states can and do pursue proactive strategies to extend central control by guiding the development of the medium to promote their own interests and priorities. Thus, through a combination of reactive and proactive strategies, authoritarian regimes have learned to counter the challenge posed by the potential democratizing influences of the Internet (Kalathil and Boas 2001). Corrales and Westhoff (2006) find that high-income, market-oriented autocratic states are less draconian because, although they fear the political consequences of Internet expansion, they welcome its economic payoffs.

An alternative line of reasoning is based in the "authoritarian bargain" between rulers and citizens by which citizens relinquish political rights to gain economic security. Desai, Olofsgård, and Yousef's (2009) empirical evidence from 45 non-democratic states confirms the trade-off of bundles of welfare benefits to secure public support. Jiang and Xu (2009) show that by providing limited improvement in administrative efficiency and transparency, Chinese provincial governments are able to deflate social tension and reestablish party legitimacy. Manipulating online structures through information delivery, agenda setting, and containment of public dissent is a more subtle form of online social control. Kardan and Sadeghiani's longitudinal study of Iran finds a positive but low trend of change in e-government initiatives from 2008 to 2009, but a substantial negative trend of change in e-democracy during the same period. Thus, they recommend against an

Table 5.3 Summary of Support for Hypotheses

Hypothesis **[Expected direction +/–]**	*Online Services (OSI) [Actual direction +/–/0]*	*E-Participation Capabilities (EPI) [Actual direction +/–/0]*
H_1: More democratic political structure [+]	Not supported, opposite direction [–]	Not supported, opposite direction [–]
H_2: Public administration performance [+]	Supported [+]	Supported [+]
H_3: Financial investment in ICTs [+]	Not supported, not significant [0]	Not supported, not significant [0]
H_4: Competitive telecom sector [+]	Supported [+]	Supported [+]

emphasis on e-government as a path to e-democracy in authoritarian countries such as Iran (Kardan and Sadeghiani 2011).

More detailed case studies are needed to understand the causal relationship between democratic institutions and e-government capabilities, including those intended for e-participation (Åström et al. 2012; Gulati, Williams, and Yates 2014; Rose 2005). Quite possibly the centralization that characterizes authoritarian regimes makes them more efficient than democracies in implementing policy decisions and directives to provide e-government services. Also, it may be the case that the absence of democratic institutions implies a lack of transparency and citizen engagement and what these governments are placing online is superficial in nature and meant to provide an appearance of openness. And in countries where there is a long-standing culture of openness, there is much more emphasis placed on maintaining existing means of government-to-citizen communication.

5. Conclusions and Limitations

Our research assesses the relationship of political factors and policy initiatives to improved e-government services and e-participation capabilities. We showed that countries that have more effective public sector governance, an independent national regulatory authority, and encourage competition in the telecommunications sector also have more extensive provision of e-government services. Specifically, these three factors have a positive and significant relationship with both the OSI and EPI (United Nations 2014). We also provide evidence suggesting that the provision of more e-government services is not associated with a more democratic political system, but rather seems to be associated with a more authoritarian political system.

Like almost all regression models developed for cross-national studies, the variables for which there is reliable national-level data are rarely comprehensive. Thus, there may be other activities or norms that explain the greater development of e-government that also can be measured. As a regional example, Wilson and Wong (2006), demonstrate the importance of "information champions" in explaining variation in Internet diffusion across African states. The appointment of a chief information officer could indicate an even stronger commitment by a nation to advancing democratic governance through technology. External leadership also may influence transformations in the public sector. For example, nations can learn from each other. There are studies to suggest that governments adopt policy innovations from nations seen as their socio-cultural peers and from neighbor states that have demonstrated past success with new policies (Simmons and Elkins 2004; Pick and Sarkar 2015). Research by Lee, Chang, and Berry (2011) showing that international organizations and neighboring countries positively influence online service provision offers support for these additional paths to e-government capability.

There are of course other factors that contribute to understanding the extent of online services and e-participation across the globe. In the regression models we estimated, more than half of the variance in e-government capability was not explained by our independent variables. These models therefore serve as a reminder of how inherently complex it is to understand and measure the extent and quality of online services at scale.

The unit of analysis of e-government in this chapter is the country. While appropriate for the OSI and EPI dependent variables in this study, interesting regional differences in countries with large populations like China and India, or large land mass, such as Russia and Australia, can only be explored as part of a more in-depth study. Even so, we hope that our findings will guide decision makers in capitals across the globe to take an active role in strengthening democracy, improving public sector performance, and developing a healthy ICT sector. If properly guided, such changes should reduce inequalities in e-government and thereby allow the ongoing information and communication revolution to improve the lives of those in countries who have yet to benefit.

Notes

1 Consistent with the literature, this study understands e-government as encompassing two dimensions: the delivery of online services and e-participation.

2 The Income Index control variable was excluded because it was not statistically significant in the EPI regression model and again was highly correlated with government performance, the Education Index, and urbanization.

References

Åström, Joachim, Martin Karlsson, Jonas Linde, and Ali Pirannejad. Understanding the Rise of e-Participation in Non-Democracies: Domestic and International Factors. *Government Information Quarterly* 29, no. 2 (2012): 142–50.

Azad, Bijan, Samer Faraj, Jie Mein Goh, and Tony Feghali. "What Shapes Global Diffusion of e-Government: Comparing the Influence of National Governance Institutions." *Journal of Global Information Management* 18, no. 2 (2010): 85–104.

Bertot, John C., Elsa Estevez, and Tomasz Janowski. Digital Public Service Innovation: Framework Proposal. In *Proceedings of International Conference on Theory and Practice of Electronic Governance (ICEGOV)*, 2016, 113–22.

Boudjelida, Abdelhamid, Sehl Mellouli, and Jungwoo Lee. Electronic Citizens Participation: Systematic Review. In *Proceedings of International Conference on Theory and Practice of Electronic Governance (ICEGOV)*, 2016, 31–9.

Brown, David. "Electronic Government and Public Administration." *International Review of Administrative Sciences* 71, no. 2 (2005): 241–54.

Chatfield, Akemi T., and Omar Alhujran. "A Cross-Country Comparative Analysis of e-Government Service Delivery Among Arab Countries." *Information Technology for Development* 15, no. 3 (2009): 151–70.

CIA. "The CIA World Factbook." 2009. Accessed October 1, 2011. www.cia.gov/library/publications/the-world-factbook/.

Corrales, Javier, and Frank Westhoff. "Information Technology Adoption and Political Regimes." *International Studies Quarterly* 50, no. 4 (2006): 911–33.

Desai, Raj M., Anders Olofsgård, and Tarik Yousef. "The Logic of Authoritarian Bargains." *Economics & Politics* 21, no. 1 (2009): 93–125.

Dwivedi, Yogesh K., Vishanth Weerakkody, and Marijn Janssen. "Moving Towards Maturity: Challenges to Successful e-Government Implementation and Diffusion." *ACM SIGMIS Database* 42, no. 4 (2011): 11–22.

Fountain, Jane E. *Building the Virtual State: Information Technology and Institutional Change*. Washington, DC: Brookings Institution Press, 2001.

Gainous, Jason, Kevin Wagner, and Jason Abbott. "Civic Disobedience: Does Internet Use Stimulate Political Unrest in East Asia?" *Journal of Information Technology & Politics* 12, no. 2 (2015): 219–36.

Gascó, Mila. "Exploring the e-Government Gap in South America." *International Journal of Public Administration* 28, nos. 7–8 (2005): 683–701.

Guillén, Mauro F., and Sandra Suárez. "Explaining the Global Digital Divide: Economic, Political and Sociological Drivers of Cross-National Internet Use." *Social Forces* 84, no. 2 (2005): 681–708.

Gulati, Girish J., Christine Williams, and David Yates. "Predictors of On-Line Services and e-Participation: A Cross-National Comparison." *Government Information Quarterly* 31, no. 4 (2014): 526–33.

Gulati, Girish J., and David Yates. "Strategy, Competition, and Investment: Explaining the Global Divide in e-Government Implementation with Policy Variables." *Electronic Government: An International Journal* 8, no. 2/3 (2011): 124–43.

Helbig, Natalie, J. Ramón Gil-García, and Enrico Ferro. "Understanding the Complexity of Electronic Government: Implications from the Digital Divide Literature." *Government Information Quarterly* 26, no. 1 (2009): 89–97.

Hindman, Matthew. *The Myth of Digital Democracy*. Princeton, NJ: Princeton University Press, 2008.

Hoffmann, Bert. *The Politics of the Internet in Third World Development: Challenges in Contrasting Regimes with Case Studies of Costa Rica and Cuba*. New York: Routledge, 2004.

ITU. *International Telecommunications Union ICT Eye Database*, 2011. Accessed October 1, 2011. www.itu.int/ITU-D/icteye/Default.aspx.

ITU. *Measuring the Information Society 2015*. International Telecommunication Union Report. Geneva, Switzerland: United Nations, 2015.

Jiang, Min, and Heng Xu. "Exploring Online Structures on Chinese Government Portals: Citizen Political Participation and Government Legitimation." *Social Science Computer Review* 27, no. 2 (2009): 174–95.

Kalathil, Shanthi, and Taylor Boas. "The Internet and State Control in Authoritarian Regimes: China, Cuba, and the Counterrevolution." *First Monday* 6, no. 8 (2001, August 6).

Kardan, Ahmad A., and Ayoob Sadeghiani. "Is e-Government a Way to e-Democracy? A Longitudinal Study of the Iranian Situation." *Government Information Quarterly* 28, no. 4 (2011): 439–562.

Kaufmann, Daniel, Aart Kraay, and Massimo Mastruzzi. "Governance Matters VIII: Aggregate and Individual Governance Indicators, 1996–2008." *Policy Research Working Paper 4978*, The World Bank, Washington, DC.

King, John L., Vijay Gurbaxani, Kenneth Kraemer, F. Warren McFarlan, K. S. Raman, and C. S. Yap. "Institutional Factors in Information Technology Innovation." *Information Systems Research* 5, no. 2 (1994): 139–69.

Krishnan, Satish, Thompson Teo, and Vivien Lim. "Examining the Relationships Among e-Government Maturity, Corruption, Economic Prosperity and Environmental Degradation: A Cross-Country Analysis." *Information & Management* 50, no. 8 (2013): 638–49.

Layne, Karen, and Jungwoo Lee. "Developing Fully Functional e-Government: A Four Stage Model." *Government Information Quarterly* 18, no. 2 (2001): 122–36.

Lee, Chung-Pin, Kaiju Chang, and Frances Berry. "Testing the Development and Diffusion of e-Government and e-Democracy: A Global Perspective." *Public Administration Review* 71, no. 3 (2011): 444–54.

Lijphart, Arend. *Patterns of Democracy: Government Forms and Performance in Thirty-Six Countries*. New Haven, CT: Yale University Press, 1999.

Ma, Lianjie, Jongpil Chung, and Stuart Thorson. "-Government in China: Bringing Economic Development Through Administrative Reform." *Government Information Quarterly* 22, no. 1 (2005): 20–37.

Macintosh, Ann, and Angus Whyte. "Towards an Evaluation Framework for eParticipation." *Transforming Government: People, Process and Policy* 2, no. 1: 16–30.

Medaglia, Rony. "eParticipation Research: Moving Characterization Forward (2006–2011)." *Government Information Quarterly* 29, no. 3 (2012): 346–60.

Moon, M. Jae, and Norris, Donald. "Does Managerial Orientation Matter? The Adoption of Reinventing Government and e-Government at the Municipal Level." *Information Systems Journal* 15, no. 1 (2005): 43–60.

Norris, Donald F., and M. Jae Moon. "Advancing e-Government at the Grassroots: Tortoise or Hare?" *Public Administration Review* 65, no. 1 (2005): 64–75.

Norris, Pippa. *Digital Divide: Civic Engagement, Information Poverty, and the Internet Worldwide*. New York: Cambridge University Press, 2001.

Osborne, Stephen P., Zoe Radnor, and Greta Nasi. "A New Theory for Public Service Management? Toward a (Public) Service-Dominant Approach." *The American Review of Public Administration* 43, no. 2 (2013): 135–58.

Pemstein, Daniel, Stephen Meserve, and James Melton. "Democratic Compromise: A Latent Variable Analysis of Ten Measures of Regime Type." *Political Analysis* 18, no. 4 (2010): 426–49.
Pick, James B., and Avijit Sarkar. *The Global Digital Divides: Explaining Change*. Berlin and Heidelberg, Germany: Springer, 2015.
Rose, Richard. "A Global Diffusion Model of e-Governance." *Journal of Public Policy* 25, no. 1 (2005): 5–27.
Shirazi, Farid. "Social Media and the Social Movements in the Middle East and North Africa: A Critical Discourse Analysis." *Information Technology & People* 26, no. 1 (2013): 28–49.
Simmons, Beth A., and Zachary Elkins. "The Globalization of Liberalization: Policy Diffusion in the International Political Economy." *American Political Science Review* 98, no. 1 (2004): 171–89.
Stier, Sebastian. "Political Determinants of e-Government Performance Revisited: Comparing Democracies and Autocracies." *Government Information Quarterly* 32, no. 3 (2015): 270–8.
Sunstein, Cass R. *Republic.com 2.0*. Princeton, NJ: Princeton University Press, 2007.
United Nations. *Human Development Report 2010–20th Anniversary Edition; the Real Wealth of Nations: Pathways to Human Development*. New York: Palgrave Macmillan, 2010.
United Nations. *UN e-Government Survey 2014: E-Government for the Future We Want*. New York: United Nations, 2014.
Weerakkody, Vishanth, Zahir Irani, Habin Lee, Ibrahim Osman, and Nitham Hindi. "e-Government Implementation: A Bird's Eye View of Issues Relating to Costs, Opportunities, Benefits and Risks." *Information Systems Frontiers* 17, no. 4 (2015): 889–915.
West, Darrell M. *Digital Government: Technology and Public Sector Performance*. Princeton, NJ: Princeton University Press, 2005.
West, Darrell M. *Improving Technology Utilization in Electronic Government Around the World*. Washington, DC: Brookings Institution, 2008.
Wilson, Ernest J., and Kelvin Wong. *Negotiating the Net: The Politics of Internet Diffusion in Africa*. Boulder, CO: Lynne Rienner, 2006.

6 ICT-Enabled E-Entertainment Services in U.S. Counties

Socio-economic Determinants and Geographic Patterns

Avijit Sarkar, James Pick, and Jessica Rosales

1. Introduction

Investigations of the digital divide have often focused on patterns of adoption, diffusion, and use of the Internet for a wide variety of purposes. The term digital divide refers to the "gap between individuals, households, businesses and geographic areas at different socio-economic levels with regard both to their opportunities to access ICTs and to their use of the Internet for a wide variety of activities" (OECD 2001: 5). Within the broader landscape of the global digital divide, recent research has been conducted based on the broad consensus among academics in various disciplines and policy makers that adoption and diffusion of ICTs including the Internet is having a profound impact on modern life in areas such as education,[1] health care, and other diverse services (Agarwal, Animesh, and Prasad 2009; Warschauer and Matuchniak 2010). Understanding the use of ICTs for accessing entertainment services, examining patterns of such use and factors that influence use as well as the implications of ICT usage for entertainment, are largely unstudied themes in the digital-divide literature at this present time. This chapter attempts to fill this void.

Generally speaking, the digital divide is an important topic because digital skills and knowledge are crucial for a person to have a satisfying and prosperous life in today's world, so knowing more about digital deficits can help organizations and governments to narrow the divide. It is also important for planners and researchers to gain better understanding of technologies in society and for people themselves to understand their digital capabilities and how to improve them. However, there are variations in understanding the digital divide. For example, it can be viewed from the standpoint of the extent of adoption and diffusion of technologies (Rogers 2003); it can be looked at from a national standpoint, considering the multidimensional digital capabilities of a country and the digital level of the country worldwide (Dutta, Mia, and Geiger 2011), or it can be evaluated on the basis of its correlates and geographic patterns for small units, such as regional districts of Europe (Vicente and Lopez 2011) or U.S. counties as in this study.

Our motivation is the fact that no research on socio-economic associations and geography of e-entertainment for continental U.S. counties exists, yet it is essential to gain knowledge of the phenomena for the world's largest economy. Specifically, the goal of this research is to understand the geographic patterns of e-entertainment use and its socio-economic associations for continental U.S. counties and recommend policies for county populations to improve their access and use of e-entertainment.

But why is entertainment a sector that requires more attention? The entertainment and media market in the United States is expected to be worth USD 600 billion in 2015 and almost 725 billion by 2019 (Statista 2015a). According to the latest U.S. Economic Census, the Internet Publishing, and Broadcasting and Web Search Portals sub-sector, which includes establishments that provide textual, audio, and/or video content of general or specific interest on the Internet exclusively, has grown significantly. Between 2007 and 2012, the number of sub-sector establishments and total employment increased by 82% and 143%, respectively (U.S. Bureau of the Census 2015).

In this chapter, we analyze the influence of social, economic, ethnic, innovation, and social capital factors on e-entertainment services for a large sample of continental U.S. counties. These factors are important because prior studies at the national, state/provincial, and individual level have consistently identified one or several of these factors as strongly associated with technology levels. Knowing what factors are highly associated with e-entertainment provides knowledge that can help interventions of varied kinds to stimulate more access and use of e-entertainment. Further, we identify and explain the nationwide geographic patterns and clustering of e-entertainment use of the Internet.

This study of e-entertainment services for the United States is novel, because it bases the research on a large sample of 3,109 continental U.S. counties, it analyzes nationwide geographic patterns of ICT-enabled e-entertainment services by counties, and examines the association of social capital with e-entertainment use of the Internet in counties. Social capital was found to facilitate Internet access and use in prior studies (Chen 2013; Pick, Sarkar, and Johnson 2015). Our empirical treatment of social capital in the form of an index (Rupasingha and Goetz 2008) differs from prior studies (Chen 2013), and to the best of our knowledge, our work represents the first attempt to study the impact of socio-economic factors on e-entertainment use at the level of U.S. counties.

The remainder of this chapter is organized into sections on a literature review of technology adoption to deliver e-services in the United States, a research model of Internet use for e-entertainment, research methodology, exploratory spatial analysis of geographical patterns of e-entertainment use, regression findings, discussion of policy implications of the findings at the county, metropolitan, state and federal levels, and conclusions.

2. Literature Review

2.1 *Digital-Divide Research and the Question of E-Entertainment*

Research on the digital divide has examined disparities and unevenness in access and use of digital technologies from varied perspectives, units of analysis, and regions. Some theory has been developed for understanding the digital divide, but the conceptual effort is somewhat disconnected and no overarching theory dominates (Pick and Sarkar 2016). Several theories are based on the adoption-diffusion theory (ADT), the unified theory of acceptance and use of technology (UTAUT), and van Dijk's theory of digital technology access and societal impact (Rogers 2003; van Dijk 2005; Venkatesh et al. 2003). ADT theory emphasizes the adoption over time of ICT innovations. UTAUT theory focuses on behavioral intention, the degree to which an individual wishes to use a technology (Niehaves and Plattfaut 2014). Van Dijk's theory proposes that personal background inequalities and positional characteristics give access to resources; this combined with ICT usage leads to participation in society, which in turn feeds back into positional attributes (van Dijk 2005).

Among the methods commonly used to examine ICT adoption and digital divides are surveys (Perrin and Duggan 2015), multivariate analysis (Khatiwada and Pigg 2010; Ono and Zavodny 2007; Vicente and Lopez 2011), structural equation modeling (Pick and Azari 2011), spatial analysis (Grubesic 2004, 2006), and qualitative methods (Kvasny and Keil 2006). Common to the methods is the goal to compare digital-divide levels of entities (i.e., nations, states/provinces, counties, organizations, individuals) and to understand how digital divides are formed, and what social, economic, political, and behavioral factors are associated with digital levels of entities.

Throughout the United States, access and use of information technologies and social media vary considerably (Azari and Pick 2005; Chen 2013; Grubesic 2004, 2006; Pick, Sarkar, and Johnson 2015; Zickuhr and Smith 2012). Most of the digital-divide research in the United States has focused at the individual level including the National Telecommunication and Information Administration (NTIA) studies in the early 2000s, Pew Center reports (Perrin and Duggan 2015; Zickuhr and Smith 2012), and academic studies (Agarwal, Animesh, and Prasad 2009; Chen 2013, Pick, Sarkar, and Johnson 2015). At the state level in the United States, adoption and use of technologies and social media have indicated the importance of urban location, social capital, and education, while social media use has been influenced by immigrant population, Asian ethnicity, and education (Pick, Sarkar, and Johnson 2015).

In the European context, a nationally representative survey series of over 1,500 individuals in the United Kingdom—a nation comparable to the United States in networked readiness, revealed the pursuit of entertainment to be the leading purpose of Internet use across age groups (Helsper and Eynon 2010). While usage declined with age cohorts, novice users with less than 6 months of Internet usage experience engaged in Internet use for entertainment more than advanced users. Helsper and Eynon added more nuance to the concept of the digital native and observed that generation alone does not define if an individual is a digital native or not. Gender, education, experience, and breadth of use also play a part. Using 2011 data from the same UK-based survey series, 44% of next generation users considered the Internet important for entertainment, by far the most for important among any category of media (Dutton and Blank 2014).

2.2 Digital Divide Research at the U.S. County Level

In this chapter, we examine the use of the Internet to pursue entertainment activities in U.S. counties. The county lies below states in the hierarchy of U.S. Census geographic entities. It has been the basis of recent ICT diffusion and digital-divide studies of the U.S. federal government (The White House 2015), as well as of studies of spatial dimensions and patterns of Internet activity and growth (Grubesic 2002; Khatiwada and Pigg 2010), e-government practices (Huang 2006), and so on. A study of the influence of social capital on economic growth of communities has identified inputs into the production of social capital at the level of U.S. counties (Rupasingha, Goetz, and Freshwater 2006). Lastly, U.S. counties are one of the country's oldest forms of government dating back to the 1600s. While the structuring of county governments is diverse and basic roles and responsibilities of county governments are established by the states, counties administer their own economy, education, justice and public safety, health care, social services, transportation, and other services that directly impact the county residents. In addition, counties are often responsible for implementing a broad array of federal, state, and local programs.

The U.S. Census officially defines metropolitan and micropolitan areas, which consist of counties associated with those areas (U.S. Bureau of the Census 2016). A metropolitan county contains one or more urban core areas of at least 50,000 population. A micropolitan county contains one or more urban areas of at least 10,000 but less than 50,000 population. All remaining counties are considered rural.

The county unit has been rarely used in U.S. digital-divide studies for several reasons: difficulty in procuring datasets for all counties and county-equivalents, especially for ICT and social media dependent variables; lack of theory to base county studies on; and the challenge of sparseness of data for rural areas. The U.S. government has collected very little data on ICT access and use by individuals at the state or county level, unlike some other nations such as China and Japan. On the other hand, the United States does collect information on information-related businesses at the county level. A study based on a 5% random sample of U.S. counties from the U.S. Economic

Census of 1997 found that important determinants of the size of technology sectors were professional/scientific workforce, services, income, federal grants, education, and Latino and Black ethnicities (Azari and Pick 2005). Khatiwada and Pigg (2010) examined socio-economic and geographic factors affecting the adoption of Internet services in U.S. counties, using Internet service providers from the U.S. Census county business patterns of 2005. Using a large sample of close to 3,000 counties, Khatiwada and Pigg determined that urban counties with more educated people and business establishments per capita and higher housing value have higher presence of Internet service providers. In other words, the market is the driving force for Internet diffusion. This study included civic organizations among the initial set of independent variables but subsequently dropped it due to multicollinearity problems. Nonetheless, civic organizations have been considered for development of social capital constructs or indices (Rupasingha, Goetz, and Freshwater 2006), affirming the importance of social capital in our model of e-entertainment adoption in U.S. counties.

While research into e-governance aspects exists for U.S. counties, most of this research analyzes county governments' adoption of e-government and related services on their websites (Huang 2006, 2007; Manoharan 2013), not individuals' use of e-government services at the nationwide level. Studies by Huang (2006, 2007) revealed that a county's population size, demographic factors such as ethnic diversity and education, economic attributes such as housing, income, and business factors are associated with adoption of e-government by counties. Huang (2007) concluded that e-government diffusion is different from e-government adoption in terms of functional sophistication. Manoharan (2013) found that predictors of e-government adoption by U.S. counties included technical capacity (IT employees), website longevity, employee support, organizational size (number of functions), external collaboration with nonprofits, private nonfarm business units, population, and education. Manoharan (2013) reiterated that research into adoption and use of the Internet by the U.S. consumer to access e-services, such as e-entertainment, e-government, and others, is presently missing in the literature. The study we present in this chapter attempts to fill this void.

3. Research Model of Internet Use for E-Entertainment

The research model of Internet use for e-entertainment in U.S. counties (Figure 6.1) is drawn from the Spatially Aware Technology Utilization Model (SATUM, Pick and Sarkar 2016), which is appropriate for examining composite influences of various social, economic, and political determinants of ICTs. SATUM was induced from extensive prior literature on the social, economic, societal, and innovation correlates of ICT access and use (Pick and Sarkar 2015, 2016). It draws on multivariate modeling and theories of geographical agglomeration, and it is relevant for data that include geographic location. In this research, we use SATUM to posit associations of 13 demographic, economic, educational, government support for education, innovation, and social capital independent factors with five e-entertainment dependent variables. While the digital-divide literature has used a number of theories and models to explain technology adoption and diffusion (Pick and Sarkar 2016), SATUM is unique in explicitly considering underlying geographic relationships as components so that a research study can assess the extent of spatial bias in standard multivariate statistical analysis, an assessment often ignored in the digital-divide literature. Screening for the presence of spatial autocorrelation, a common bias problem plaguing any geographically referenced phenomenon, is part of SATUM, as shown in Figure 6.1. Also, k-means cluster analysis is applied as an exploratory method to determine the high- and low-intensity geographical groupings of counties throughout the United States.

Demographic influences of factors such as race and ethnicity, urban location, and educational attainment on the adoption and use of ICTs have been extensively discussed and validated in prior literature (Agarwal, Animesh, and Prasad 2009; Chen 2013; Grubesic 2006; Niehaves and Plattfaut 2014; NTIA 2011; Perrin and Duggan 2015). As an education variable, college education is included due to its use in prior studies (Azari and Pick 2005; Chen 2013, Pick, Sarkar, and Johnson 2015) and because it represents a broader cumulative range of educational experiences. In our model (Figure 6.1), we introduce young dependency ratio, working age population, Asian, Black, and Hispanic as race/ethnicity variables, urban location, and college education, and local government educational expenditure as independent variables to be associated with e-entertainment use in counties. Since elderly people show greater reluctance to adopt new technologies than teenagers (Niehaves and Plattfaut 2014; Srinuan and Bohlin 2011) and age influences on ICT adoption have been well-established (Ono and Zavodny 2007; Selwyn 2006), we include the age variables of young dependency ratio and working age population in our conceptual model. We include working age population because of its importance for Internet and mobile phone use in a study of digital divide among Japanese prefectures (Nishida, Pick, and Sarkar 2014). Ethnic variables were shown to have important influences on adoption of technologies for individuals in the United States (NTIA 2010; Perrin and Duggan 2015), while proportion of urban population is known as a crucial factor for technology use for individuals in the United States (Chen 2013) and U.S. states (Pick, Sarkar, and Johnson 2015).

Income is included in the model because of the strong influence of household income on Internet use in the United States for the period 2010–2015 (Perrin and Duggan 2015), while service occupation was found to be a significant correlate of payroll/receipts for the broadcasting/telecommunications industry and motion picture/sound industry for U.S. counties (Azari and Pick 2005). We reason that construction occupation, significant in a worldwide study of nations (Quibria et al. 2003), is essential for developing modern Internet infrastructure in U.S. counties.

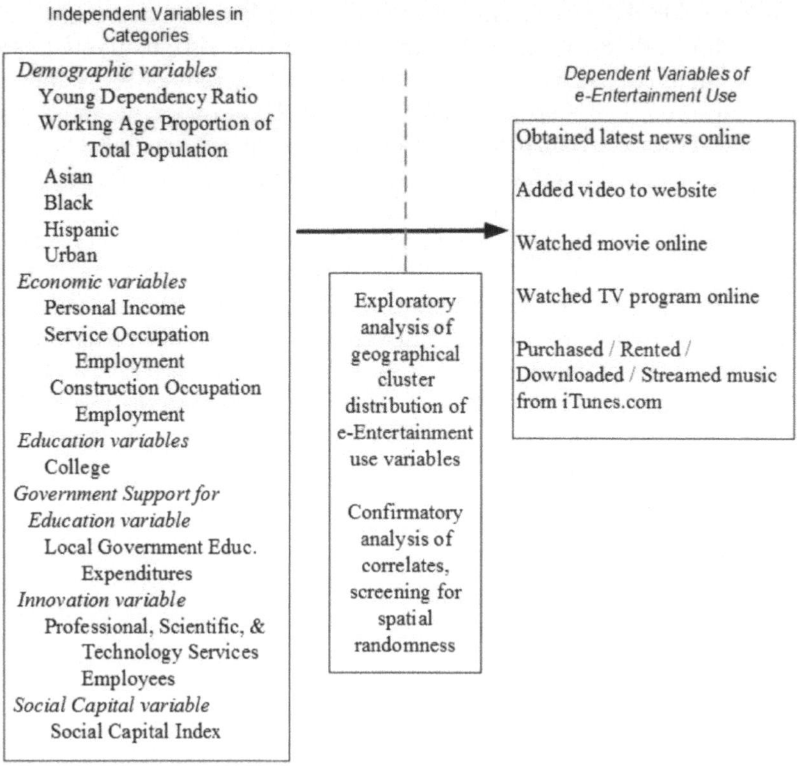

Figure 6.1 Operationalized Conceptual Model for E-Entertainment Use in U.S. Counties.

Innovation, which manifests itself in different forms in the digital-divide literature, has often been found to be associated with higher levels of ICT utilization (Azari and Pick 2005; Pick and Azari 2008; Quibria et al. 2003). In addition, we include professional, scientific, and technical services workforce as an independent variable proxy for innovation, since this segment of the workforce is more likely to use ICTs to foster innovation.

Social capital, in turn, is the scope of ties and linkages in a population through physical and communication means as well as by organizations that foster human collaboration or by bonding between people who possess resources (Chen 2013). Social capital was a key factor in a survey of the U.S. digital divide for a sample of individuals (Chen 2013), as well as in a two-stage study of the decisions by persons to go online, subject to peer influences (Agarwal, Animesh, and Prasad 2009). Chen (2013) posited that social capital might effect digital divides in two ways: through bonding social capital, namely, strengthening ties among similar, peer group people and by bridging social capital, namely, strengthening ties to heterogeneous groups outside the peer group, and the Chen study found some empirical support for both. Social capital of the bonding type also manifested itself in the form of cooperative society membership for a study of ICT adoption in Indian states (Pick, Nishida, and Sarkar 2014). In the present study, we posit social capital to enhance e-entertainment use in continental U.S. counties based upon prior literature and use a Putnam-like index (Putnam 2000) developed by Rupasingha and Goetz (2008).

The choice of the five e-entertainment dependent variables from a set of over 50 Internet use variables encompasses streaming music, TV, and movies, and keeping up with world news, which are among the top five most popular online activities of certain user groups, such as millennials in 2015 (Statista 2015b). Also, it encompasses e-entertainment service providers such as Netflix, YouTube, and iTunes, which were among the five leading Internet applications in North America in March 2015 (Statista 2015c).

Our research questions are:

1. *What are the spatial patterns of high and low levels of e-entertainment for U.S. counties?*
2. *What are the social, economic, ethnic, innovation, and social capital influences on e-entertainment services for U.S. counties?*

4. Research Methodology

4.1 Research Method and Justification

The research methodology collects data, which are analyzed descriptively, mapped, and explored by cluster analysis, and analyzed by spatial autocorrelation; this is followed by regression analysis, and, finally, diagnostic testing of residuals. The justification is that this method will provide both geographical information about the spatial patterning of e-entertainment in U.S. counties, as well as regression-based assessment of multiple correlates of e-entertainment, while evaluating the quality of the regressions by diagnostic and spatial testing of residuals. The approach has been applied to several studies conducted by the authors (Nishida, Pick, and Sarkar 2014; Pick, Nishida, and Sarkar 2014, 2015; Pick and Sarkar 2015).

4.2 Research Design

The methodological research design included (1) data collection, (2) descriptive statistics and correlation analysis, and (3) exploratory analysis of geographic patterns for dependent e-entertainment variables, which consisted of cluster analysis based on the full set of dependent variables and computation of Moran's index (Moran's I) for each dependent variable, (4) confirmatory analysis of correlates of each dependent variable using OLS stepwise multiple regression, and (5) diagnostic and spatial autocorrelation analysis of regression residuals, based on the Moran's I test.

4.3 Data Collection Techniques

Data on the independent variables such as population, ethnicity, employment, income, and governmental support for education were obtained at the county level from various census sources (see Table 6.1). Data on U.S. county-level social capital came from Rupasingha and Goetz (2008). While data for several independent variables were obtained for the period 2010–2012, social capital data were from the year 2009, which is the latest year in which such data are available. Data on the five e-entertainment dependent variables—namely, use of the Internet to obtain latest news, add a video online, watch movies online, watch TV programs online, and purchase/rent/order streaming music from iTunes—were provided by the Esri/GfK MRI DoubleBase Survey (Esri 2014). Esri computes estimates of these dependent variables; relevant data collection and estimation methodology statement for dependent variable data can be found in Esri (2015). The federal government and Esri/GfK perform data verification and validation.

Since all dependent variables were for the three-year period 2010–2012, time simultaneity is honored. The fact that the data for some of the independent variables lags by five years is acceptable since it is reasonable to assume a time-lagged impact of independent variables, such as employment on e-entertainment dependent variables. Our sample of 3,109 counties consists of all continental U.S. counties, excluding those of Alaska and Hawaii, which were not included because the data on social capital were unavailable for these two states. Henceforth, in this chapter, when our sample or data are referred to, it refers to counties in the lower 48 states of the continental United States. Variable definitions, sources, and descriptive statistics (n = 3,109 counties) of the dependent and independent variables appear in Table 6.1.

4.4 Data Analytics

Descriptive statistics, means, and standard deviations were computed for all variables in order to gauge the averages and extent of variation for all U.S. counties. Correlation analysis was applied in order to screen the independent variables for multi-collinearity. Variables were mapped using a Geographic Information System (GIS). Given our sample of 3,109 counties, map displays revealed key trends, without overloading the display. Dependent variables were mapped to gain preliminary understanding of the geographic patterning of e-entertainment for the United States. K-means cluster analysis of the dependent variables was then applied to identify groups of counties that are most similar in their use of all five e-entertainment dependent variables. K-means cluster analysis is an exploratory, not confirmatory method.

Table 6.1 Descriptive Statistics

Abbreviation	*Name of Dependent Variable*	*Source*	*Year of Data*	*Definition*	*Min*	*Max*	*Mean*	*SD*
BRDBND	Population with access to broadband	NTIA_SBI_Analyze	2012	% POP with access to any broadband technology (excluding satellite)	0.27	1.00	0.99	0.04
NEWS30R	Internet last 30 days: obtained latest news	BA	2010–2012	% Adults who obtained latest news using Internet in past 30 days	0.16	0.66	0.35	0.08
VIDAUTH30R	Internet last 30 days: added video to website	BA	2010–2012	% Adults who added video to website using Internet in past 30 days	0.02	0.10	0.04	0.01
MOVIE30R	Internet last 30 days: watched movie online	BA	2010–2012	% Adults who watched movie online in past 30 days	0.03	0.26	0.08	0.03
TVOL30R	Internet last 30 days: watched TV program online	BA	2010–2012	% Adults who watched TV program online in past 30 days	0.04	0.29	0.09	0.03
ITUNE12R	Ordered from website last 12 mo: iTunes.com	BA	2010–2012	% Adults who purchased/rented/downloaded/streamed music from itunes.com in past 12 months	0.02	0.17	0.06	0.02

Table 6.1 (Continued)

Variable	*Name of Independent Variable*	*Source*	*Year of Data*	*Definition*	*Min*	*Max*	*Mean*	*SD*
YOUNGDEPR	Young Dependency Ratio	CENDEC10_DP01	2010	POP 0–19/POP 20–64	0.17	0.88	0.45	0.07
COLLEGER	College Graduates or Higher, Age 18+ (%)	CENACS10_DP02	2008–2012	Population Estimate of College Graduates or Higher, Age 18+	0.03	0.54	0.13	0.06
PINCPC12	Personal Income Per Capita	BEA_CA1–3	2012	Personal Income Per Capita ($)	17,264.00	116843.00	36523.76	9182.78
WKAGEPOP	Working Age Pop (Pop. 20–64)/Total Population	CENDEC10_DP01	NA	POP 20–64/Total POP	0.46	0.74	0.58	0.03
SERVICER	Service Occupations (%)	CENACS10_DP03	2008–2012	Persons 16+ in Service Occupations	0.03	0.21	0.10	0.02
CONSTRUCTR	Construction Industry, Employed Persons 16+ (%)	CENACS10_DP03	2008–2012	Construction Industry, Employed Persons 16+	0.00	0.16	0.04	0.01
ASIANR	Asian Population (%)	CENDEC10_DP01	2010	Asian Population	0.00	0.36	0.01	0.02
BLACKR	Black Population (%)	CENDEC10_DP01	2010	Black or African-American Population	0.00	0.86	0.10	0.15
HISPANICR	Hispanic/Latino Population (%)	CENDEC10_DP01	2010	Hispanic or Latino Population	0.00	0.96	0.08	0.13
URBAN	Urban Population (%)	CENDEC10_DP01	2010	Urban Population	0.00	100.00	41.48	31.44
LGOVEDEXPR	Local Govt Education Expenditures	NCES_F33	2009	Sum of School District Expenditures divided by total pop for all School Districts within a County ($)	69.57	36664.34	1985.97	1144.67

(Continued)

Table 6.1 (Continued)

Variable	*Name of Independent Variable*	*Source*	*Year of Data*	*Definition*	*Min*	*Max*	*Mean*	*SD*
PSTSVCEMPR	Professional, Scientific, and Tech Services Employees (%)	CENECON07	2007	Number of Professional, Scientific, and Technical Services Employees	0.00	0.18	0.01	0.01
FACTORAVE	Transformed Social Capital	aese.psu.edu	2009		0.00	0.47	0.17	0.04
						n = 3,109		

	SOURCES	*Abbrev*	*Detailed Reference*
	U.S. Census Bureau, DEC 2010, Table DP-01	CENDEC10_DP01	U.S. Census Bureau, Decennial Census 2010, Summary File 1
	U.S. Census Bureau, ACS 2012, Table DP-02	CENACS12_DP02	U.S. Census Bureau, American Community Survey 2012, 5-year estimates, Table DP-02
	U.S. Census Bureau, ACS 2012, Table DP-03	CENACS12_DP03	U.S. Census Bureau, American Community Survey 2012, 5-year estimates, Table DP-03
	U.S. Census Bureau, ACS 2012, Table DP-04	CENACS12_DP04	U.S. Census Bureau, American Community Survey 2012, 5-year estimates, Table DP-04
	U.S. Census Bureau, ACS 2012, Table DP-05	CENACS12_DP05	U.S. Census Bureau, American Community Survey 2012, 5-year estimates, Table DP-05
	Federal Communications Commission, Form 477	FCC_477	Federal Communications Commission, Form 477, Local Telephone Competition, and Broadband Deployment
	Bureau of Economic Analysis, CA1–3	BEA_CA1–3	Commerce, Bureau of Economic Analysis, CA1–3, Personal Income, Per Capita Income
	Bureau of Economic Analysis, CA04	BEA_CA04	U.S. Dept of Commerce, Bureau of Economic Analysis, CA04, Personal Income Summary

Table 6.1 (Continued)

Variable	*Name of Independent Variable*	*Source*	*Year of Data*	*Definition*	*Min*	*Max*	*Mean*	*SD*
	SOURCES	*Abbrev*	*Detailed Reference*					
	U.S. Census Bureau, Economic Census 2007	CENECON07	U.S. Census Bureau, Economic Census 2007					
	U.S. Department of Commerce, NTIA, State Broadband Initiative, Analyze Table	NTIA_SBI_ Analyze	U.S. Dept of Commerce, National Telecommunications, and Information Administration, State Broadband Initiative (CSV format December 31, 2012).					
	National Center for Education Statistics, F-33	NCES_F33	U.S. Census Bureau, Governments Division, Local Education Agency (School District) Finance Survey (F-33), National Center for Education Statistics, Common Core of Data					
	Esri Business Analyst Data	BA	Esri Business Analyst Data, GfK MRI DoubleBase Survey 2012					
	Rupasingha and Goetz, 2008	—	Pennsylvania State University's Northeast Regional Center for Rural Development					

Cluster analysis results were mapped to provide a holistic perspective on clusters of counties with high, moderate, and low use of the Internet for e-entertainment along with important visual cues.

A key question is this study is whether e-entertainment usage in continental U.S. counties shows statistically significant patterns of agglomeration of high and low values, or is it spatially randomly distributed. We diagnose spatial autocorrelation for each dependent variable by using Moran's I test statistic. Moran's I test is inferential; the null hypothesis is that the values of a variable are randomly distributed spatially (Openshaw 1984). The Moran's I statistics for the dependent variables indicate how agglomerated counties are (positive value of Moran's I); if they are randomly distributed (Moran's I equals 0) or if a county with high e-entertainment is surrounded by low valued counties and vice versa (negative value of Moran's I).

OLS regressions were subsequently performed for each e-entertainment usage dependent variable, in stepwise order, allowing in only those independent variables with significance levels equal or less than 0.05. As an additional test of multi-collinearity, the variance inflation factor (VIF) was computed for each independent variable. We utilize the common cut-off of five or greater for VIF to be of concern and no multi-collinearity problems were detected. Three diagnostic test statistics—Joint Wald, Koenker (BP), and Jarque-Bera were administered to ensure that regression assumptions are met. Model relationships that result in spatially random errors were regarded as valid. If errors in the model fit are spatially autocorrelated, it implies that the geographic forces are exogenous to the conceptual model. If Moran's I testing indicates that regression residuals are not spatially randomly distributed, then regressions results have to be treated with caution.

5. Findings: Geographic Patterns of Internet Use for E-Entertainment in U.S. Counties

Geographic findings are presented first in a section on the national geographic distribution of individual e-entertainment variables, then in a section on the agglomeration of the individual variables, and finally in a section on the cluster analysis that identifies clusters based on all five dependent variables together.

5.1 Geographic Distribution of Individual Variables

The geographic distribution of use of Internet for watching movies, seen in Figure 6.2, shows higher per capita usage in the western U.S. counties, in the counties of the Boston-Washington megalopolitan area, and in other parts of the country—often centered around metropolitan areas. Use of the Internet for watching movies is comparatively lower along a north-south axis east of the Rocky Mountains extending in the rural and prairie areas from the Dakotas to Texas, in the non-metropolitan areas of the Southern

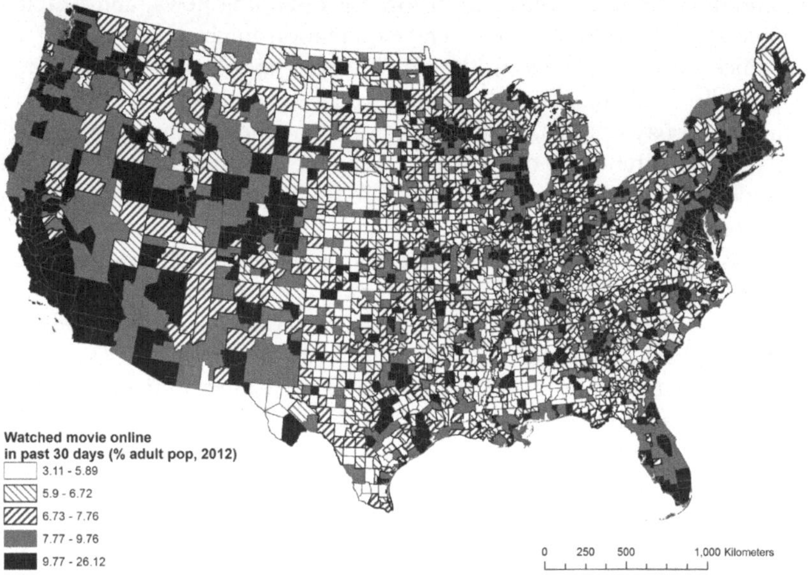

Figure 6.2 Internet Use for Watching Movies, U.S. Counties, 2012.

United States, excluding Florida and metropolitan parts of Texas, and in the Appalachian region.

The nationwide spatial distributions for the five e-entertainment factors may be summarized as having the large and intensive concentrations of high e-entertainment levels in the Boston to Washington megalopolis; West Coast counties extending from San Diego through Los Angeles to San Francisco and up through the Oregon coast to Seattle; an intensive high level in greater Denver stretching west to Salt Lake City, Chicago, and Milwaukee metro areas; Atlanta; and mid, and southern parts of Florida. They bear resemblance to the core of broadband providers identified by Grubesic (2006) at the zip-code level, which emphasizes the northeastern megalopolis, Atlanta, and Central/South Florida. A difference, however, is that the entire western coastal counties are not highlighted in Grubesic's study. Grubesic's Chicago-Milwaukee high intensity area is much smaller consisting of Chicago city and inner suburbs extending only to Milwaukee. In general, the combined factors of expanded mobile display capability, extension of wireless signal strength, consumer awareness, and growth in the metro area fringes would account for e-entertainment's leapfrogging outside of traditionally defined large metro areas.

Two prominent exceptions to the overall similarity of high-level geographic patterns apply for particular regions of the nation. First, the South

Texas counties bordering Mexico are very high for ordering from iTunes and adding video to a website, but low for obtaining news, and watching a movie or TV program online. The explanation might relate to cost factors, since this area is among the most poverty-stricken in the nation, so the periodic "Free on iTunes" segments and free uploads would be more popular than pay options. Also, the region's low-income households might have reduced interest in online news. Second, we posit that online news has a more intensive and broader reach in the upper Midwest—namely, in Minnesota, northern Illinois, Indiana, and, Ohio than online movies, TV programs, and adding videos. The reason is unclear, but possible factors requiring further research include greater intensity of online news and providers, heightened reduction in print news, elevated educational levels, and expanded presence of public TV and radio news websites.

Common nationwide concentrations of low e-entertainment levels are located along a north-south axis east of the Rocky Mountains and extending in the rural and prairie areas from the Dakotas to Texas, as well as in the nonmetropolitan South, excluding Florida and Texas, and in the Appalachian region. Most of these predominantly rural counties also have low average incomes, especially in Appalachia and the South, so the lack of affordability of e-entertainment services can reduce usage. These areas are characterized by fewer Internet service providers (Khatiwada and Pigg 2010) and lower Internet speeds. Another reason for low e-entertainment usage in Appalachian counties and scattered sections of the South is somewhat lower access to broadband, in the range of 50%–95% of the national average. The distribution pattern of low use is very consistent across e-entertainment types, with the exception of watching a TV program online for the N-S prairie area, which rises from typically low level for the other four types to a medium level.

The county-level mapping provides a much richer and more detailed view of the geographic textures of e-entertainment nationwide. For instance, the areas of high use and low use do not correspond to state boundaries, which points to major weaknesses in relying on and interpreting state- or provincial units of analysis of ICT/Internet variables (Pick, Sarkar, and Johnson 2015). The value of cluster maps and characterizations is that profiles of usage patterns become available providing a more refined view of the multiple dimensions of user experience. The cluster characterizations could be further detailed by producing clusters and characterizations for smaller geographies of interest, such as zip codes in populous metropolitan areas, census tracts in university towns, or a sample of impoverished areas.

5.2 Spatial Autocorrelation Findings for E-Entertainment Variables

Findings for Moran's I indicate that significant positive autocorrelation is present for all the e-entertainment dependent variables, which is reflected in the agglomeration of high values and low values (see Table 6.3). In other words, there are "hot spots" where many high valued counties neighbor each other and "cold spots," where many low valued counties neighbor each other. This confirms that for the nation as a whole, there are "hot" and "cold" spots for each of the e-entertainment variables, which is important in recognizing that neighboring counties are usually at the same high or low e-entertainment level.

The dependent variables were also mapped using the Local Moran's I statistic. This Local Indicator of Spatial Association (LISA)-based method shows local patterns of spatial association that help to identify a single "outlier" county relative to its neighbors. LISA analysis of e-entertainment variables shows combinations of "high-high" counties (i.e., a high e-entertainment use county surrounded by other high e-entertainment use peers) in U.S. metro areas and "low-low" counties in rural areas. Interestingly, a handful of "high-low" outliers (i.e., a high e-entertainment use county surrounded by low e-entertainment use neighboring counties) are also observed. Upon closer examination, we see that almost every "high" county in a "high-low" combination is home to at least one college or university, or contains a military reservation or government laboratory. A case in point is Brazos County, Texas, which is a "high" e-entertainment "outlier" county surrounded by other low e-entertainment counties in a rural part of Texas. Unsurprisingly, Brazos County is home to Texas Agricultural and Mechanical University (TAMU), a research-intensive flagship university.

5.3 Cluster Analysis of the Five E-Entertainment Variables

The trends noted for individual e-entertainment variables are summarized and characterized through applying k-means cluster analysis, and mapping and characterizing the clusters. As seen in Figure 6.3, the very high-usage areas appear in clusters 1 and 2. For the West Coast, rather than a solidly high-usage coastal region, only selected concentrations appear in clusters 1 and 2, in particular San Diego, Orange, Ventura, Santa Barbara, and San Luis Obispo counties, most of the San Francisco Bay Area, and Portland and Seattle metro areas.

The characterization of the clusters in Table 6.2 is helpful in understanding the actual meaning of high versus low e-entertainment penetration. For the high-level clusters 1 and 2, half of the county populations on average use online news, compared to only 6% that adds video to a website and 11% to 17% watches movies or TV online or accesses iTunes. Partly the differences reflect the decline in print news, but also that accessing news is a less time consuming and more user-friendly activity than say uploading videos. At the low end for clusters 7 and 8, about 30% of county populations access online news, compared to 3%–4% who add video to a website, and 5%–8% for the other variables. For people in the latter clusters, online news stands out as the only prevalent form of e-entertainment. Reasons for low usage of e-entertainment include higher relative cost (versus household income), lower bandwidth, and less leisurely lifestyles.

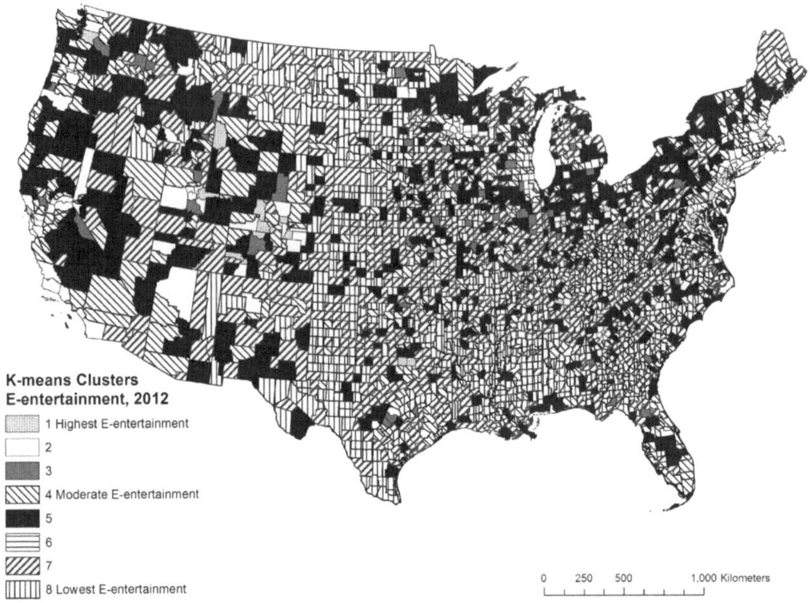

Figure 6.3 K-means Clusters of Internet Use for E-Entertainment, Continental U.S. Counties, 2010–.

6. Findings on Correlates of E-Entertainment Use

The OLS regression findings identify the determinants of e-entertainment uses for all the counties in the lower 48 states of the United States, as well as for the subsamples of metropolitan, micropolitan, and rural counties. For the entire lower-48 sample of 3,109 counties, the results in Table 3 reveal that the most important e-entertainment determinant is college graduation, followed by young dependency ratio and working age population. The education finding is in concert with prior research (Perrin and Duggan 2015; Pick, Sarkar, and Johnson 2015), which concerns larger geographic units. At the county level, this finding is in concert with strong association of college graduation with receipts and payrolls in the broadcasting/telecommunications sector, but is not consistent with the lack of association of college education with the motion picture-sound sector, in a prior study of counties (Azari and Pick 2005). It makes sense that counties with high levels of education would have more e-entertainment use, since more educated citizenry would tend to seek more news, information, video, or TV information. This was corroborated by multiple Pew Research Center findings including that 47% of a random sample of 18+ college-educated population in the United States played video games, compared to 40% of less-than-high-school educated (Duggan and Page 2015), and that 44% of

Table 6.2 K-means Cluster Characteristics for E-Entertainment, Continental U.S. Counties, 2010–2012

Dependent Variable	*Cluster Number* 1	2	3	4	5	6	7	8	*MAX*	*MIN*	*RATIO of Max to Min*
Internet last 30 days: obtained latest news	0.56	0.49	0.50	0.42	0.37	0.23	0.32	0.26	0.37	0.23	1.65
Internet last 30 days: added video to website	0.06	0.06	0.07	0.05	0.04	0.10	0.03	0.04	0.10	0.03	2.89
Internet last 30 days: watched movie online	0.14	0.12	0.18	0.11	0.08	0.18	0.07	0.05	0.18	0.05	3.28
Internet last 30 days: watched TV program online	0.17	0.14	0.21	0.12	0.09	0.25	0.08	0.07	0.25	0.07	3.38
Ordered from website last 12 mo: iTunes.com	0.14	0.11	0.11	0.09	0.07	0.13	0.05	0.05	0.13	0.05	2.62
Number of Counties	43	185	63	426	715	1	903	773			

college-educated respondents obtained daily news on a smartphone versus 31% of high school educated (Pew Research Center 2012).

Results show that young people are more likely to utilize the five forms of e-entertainment. Young persons under age 20 in 2010–2012, the period of the study, were nearly entirely raised in the era of the web, which commenced in 1991, so they grew up exposed to it. As new forms of e-entertainment, such as iTunes, obtaining the latest news, online TV, online movies, and video uploading became available, the young user adopted the web-based delivery mode much more rapidly than older people. This finding for older people was corroborated by a study of small regions in Europe, in which greater prevalence of older people reduced ICT use (Vicente and López 2011). Likewise, counties with high proportion of working age population (20–64 years old) have significant use of e-entertainment. This age group consists of a combination of Generation X and the parents of millennials. Generation X had exposure as teens and young adults to web-based e-entertainment, while millennial parents have had indirect effects of learning to use it from their children.

The proportion of adults who are service workers is a secondary correlate of all dependent variables except for iTunes and corresponds to findings at the U.S. county level for the effects of service workforce on receipts/payroll for motion picture/sound and broadcasting/telecommunications sectors (Azari and Pick 2005).

Regarding ethnicity, counties with high Asian proportion have significant association with e-entertainment, although at a more moderate level compared to education or age structure. This effect is related to generally high levels of Asian technology usage across the United States, including that 97% of Asians reported using the Internet (Perrin and Duggan 2015). Another recent Pew finding showed that the Asian advantage holds also for "having broadband at home," in particular 84% of English-speaking Asians, versus 50% of Hispanics. Likewise, 91% of Asians own a smartphone versus 65% of Hispanics (Perrin 2016). On the other hand, counties with high levels of Black or Hispanic population tend to have reduced usage of obtaining the latest web-based news, a finding similar to associations noted between Black and Hispanic population groups and Internet usage in the nation as a whole (Perrin and Duggan 2015). The finding for African-American race being associated with reduced Internet use is similar to a prior finding (Chen 2013). However, the finding for Latinos differs from prior studies in which Hispanic ethnicity was associated with increased receipts/payrolls for motion picture/sound industry (Azari and Pick 2005) and positively related to Facebook usage in U.S. states (Pick, Sarkar, and Johnson 2015).

The positive influence of urban location on iTunes, obtaining the latest news, and watching movies online is due to the greater entertainment content provision in urban areas, as well as the generally expanded Internet usage in the more sophisticated urban economies, and the associated presence of younger average age, higher income, and greater educational attainment (Perrin and Duggan 2015).

Although social capital in prior studies has partially been related to higher use of technologies at the individual level for the United States (Chen 2013) and the state level for India (Pick, Nishida, and Sarkar 2014), our findings at the county-level show that social capital reduces both adding video to a website and iTunes usage. The explanation is a substitution effect—namely, the lack of social capital or neighborhood social connections—such as clubs, interest groups, community organizations, might encourage the more physically isolated citizens to make up for social connections virtually through substituting e-entertainment, at least for adding video to a website and iTunes usage. The difference from the study of India (Pick, Nishida, and Sarkar 2014) could be explained by the low level of social media use in India in 2010 (eMarketer 2012), so there would not be a substitution effect, whereas the difference from the United States (Chen 2013) is unexplained. Regarding the lack of social capital's effect on obtaining news, adding videos, and watching TV we speculate that those forms of e-entertainment are more individually directed and not tied to physical social networking.

For the national sample, adjusted R^2 is very high (i.e., over 0.63). Although the Joint Wald Statistic is consistently significant confirming lack of joint influences from independent variables, the significant Koenker and Jarque-Bera statistics indicate that the regression model has substantial heteroscedasticity—namely, inconsistent variance of residuals, and that the residuals do not conform to the normal distribution in skewness and kurtosis.

Such weaknesses may be due to presence of outliers in the very large sample, and hence the findings should be viewed somewhat cautiously. Moran's I indicates that significant autocorrelation is present for all the dependent variables, which is reflected visually in the agglomeration of high values and low values present for the map of watching movies in Figure 6.2.

Although the findings for the metropolitan and micropolitan subsamples largely correspond to those for the nation, the rural subsample has distinctive features. For the metropolitan sample (Table 6.3), the three most dominant independent factors again are college education, young dependency ratio, and working age population. However, college education has stronger effect in metro areas versus the nation, which reflects that metropolitan counties' e-entertainment is even more closely related to high educational attainment. Service occupation is significant for watching TV online, adding video to website, and watching a movie online, while construction occupation has inverse effect for the same three variables. Construction workers' reduced levels of usage may be due to their lesser education and low work exposure to web-based e-entertainment. Ethnicity has only minor influences such as the inverse association of Blacks with obtaining the latest news, similar to the nation, and the association of Hispanics with adding video to a website, which is unexplained. Social capital has no relationship to e-entertainment

Table 6.3 OLS Regression Findings for Socio-Economic Determinants of E-Entertainment Variables, 2010–2012, Entire Country (Lower 48 States), Metropolitan, Micropolitan, and Rural Samp

		Lower 48 States					*Metropolitan*			
*Independent Variable**	*Indep. Var. Definition*	*ITUNE 12R*	*NEWS 30R*	*TVOL 30R*	*VIDAUTH 30R*	*MOVIE 30R*	*ITUNE 12R*	*NEWS 30R*	*TVOL 30R*	*VIDA 30R*
YOUNGDEPR	Young Dependency Ratio	0.303 ***	0.164 ***	0.208 ***	0.303 ***	0.261 ***	0.367 ***	0.233 ***	0.285 ***	0.315 ***
COLLEGER	College Graduates or Higher, Age 18+	0.671 ***	0.626 ***	0.555 ***	0.466	0.440 ***	0.862 ***	0.810 ***	0.612 ***	0.683 ***
PINCPC12	Personal Income Per Capita									−0.17 ***
WKAGEPOP	Working Age Pop (Pop. 20–64)/Total Population	0.295 ***	0.214 ***	0.295 ***	0.380 ***	0.337 ***	0.282 ***	0.207 ***	0.364 ***	0.433 ***
SERVICER	Persons 16+ in Service Occupations		0.079 ***	0.138 ***	0.125 ***	0.145 ***			0.161 ***	0.124 ***
CONSTRUCTR	Persons 16+ in Construction Occupations			−0.113 ***					−0.169 ***	−0.15 ***
ASIANR	Asian	0.132 ***	0.109 ***	0.145 ***	0.198 ***	0.208 ***				
BLACKR	Black		−0.204 ***					−0.170 ***		
HISPANICR	Hispanic		−0.107 ***							0.161 ***
URBAN	Urban	0.126 ***	0.216 ***			0.154 ***	N/A	N/A	N/A	N/A
LGOVEDEXPR	Local Govt Education Expenditures									
PSTSVCEMPR	Prof, Scientific, Tech Services									
FACTORAVE	Social Capital	−0.096 ***			−0.148 ***					
Joint Wald Statistic		6867.020 ***	7438.274 ***	2489.601 ***	2832.429 ***	3565.527 ***	3192.974 ***	2445.935 ***	2097.008 ***	2362. ***
Koenker (BP)		219.540 ***	141.964 ***	206.530 ***	240.588 ***	235.045 ***	23.158 ***	2.786	55.076 ***	29.14 ***
Jarque-Bera		463.499 ***	130.998 ***	17578.058 ***	9084.632 ***	11893.920 ***	92.258 ***	53.486 ***	3596.810 ***	2689. ***
Spatial Autocorrelation of Dep Variable (Moran's I)		0.499 ***	0.547 ***	0.309 ***	0.343 ***	0.353 ***	0.631 ***	0.648 ***	0.452 ***	0.472 ***
Spatial Autocorrelation of Regression Residuals (Moran's I)		0.244 ***	0.357 ***	0.065	0.113	0.092	0.501 ***	0.567 ***	0.119	0.111
Adjusted R^2		0.785 ***	0.764 ***	0.631 ***	0.653 ***	0.676 ***	0.812 ***	0.775 ***	0.700 ***	0.728 ***
Sample Size		3109	3109	3109	3109	3109	1161	1161	1161	1161

* Signif. At 0.05, ** Signif. At 0.01, *** Signif. At 0.001
* Refer to Table 6.1 for the variable name abbreviations

Table 6.3 (Continued)

	Micropolitan					*Rural*				
MOVIE 30R	*ITUNE 12R*	*NEWS 30R*	*TVOL 30R*	*VIDAUTH 30R*	*MOVIE 30R*	*ITUNE 12R*	*NEWS 30R*	*TVOL 30R*	*VIDAUTH 30R*	*MOVIE 30R*
0.345 ***	0.407 ***		0.311 ***	0.394 ***	0.439 ***					
0.552 ***	0.644 ***	0.522 ***	0.478 ***	0.511 ***	0.425 ***	0.392 ***	0.466 ***	0.421 ***	0.313 ***	0.296 ***
				−0.178 ***			−0.148 ***			
0.445 ***	0.386 ***		0.393 ***	0.416 ***	0.494 ***					
0.190 ***	0.163 ***	0.206 ***	0.172 ***	0.164 ***	0.183 ***	0.141 ***		0.108 ***		0.184 ***
−0.181 ***			−0.159 ***		−0.129 ***		0.120 ***			
	0.126 ***	0.211 ***	0.234 ***	0.225 ***	0.238 ***	0.229	0.276 ***	0.224 ***	0.221 ***	0.270 ***
		−0.243 ***					−0.294 ***		0.146 ***	−0.153 ***
		−0.184 ***				0.127 ***			0.248 ***	
N/A	N/A	N/A	N/A	N/A	N/A	N/A	N/A	N/A	N/A	N/A
					-0.148 ***					
	−0.080 ***			−0.112 ***		−0.250 ***			−0.299 ***	−0.188 ***
1931.894 ***	1196.073 ***	1656.720 ***	474.021 ***	380.346 ***	535.457 ***	139.277 ***	1079.110 ***	147.743 ***	242.782 ***	210.342 ***
56.722 ***	54.563 ***	28.271 ***	120.990 ***	104.520 ***	121.519 ***	195.855 ***	129.130 ***	108.680 ***	126.223 ***	143.785 ***
1709.520 ***	80.259 ***	85.111 ***	2339.509 ***	1082.878 ***	1585.851 ***	6193.948 ***	30.271 ***	5892.443 ***	22834.132 ***	7393.511 ***
0.523 ***	0.242	0.233	-0.052	0.046	0.017	0.186*	0.480 ***	0.157*	0.217**	0.108
0.078	0.584	0.346	0.051	0.773	0.147	0.082	0.358 ***	-0.069	0.040	-0.065
0.710 ***	0.730 ***	0.701 ***	0.604 ***	0.600 ***	0.634 ***	0.316 ***	0.520 ***	0.324 ***	0.282 ***	0.315 ***
1161	637	637	637	637	637	1311	1311	1311	1311	1311

in metropolitan counties, which may be due to the ease in the city of compensating for low social capital with widespread social media.

For the micropolitan sample, we find again consistently positive effects of college education, young dependency ratio, working age population, and service occupation. Construction work relates inversely to watching TV online and watching a movie online. For ethnicities, a high Asian percent is again related to e-entertainment, as with the nation and metropolitan samples, while counties with high Black or Hispanic populations have comparatively lower e-entertainment level. Social capital is consistent with the national sample in reducing iTunes usage and adding video to a website.

For the rural sample, the effect of a county's percentage of college education is the most important factor, followed by percentage of Asian population, and service occupations. The proportion of Black race in rural counties is mixed in effect, while being Hispanic is positive for iTunes and adding video to a website, findings which are unexplained.

For the subsamples, the adjusted R^2 is highly significant in all regressions. The Joint Wald diagnostic statistic reveals a lack of association among independent variables. However, the Koenker and Jarque-Bera tests indicate uneven residuals and violation of skewness and kurtosis of the residual distribution, connoting some need for caution in relation to the findings. Moran's I measure for residuals is mostly random, demonstrating overall a good chance that the regression errors are not spatially biased. The next section discusses important findings in relation to the existing literature.

7. Discussion on Correlates of E-Entertainment Use

As seen in the previous section, the consistent correlates of e-entertainment in the continental U.S. counties are education, age structure, working age population, and service occupations, as well as urban location for the countrywide sample. These are well known and correspond to findings in prior studies (Azari and Pick 2005; Pick and Azari 2008; Perrin and Duggan 2015; Pick, Nishida, and Sarkar 2015). For instance, an investigation of 160 metropolitan and micropolitan U.S counties in 2000 showed that college graduation per capita was related strongly to receipts and payroll for the broadcasting and telecommunications industry sector, but not so for the motion picture and sound industry sector (Azari and Pick 2005). For a recent sample of the 50 U.S. states, higher education was consistently positive for contemporary ICT variables, except for its inverse effect on Twitter use (Pick, Nishida, and Sarkar 2015). It can be reasoned that Twitter is less appealing to the highly educated, given its succinct limit on length of message and often fleeting content.

Our findings on e-entertainment being associated with young and working age population are supported by results of a factor-analytic study of the dependent factor of ICT use for 164 subnational regions extending across the member states of the European Union (Vicente and López

2011)—findings that indicated that one of the three correlates of the ICT factor was the inverse effect of population over 65, which supports the positive effects of youth and working age. In concert with the present study, an earlier county-based study (Azari and Pick 2005) found that service occupation was a consistent positive influence on receipts and payrolls in both the broadcasting/telecommunications and motion picture/sound sub-industries, while other studies have confirmed that urban location is related to ICT uses (for example, Malecki 2003; Perrin and Duggan 2015; Pick, Nishida, and Sarkar 2015).

Ethnicity and race have varied association with e-entertainment use at the county level. Being Asian is the most prevalent influence, although not significant in the metropolitan subsample. Its high impact overall was seen previously in less-urban U.S. states (Pick, Sarkar, and Johnson 2015), and for the whole nation at the household level (Perrin and Duggan 2015). By contrast, nationally, Black race is significant and inverse only for the online news variable. In rural counties, it is additionally inverse for watching an online movie and positive for adding video to a website. Its overall inverse effect was strongly corroborated in a national survey by Perrin and Duggan (2015), but that study pointed to a narrowing gap among different racial and ethnic groups: "African Americans have seen the greatest growth rate between 2000 and today." It may be that the historical lower ICT use by Blacks and, in some instances, the long-term failure of ICT training (Kvasny and Keil 2006) is lessening today.

For Hispanics, the present results are less significant and inconsistent in directionality, so we recommend deeper research, perhaps with a design that does not aggregate all Hispanic populations together, but sorts them by country of origin, generational time in the United States and other refinements. Ethnic and racial differences in digital divides are not limited to U.S. counties; for example, in a study of 164 European subnational units, residence in a Nordic nation was a significant positive factor (Vicente and López 2011). Such comparative digital-divide studies imply that ethnic/racial influences should be studied for smaller geographic units in other parts of the world, which would shed more light on the complex dimensions of such influences.

Social capital and presence of professional/scientific/technical (PST) employees have no association with e-entertainment factors, in considerable contrast with other studies of the United States and Europe. Social capital was highly associated with ICT and social media variables in a study of U.S. states (Pick, Sarkar, and Johnson 2015) and in a nationwide survey study (Chen 2013). However, we argue that its absence of association in our study is due to the competing effects of social media, available throughout the United States. In particular, by 2012, the percent of online U.S. adults using Facebook, LinkedIn, and Twitter reached 67%, 20%, and 16%, respectively (Duggan et al. 2015), so the recent upsurge in social media use has substituted for physical social capital, which reduced to insignificance many of the social capital effects noted in prior studies with 5-year-old or even older data. Another possible explanation of the absence for association is the lack of explicit inclusion and modeling of "bonding" social capital in the social capital index used in the present study. Bridging social capital is helpful to promote Internet access but not to foster actual use of the Internet. To promote Internet use and online communication, resource-rich "bonding" social capital is most effective (Chen 2013). The index developed by Rupasingha and Goetz (2008) used in our study factors in the density of religious, professional, and labor associations, bowling alleys, and civic and social organizations in U.S. counties; however, explicit inclusion of "bonding" social capital based on blood, marriage, and related strong ties is missing from the index, thus possibly causing the lack of association of social capital with e-entertainment indicators for the most part.

The lack of association for PST employees is surprising, given this variable's major importance for revenues and payrolls of broadcasting/telecomm and motion picture/sound sectors in U.S. counties noted in Azari and Pick (2005). We reason that the lack of association might be explained by increases in outsourcing and telecommuting so that PST workers who influence e-entertainment usage are less likely than in the past to be located in the same county as the usage, but rather, in an online/outsourcing milieu, many of them can be located anywhere in the nation or beyond. For instance, one outsourcing survey of businesses indicated U.S. legal and accounting professional services are 40%–65% and 30%–50% outsourced, respectively (Deloitte 2014). The trend toward outsourcing would result in a more random association of PST and e-entertainment for counties.

8. Policy Implications and Limitations

This study informs the formation of policies by county governments and by state and federal governments affecting county citizens to improve their access to e-entertainment. A starting point is to consider the suppliers of e-entertainment, and then turn to policies that can influence supply and demand. E-entertainment is supplied mostly by businesses and somewhat by nonprofits. Businesses providing e-entertainment are competitive and adjusting to the fast-moving changes in the content, delivery mode, and coordination of e-entertainment (Loebbecke and Powell 2002; The Economist 2013a, b, c). All aspects of e-entertainment are growing in this decade (The Economist 2013c) including e-music, e-videos, e-television, e-publishing, and e-news. Estimates indicate that although in 2008 one-tenth of consumer buying of e-entertainment was in digital form, by 2017 it will be about one-half (The Economist 2013c). In addition to commercial e-entertainment services, public media nonprofits are expanding quickly into the web and mobile spaces to supply news, interviews, videos, and publications in digital form.

Given this backdrop, what roles do county and metropolitan government policies play, as well as those of state and federal governments? We divide our recommendations, based on the present research, into state and federal policies and policies for populous counties and metropolitan areas. Small counties are unlikely to have e-entertainment policies but rather rely on the state and federal governments.

Recommendations for federal and state government policies affecting county citizens:

- The federal and state governments can establish policies favorable to college education, which has many implications, one being that it would positively affect e-entertainment use, based on our findings. Furthermore, a more direct effect would be increased federal and state government policies to provide support to publicly supported entertainment by television, radio, and online entities that are often associated with state universities, colleges, and community colleges.

Recommendations for metropolitan and county policies affecting county citizens:

- Metropolitan and city government policies can seek to establish or expand free public broadband capability throughout the city or metropolitan areas, providing a broader base of e-entertainment access for citizenry. The present study supports such initiatives, since they can reach geographic parts of these city administrative districts that have ICT-deprived population.
- Metropolitan and city government policies can provide funding for, and/or operate city and county training programs that help deprived citizens to gain access to and make usage of e-entertainment, such as the non-college-educated, older people, ones in manufacturing and other non-service occupations, and those located outside of the urbanized areas. It is essential that such policies not only provide one-time training but also include further pathways that sustain the training and/or lead to further formal education. This has been shown in failures of short-term city technology programs to serve the ICT-disadvantaged populations such as technology training in deprived zones of Atlanta and the enhanced Internet access in Lagrange, Georgia, programs that collapsed in the long-term because of the lack of follow-up initiatives (Kvasny and Keil 2006).

Government policies at several levels can lead to actions that favor the narrowing of the digital divide present among U.S. counties, while bringing benefits to all citizens.

This study has several limitations that include occasional spatial bias, exclusion of Alaska and Hawaii counties, and lack of democracy/societal openness variables. The spatial bias in the residuals of some regressions could be addressed in future studies by introduction of additional independent variables, transformations, more elaborate models, or a non-linear approach in the regressions that would encompass location sufficiently to eliminate the bias. Structural equation modeling might replace regression analysis given the large sample size. The exclusion of Alaska and Hawaii was due to absence of social capital data at the county level for those states. A set of consistent variables on democracy and societal openness for counties nationwide and reliable data for the same is presently unavailable; therefore, that dimension, often important in other studies at the state and national levels, is excluded. Another challenge to explain in detail, based mostly on secondary sources, the e-entertainment patterns for 3,109 continental U.S. counties that are a novel finding of this study.

The limitations might have some partial effect on the detailed findings, but we have no reason to anticipate a major effect on the answers to the research questions and resulting policy recommendations. The limitation in explanations, although not diminishing the answers to the research questions, might weaken the broader picture of the resultant policy recommendations. While beyond the scope of one chapter, a series of future articles on each e-entertainment dependent variable or a lengthy report might enhance this work.

Future directions of this research could, first of all, introduce designs to improve or eliminate all of these limitations. In addition, since e-entertainment is growing rapidly among audiences, longitudinal studies, such as fixed effects modeling, could demonstrate the patterns of e-entertainment change over time and how change relates to social, economic, ethnic, and political factors. Other investigations could seek to support explanations of detailed geographic patterns for individual e-entertainment attributes. With accurate data, studies could be conducted at country-equivalent geographies in other parts of the world to gain comparative data on spatial patterns and factors influencing change. Additionally, case studies can illuminate complex organizational and behavior aspects of e-entertainment change.

9. Conclusion

This research is the first attempt at the U.S. county level to systematically investigate use of the Internet for e-entertainment as part of ICT-enabled digital services of value to the society. Adopting the SATUM developed by the authors, associations of 13 demographic, socio-economic, innovation, and social capital independent variables with five e-entertainment dependent variables are posited and empirically tested. The research is unique due to its geographic scope of over 3,000 counties of the United States, and to novel indicators of e-entertainment. Employing spatial analytical methods, we determine clusters of high and low e-entertainment in continental U.S. counties, which are for the most part aligned with the urban-rural divide with some interesting nuances. Youth and working age population are found to be significantly associated with Internet use for e-entertainment—in counties nationwide, in metropolitan, as well as micropolitan areas. The importance of college education, irrespective of the urban-rural divide, is also reinforced. Contrary to findings in prior literature, social capital is found to reduce adding videos online and iTunes usage. Explanations are provided along with implications of our findings, and policies at the county, metropolitan, state, and federal state levels are recommended in concert with the findings.

Note

1 Education in this chapter refers to the demographic variable of educational attainment, not to ICT education.

References

Agarwal, Ritu, Animesh Animesh, and Kisalaya Prasad. "Social Interactions and the 'Digital Divide': Explaining Variations in Internet Use." *Information Systems Research* 20, no. 2 (2009): 277–94.

Azari, R., and J. B. Pick. "Technology and Society: Socioeconomic Influences on Technological Sectors for United States Counties." *International Journal of Information Management* 25, no. 1 (2005): 25–37.

Chen, W. "The Implications of Social Capital for the Digital Divides in America." *The Information Society* 29 (2013): 13–25.

Deloitte. *Deloitte's 2014 Global Outsourcing and winsourcing survey*. New York: Deloitte. December 2014. http://www2.deloitte.com/content/dam/Deloitte/us/Documents/strategy/us-2014-global-outsourcing-insourcing-survey-report-123114.pdf.

Duggan, M., N. B. Elison, C. Lampe, A. Lenhard, and M. Madden. *Social Media Update 2014*. Washington, DC: Pew Research Center, January 9, 2015.

Duggan, M., and D. Page. *Gaming and Gamers*. Washington, DC: Pew Research Center, December 2015.

Dutta, S., I. Mia, and T. Geiger. "The Networked Readiness Index 2010–2011: Celebrating 10 Years of Assessing Networked Readiness." In *The Global Information Technology Report 2010–2011*, edited by S. Dutta and I. Mia, 3–32. Geneva, Switzerland: World Economic Forum, 2011.

Dutton, W. H., and G. Blank. "There Emergence of Next Generation Internet Users." *International Economics and Economic Policy* 11 (2014): 29–47.

The Economist. "The Travel Channels; TV Everywhere." July 20, 2013a, 59.

The Economist. "Pennies Streaming from Heaven; the Entertainment Industry and Online Media." August 17, 2013b, 14.

The Economist. "Counting the Change: Digital Media." August 17, 2013c, 53–4.

eMarketer. "Social Networking in India Rises Quickly Among Web Users, But Just 6.3% of All Residents Use Social Sites."*eMarketer*, August 31, 2012. www.emarketer.com

Esri. "Business Analyst." May 1, 2014. www.esri.com/software/businessanalyst.

Esri. "2013 U.S. Market Pontential Data Methodology Statement." 2015. http://support.esri.com/en/knowledgebase/whitepapers/view/productid/153/metaid/2037.

Grubesic, T. H. "Spatial Dimensions of Internet Activity." *Telecommunications Policy* 26 (2002): 363–87.
Grubesic, T. H. "The Geodemographic Correlates of Broadband Access and Availability in the United States." *Telematics and Informatics* 21 (2004): 335–58.
Grubesic, T. H. "A Spatial Taxonomy of Broadband Regions in the United States." *Information Economics and Policy* 18 (2006): 423–48.
Helsper, E. J., and R. Eynon. "Digital Natives: Where Is the Evidence?" *British Educational Research Journal* 36, no. 3 (2010): 503–20.
Huang, Z. "E-Government Practices at Local Levels: An Analysis of U.S. Counties' Websites." *Issues in Information Systems* 7, no. 2 (2006): 165–70.
Huang, Z. "A Comprehensive Analysis of U.S. Counties' e-Government Portals: Development Status and Functionalities." *European Journal of Information Systems* 16 (2007): 149–64.
Khatiwada, L. K., and K. E. Pigg. "Internet Service Provision in the U.S. Counties: Is Spatial Pattern a Function of Demand." *American Behavioral Scientist* 53, no. 9 (2010): 1326–43.
Kvasny, L., and M. Keil. "The Challenges of Redressing the Digital Divide: A Tale of Two U.S. Cities." *Information Systems Journal* 16, no. 1 (2006): 22–53.
Loebbecke, C., and P. Powell. "e-Business in the Entertainment Sector: The Egmont Case." *International Journal of Information Management* 22 (2002): 307–22.
Malecki, E. J. "Digital Development in Rural Areas: Potentials and Pitfalls." *Journal of Rural Studies* 19 (2003): 201–14.
Manoharan, A. "A Study of the Determinants of County e-Government in the United States." *The American Review of Public Administration* 43, no. 2 (2013): 159–78.
Niehaves, B., and R. Plattfaut. "Internet Adoption by the Elderly: Employing IS Technology Acceptance Theories for Understanding the Age-Related Digital Divide." *European Journal of Information Systems* 25 (2014): 708–26.
Nishida, T., J. B. Pick, and A. Sarkar. "Japan's Prefectural Digital Divide: Multivariate and Spatial Analysis." *Telecommunications Policy* 38, no. 11 (2014): 992–1110.
NTIA. *Digital Nation: Expanding Internet Usage*. Washington, DC: National Telecommunications and Information Administration, U.S. Department of Commerce, 2010.
NTIA. *Exploring the Digital Nation: Computer and Internet Usage at Home*. Washington, DC: National Telecommunications and Information Administration, U.S. Department of Commerce, 2011.
OECD. *Understanding the Digital Divide*. Paris, France: Organization for Economic Cooperation and Development, 2001.
Ono, H., and M. Zavodny. "Digital Inequality: A Five Country Comparison Using Microdata." Social *Science Research* 363 (2007): 116–39.
Openshaw, S. *The Modifiable Areal Unit Problem*. Norwich: Geobooks, 1984.
Perrin, A. *English-Speaking Asian Americans Stand Out for Their Technology Use: FacTank*. Washington, DC: Pew Research Center, February 18, 2016.
Perrin, A., and M. Duggan. *Americans' Internet Access: 2000–2015*. Washington, DC: Pew Research Center, June 2015.
Pew Resarch Center. *The Demographics of Mobile News Habits*. Washington, DC: Pew Research Center, December 11, 2012.
Pick, J. B., and R. Azari. "Global Digital Divide: Influence of Socioeconomic, Governmental, and Accessibility Factors on Information Technology." *Information Technology for Development* 14, no. 2 (2008): 91–115.

Pick, J. B., and R. Azari. "Global Model of Utilization of Technology Based on Governmental, Social, Economic, and Business Investment Factors." *Journal of Management Information Systems* 28, no. 1 (2011): 51–85.

Pick, J. B., and A. Sarkar. *The Global Digital Divides: Explaining Change*. Heidelberg, Germany: Springer-Verlag, 2015.

Pick, J. B., and A. Sarkar. "Theories of the Digital Divide: Critical Comparison." In *Proceedings of the 49th Hawaii International Conference on System Sciences*, IEEE Computer Society, Hawaii, 2016.

Pick, J. B., T. Nishida, and A. Sarkar. "Broadband Utilization in the Indian States: Socio-Economic Correlates and Geographic Aspects." In *Management of Broadband Technology Innovation*, edited by Jyoti Choudrie and Catherine Middleton, 269–96. Oxford: Routledge, 2014.

Pick, J. B., A. Sarkar, and J. Johnson. "United States Digital Divide: State Level Analysis of Spatial Clustering and Multivariate Determinants of ICT Utilization." *Socio-Economic Planning Sciences* 49 (2015): 16–32.

Putnam, R. D. *Bowling Alone: The Collapse and Revival of American Community*. New York: Simon and Schuster, 2000.

Quibria, M. G., S. N. Ahmed, T. Tschang, and M.-L. Reyes-Macasaquit. "Digital Divide: Determinants and Policies with Special Reference to Asia." *Journal of Asian Economics* 13 (2003): 188–825.

Rogers, E. *Diffusion of Innovations*, 5th ed. New York: Free Press, 2003.

Rupasingha, A., and S. J. Goetz. *U.S. County-Level Social Capital Data, 1990–2005*. University Park, PA: Penn State University, The Northeast Regional Center for Rural Development, 2008.

Rupasingha, A., S. J. Goetz, and D. Freshwater. "The Production of Social Capital in U.S. Counties." *Journal of Socio-Economics* 35 (2006): 83–101.

Selwyn, N. "Digital Division or Digital Decision? A Study of Non-Users and Low-Users of Computers." *Poetics* 34, nos. 4–5 (2006): 273–92.

Srinuan, C., and E. Bohlin. "Understanding the Digital Divide: A Literature Survey and Ways Forward." In *Proceedings of 22nd European Regional Conference of the International Telecommunications Society*, Innovative ICT Applications—Emerging Regulatory, Economic and Policy Issues, Budapest, 2011.

Statista. "Value of the U.S. Entertainment and Media Market 2015–2019." 2015a. Accessed November 1, 2015. www.statista.com/statistics/237769/value-of-the-us-entertainment-and-media-market/

Statista. "Leading U.S. Millennial Digital Activities 2015." 2015b. Accessed November 1, 2015. www.statista.com/statistics/470808/us-millennials-Internet-usage-activities/

Statista. "Fixed Internet Applications: Share of Aggregate Traffic in North America 2015." 2015c. Accessed November 1, 2015. www.statista.com/statistics/222337/fixed-access-Internet-services-by-daily-aggregate-traffic-volume/

U.S. Bureau of the Census. *Information: Geographic Area Series: Comparative Statistics for the U.S.: 2012 and 2007. EC1251A2. 2012 Economic Census of the United States*. Washington, DC: U.S. Bureau of the Census, 2015.

U.S. Bureau of the Census. *Metropolitan and Micropolitan*. Washington, DC: U.S. Bureau of the Census, 2016. www.census.gov/population/metro.

Van Dijk, J. A. G. M. *The Deepening Divide: Inequality in the Information Society*. Thousand Oaks, CA: Sage Publications, 2005.

Venkatesh, V., M. G. Morris, G. B. Davis, and F. D. Davis. "User Acceptance of Information Technology: Toward a Unified View." *MIS Quarterly* 27, no. 3 (2003): 425–78.

Vicente, M. R., and A. J. Lopez. "Assessing the Regional Digital Divide Across the European Union-27." *Telecommunications Policy* 13 (2011): 220–37.

Warschauer, M., and T. Matuchniak. "New Technology and Digital Worlds: Analyzing Evidence of Equity in Access, Use, and Outcomes." *Review of Research in Education* 34, no. 1 (2010): 179–225.

The White House. "Mapping the Digital Divide." 2015. Accessed November 1, 2015. www.whitehouse.gov/sites/default/files/wh_digital_divide_issue_brief.pdf.

Zickuhr, K., and A. Smith. *Digital Differences*. Washington, DC: Pew Research Center, 2012.

7 E-Health as an Enabler of Social Inclusion

Ken Clarke, Adam Lodders, Robyn Garnett, Anne Holland, Rodrigo Mariño, and Zaher Joukhadar

1. Introduction

Social isolation kills (House 2001). This is the ultimate health risk. Health inequality in general is caused by the asymmetric distribution of income and services. It is the direct result of a combination of bad social and economic policy and planning. These factors drive the social determinants of health and cause much of the health inequality around the globe (Marmot et al. 2008). When it comes to accessing traditional health services, some sections of the community can experience difficulties as a direct result of exclusion factors, such as social isolation, transportation difficulties, limited health-care supply (Goins et al. 2005), remote location (Shucksmith 2003), and having English as a second language (Webb 2006).

ICTs can improve access and reach of health services throughout the community. Digital innovation can even make entirely new types of health services possible. Health services that are delivered via ICTs are known as e-health services (Blobel 2010). However, for disadvantaged community members to be able to benefit from e-health programs, the "digital divide" must be addressed (Norris 2001). The digital divide is another socially excluding factor that distinguishes the "digital haves" from the "digital have-nots" (Van Dijk 2006). The latter are those people with little or no access to, or skills to use digital tools such as smartphones, personal computers, and the Internet (Selwyn 2006).

This chapter demonstrates by practical examples how properly designed and implemented innovative e-health applications can also bridge the digital divide in many health-care scenarios, overcoming the barriers of social isolation. The intention of these implementations is to improve equality in access to health care and hence deliver better health outcomes across the population. This chapter addresses the role of e-health as it relates to social inclusion through four distinct case studies across the entire health-care spectrum: education and prevention, early intervention, acute care, and chronic disease management. Analysis of the common themes arising from these usages of e-health will demonstrate the potential of ICTs to overcome many barriers faced by socially isolated and underserved individuals to improve their well-being and quality of life. As we will also demonstrate, innovations in delivery can also have profound impacts by allowing people to remain in their homes and local community, receive specialist services and, by careful design or sometimes as a very welcome side effect, ameliorate their feelings of social isolation.

2. An Interdisciplinary Approach

The case studies presented next are unique in that they are the result of original research supported by the Melbourne Networked Society Institute based at the University of Melbourne in Australia (The University of Melbourne 2016). The Institute funds innovative interdisciplinary research that seeks to understand the impact, challenges, and benefits of increased connectivity for society. This research is driven by a connection of key societal challenges with ICTs solutions. Such an approach eschews traditional conceptions of methodology, which is often instilled from within discipline-specific histories and cultures (Huber 1990). Instead, institute-based research projects are developed in an iterative manner based upon continual feedback, development and engagement between project partners that seek to transcend the traditional organizing structures for knowledge generation within universities (Tucker and Lodders 2011: 2). For example, in the Institute's health research domain, the connection of researchers from clinical, engineering and socio-cultural research backgrounds is essential in furthering innovation (Tucker and Lodders 2011: 1).

Interdisciplinary research enables "collaboration transcending discipline boundaries to create new knowledge" (Tucker and Lodders 2011: 2). Interdisciplinary research denotes a specific form of inquiry and is distinct from multidisciplinary and cross-disciplinary approaches (Davies and Devlin 2007). The latter two forms of disciplinary collaboration retain much of their disciplinary domains. On the other hand, interdisciplinary research actively seeks to blend knowledge, traditions, methodologies, and practices from across disciplines to create new ideas and solutions to pressing matters of social concern. Therefore, the methodology adopted by the Melbourne Networked Society Institute cannot be defined within a specific discipline or disciplinary tradition, as projects actively select the best research methods available to deliver innovative, practical, and creative solutions to society's problems.

Each of the case studies presented next was the result of research funded by the Melbourne Networked Society Institute. Projects were selected on the grounds of their interdisciplinarity, novelty, and reliance on connectivity using ICTs as a core component. Interdisciplinary project teams identified the key social problem to be addressed—for example how to support early intervention for autism in regional and rural settings—as well as the best technical solutions to the problem. This was followed by iterative testing and validation in a laboratory setting before subsequent use in real-world scenarios. This agile approach typically adopted by Institute projects enabled the rapid prototyping of proof-of-concepts that have the potential to be rolled out as new and improved ICTs services for society. Each of the case studies that follow adopts this methodology.

3. Four Case Studies

3.1 Improving Health Literacy for Diabetes Patients

The initial case study in this chapter centers on health literacy, because the first step in the road to good health is having the knowledge to take care of oneself. Health literacy is "the degree to which individuals have the capacity to obtain, process, and understand basic health information and services needed to make appropriate health decisions" (U.S. Department of Health and Human Services 2000). Better levels of health literacy can increase quality of life, optimize utilization of health-care services, and reduce the burden of disease (Berkman et al. 2011). It has also long been noted that socially isolated members of the community are less healthy, both physically and psychologically, and have higher mortality rates (House, Landis, and Umberson 1988). Thus higher levels of both health literacy and social engagement can help drive generally better health outcomes for the community. However, health-literacy levels across the community are generally low (House, Landis, and Umberson 1988), and there is a complex interdependency among different kinds of literacies that contribute to health and e-health literacy (Norman and Skinner 2006). To improve health-literacy levels, experts have recommended more personal forms of communication and educational outreach with "significant widening of the content and methods used" (Nutbeam 2000).

This case study is the result of innovative interdisciplinary work that involved the collaboration of engineers, clinicians, and community representatives and focused on developing a proof-of-concept system to increase health literacy for type 2 diabetes. Type 2 diabetes is a major chronic health issue and health literacy is a factor in its prevention and management (Zimmet 2011). The Internet has increased access to volumes of information, and while beneficial, it also creates the risk of misinformation. This is a challenge for diabetes educators, as the best approach is to "push situation and user-specific quality knowledge to users based on their actual individual needs, circumstances and profiles at any given time" (Boulos et al. 2006). Meeting this challenge will result in people with diabetes obtaining access to timely and tailored information. Therefore, Internet access is vital in ensuring that these people are not socially excluded from essential medical information.

Broadcast television has almost universal presence and pervasive influence in homes. Unfortunately, for consumer health information and patient education, broadcast television does not provide health information that is reliably understandable and appropriate for any particular viewer. Therefore, the viewer is often left to rely upon non-tailored content that further isolates them from their peers in the community. Internet protocol television (IPTV) offers services not provided by broadcast television, such as replays of television shows at viewer-selected times, interaction with live television shows or video on-demand. IPTV was defined by the Open IPTV Forum in 2013 as the delivery of television services "using Internet Protocol over a managed broadband network" and is thus distinct from simple Internet-connected TV, unmediated access to Internet content or a hybrid of these (Open IPTV Forum 2013). It has become technically possible to integrate web technologies with IPTV to build systems that enable intelligently personalized recommendations and selections of television content (Blanco-Fernández et al. 2011; Chang, Irvan, and Terano 2013). Such a system can enable diabetes educators to deliver personalized health information directly to people with diabetes in their homes, encouraging people with diabetes to access relevant and tailored health information.

The Melbourne Networked Society Institute developed (using the agile interdisciplinary method) and trialed a health-literacy IPTV system for diabetes. Nine of the 13 participants with type 2 diabetes were aged 65–80 (Gray et al. 2014). They also had low usage of ICTs, with eight participants using a computer no more than twice a month and having no broadband connection at home. The health education video IPTV content consisted of filmed interviews of people with diabetes and their carers. Each interview communicated experiences of this health condition, such as symptoms, lifestyle choices, exercise, diet, mental health. Each interview was segmented into several 5- to 10-minute videos, and each short video was identified with one or more health-condition-related categories using keywords such as "depression" or "medication." The videos were loaded onto an Ericsson IPTV server and allocated an ID tag (Arberg et al. 2007). The system, now called LivBetter, can match a particular video's ID tag automatically with a health-care consumer's needs, as specified by the health-condition-related keywords they have in their account profile (LivBetter 2016). Alternatively, a health educator logged into the system at their workstation, and with permission to access a client's IPTV account, can manually allocate appropriate videos to their client based on their expert opinion of the client's information needs and wants based on previous interviews. Note that via this process, the need for ongoing face-to-face interaction with an educator is reduced, lightening the burden of travel for users and allowing education service delivery irrespective of location as long as a broadband connection is available.

The consumer interface consists of a large screen television connected to the Internet via a set top box as shown in Figure 7.1, which decodes the incoming broadband data from the remote IPTV video server and provides a high definition picture. The client must login to the IPTV system via their television screen using a remote control before they can watch their assigned videos. This is not a technical limitation but part of the necessary

Figure 7.1 Health-Literacy Internet Protocol Television (IPTV) System in Use.

privacy requirements to ensure no one else can access a user's account without authorization. The login process is as close as a consumer gets to requiring computer skills and the passwords can be made to be relatively simple to remember and type, such as four digit numbers. Users have three different remote controls they can choose: a standard TV remote as well as two different types for the set top box. The first of the latter is a standard handheld device with alpha-numeric buttons, and the second is a special keyboard-style remote (as seen in Figure 7.1) to make typing of usernames and passwords easier for users who specifically need this, such as those with impaired eyesight. Lowering the barriers to usage of technology via familiar interfaces and suitable ergonomic design helps reduce the so-called digital divide.

After logging in, the user can see a simple and clear menu of health-condition-related categories down the left side of the screen with thumbnails and text descriptions of each video title to the right, as seen in Figure 7.1. The remote control can be used to scroll down and across to the desired selection and the video can be played, paused, and fast forwarded as if the television were connected to a DVD player in the room, even though the content is being streamed over a broadband network from a distant server. This is important for making the user immediately comfortable with an interface that looks and feels just like the technology available in their home environment.

The health-literacy IPTV system was found to enable health information access among people with diabetes and their carers. Usability factors included familiarity with television technology, simplicity of screen layout and easy-to-launch video content. Major usefulness factors were a sense of affirmation, a sense of affinity, and a way to talk about the condition with family and friends. Despite general familiarity with the technology, some participants still identified usability barriers to health information access—namely, the complexity of the remote controls, glitches in menus, and problems loading video content. Some also found usefulness barriers in terms of content that was unsettling, not relevant to their situation or contrary to their idea of independently managing their condition. Their suggestions for improving the health information consumer experience of IPTV focused on the manageability of remote control devices, screen readability, content metadata, richer types of content, follow-up features, and finding pathways to additional community support.

This proof-of-concept demonstrated that it is possible to use home television to communicate person-specific professionally selected health information, and thus extend the reach and scale of home-based health services. This is particularly important in addressing the impact of isolation from health information because of location, age, health, and mobility. It can help close the digital divide, which otherwise has the potential for adverse impacts on the long-term health of those with limited ICT skills and opportunity. The technology is applicable to other chronic diseases and health conditions, and as such, it provides a convenient vehicle for the delivery of tailored health information across the community. It also overcomes a major barrier for those with chronic diseases who are very likely to be elderly and either unable or unwilling to travel. It is important to note that this also encompasses metro-based populations, not just those in rural and regional areas who are generally underserved by most community services. As a result, this enhances community connection. While this technology increases health literacy in the home, other e-health interventions can increase access to more specialized health services. The next section discusses the impact of telepractice for delivery of clinical interventions.

4. Enhancing Social Inclusion for Children With Autism Spectrum Disorder

E-health platforms can enhance social inclusion by providing support for the development of social communication, a core deficit in children with Autism Spectrum Disorder (ASD). ASD is a diagnostic term used to describe individuals with ongoing, complex neurodevelopmental disorders characterized by persistent difficulty in social communication and interaction, as well as encompassing restricted, repetitive patterns of behavior and interests (American Psychiatric Association 2013).

According to the Australian Bureau of Statistics 2009, 64,600 Australians were reported as having ASD, and Australian Centrelink data from 2012 represented a figure of one in 61 school-aged children with an Autism

Spectrum diagnosis (Buckley 2014). These Australian statistics are consistent with the Autism and Developmental Disabilities Monitoring Network in the United States, which estimates that one in 68 children is diagnosed with ASD (Centers for Disease Control and Prevention 2014).

Many experts believe that accessing intervention as early as possible may help minimize the impact of ASD related disorders on later life functioning, and accumulating evidence is demonstrating the effectiveness of early intervention for children with ASD (National Autism Centre 2015; Prior and Roberts 2012). To ensure inclusive access to timely early intervention services, however, there is a need for innovation in service-delivery models and individualization of interventions so as to more fully meet the needs of this heterogeneous group of children living within diverse family structures and environments (Stahmer, Schreibman, and Cunningham 2011; Trembath and Vivanti 2014). Such models underpin improvements in social inclusion for this population.

The use of ICTs for medical service delivery has increased over the last decade in alignment with improved technical capacity, such as increased data processing speeds and greater telecommunications bandwidth (Kumar and Cohn 2013). In the medical field, these have resulted in a telepractice that is the connection of health professionals with clients via a broadband network and ICTs in order to deliver remote consultations, assessments, and health education, for example. Such approaches are well suited to the autism field, due to the difficulty and inconsistency with access to timely and adequate assessment and intervention services for children with ASD (Braddock and Twyman 2014; Ruggero et al. 2012). A telepractice approach enables families to overcome the barriers to services for their children, including access to experienced clinicians and services, lack of social support, and financial constraints, especially when considering the common recommendations for 25 hours of intervention per week (Carlon, Carter, and Stephenson 2013; Mackintosh, Goin-Kochel and Myers 2012; Payakachat et al. 2012).

According to the Australian Bureau of Statistics for 2012-13, in 2013 83% of Australian households were connected to the Internet. This presents a great opportunity for ICTs to increase service availability, improve access and reduce social isolation (Meredith, Firmin, and McAllister 2013). Research to date has revealed positive indicators that successful ASD specific assessment and intervention can occur via telepractice based methods (Aresti-Bartolome and Garcia-Zapirain 2014; Boisvert et al. 2010).

The provision of the intervention must be flexible to meet the diverse needs of families. For this reason, the pilot project was developed at the Melbourne Networked Society Institute to determine the feasibility and value of migrating ASD specific, parent intervention training from traditional face-to-face delivery to an ICT-enabled format, using readily available technologies. The viability of this telepractice delivery model was tested via real time, online provision of eight, two-and-a-half-hour group training sessions to two parents simultaneously. Additionally, three, 1.5-hour-telepractice-based individual skills coaching sessions were provided, involving parent-child interactions in participants' homes.

To enhance the accessibility of this intervention, potentially suitable telepractice hardware and software configurations were identified and tested in relation to their general availability to the broader public, ease of use, effectiveness, and range of realistic conditions that might exist in home and clinical environments. Laboratory testing was undertaken at the Institute so as to select technologies that would maximize both accessibility and optimization of the teleconferencing experience, for both the service provider and participants. The findings from this testing show that there is one effective solution that involves using PCs, iPads, and videoconferencing software such as, Vsee and Zoom (Vsee 2016; Zoom 2016). This solution was effective in that it used off the shelf technologies that were readably accessible to consumers and clinicians.

The pilot ICTs-delivered intervention package was evaluated in four areas: design implications, effectiveness, participant satisfaction, and potential cost-benefits. Each of these areas has an impact on the subsequent support required to increase social inclusion. The intervention migrated traditional "face-to-face" classroom delivery successfully. The user experience was impacted in line with audio and visual quality, interaction with the computer screen, and interaction with group members. The intervention was effective, measured by the child's social-communication skills and parental use of communication facilitating strategies, prior to intervention, and again at six weeks post-intervention. As a result, the increased social-communications skills brought about by the technology improved the social skills of the child, which had a positive impact on social inclusion. Through continued use, parents were able to increase their strategic engagement with the child, while the child improved their communications skills. Participant satisfaction was measured via qualitative interviews and both parents and clinicians reported satisfaction with the ICTs-enabled service-delivery method. One parent stated,

> Usually I do not share my opinion when I am in group. I have anxiety to speak up in groups, but it's easier to speak when you are at home, just on your computer. It's not like the whole 20 people are staring at you when you speak.

Finally, simple cost comparisons between the service and equivalent face-to-face delivery revealed that savings were made using telepractice delivery for both parents and clinicians and in relation to travel costs and time. In addition, equipment and venue costs were reduced for the clinicians using ICTs.

However, such an approach was limited, relying on active technological intervention and review. As mentioned in the introduction, digital innovation can also make entirely new health services possible. Thus, in addition to increasing the availability of assessment for families of young children with ASD, the Microsoft Kinect sensor (Zhang 2012) was identified as a tool to enable the remote analysis of social communication in young children with ASD. The model afforded by the Microsoft Kinect, originally designed to track human motion for video-game control, effectively takes the "patient" to the "doctor" via transfer of skeleton tracking data from the child's home to the clinician's office. A screenshot of the skeleton tracking and analysis appears in Figure 7.2.

The software development served as a "proof of concept" involving collaboration between speech pathologists and software engineers in three stages. The first stage examined extraction of key physical parameters regarding parent and child interaction and involved recognition of particular actions, not just static poses, as a requirement. A series of consultations between the speech pathologists and software engineers on the project produced a list of desirable measurements, including head level (height offset between parent and child), proximity, hyperactivity level, touch, imitation, and response delay time. The second stage concerned feasibility testing and software calibration using data collected via simulated interactions in the laboratory. Software training sets, comprising 3D skeleton joint positions were developed. The simulated scenarios were used to test the capability of the system to accurately track and measure key elements of the interaction in real time. Finally, a proof of concept was tested using the Kinect software toolbox to assess interactions between a small number of parents and children with ASD. This occurred for the purpose of validating software outputs in realistic situations and in order to identify factors that may negatively impact the effective operation of the system.

This Kinect project resulted in the development of an automated tracking and analysis system, providing meaningful statistics on the quality of

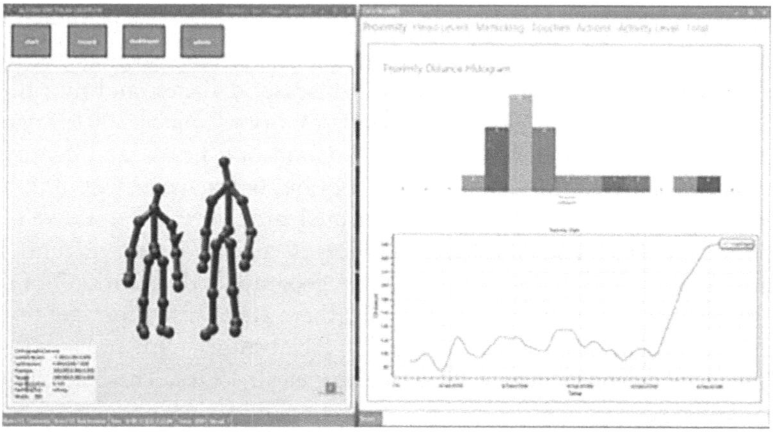

Figure 7.2 Kinect Skeleton Tracking and Analysis.

parent-child interaction via a highly automated real-time user interface. A prototype dashboard displays Kinect sensor output, presenting both real-time and cumulative interaction measurements alongside skeleton avatars. It is envisaged that the Kinect system has the potential to support the remote assessment of children with ASD for the purpose of intervention planning and evaluation of progress. The toolbox can be developed for future iterations to create a portable and readily available "expert system" providing early intervention support to families who are typically underserved and in areas where services are scarce. This system could not only benefit children with ASD, but be applied more broadly to other client groups requiring some form of rehabilitation or intervention therapy.

The project has provided promising results for enhancing the social inclusion of these families. Such innovative ICT-enabled services may allow more children with ASD to access intervention earlier, or may allow access to intervention for families that would not traditionally have been able to participate due to factors, such as time, transport, cost, care for other children, or lack of suitably trained professionals in the area (Houston 2013). Additional benefits of ICT-enabled services may include increased service choice for families, potential for continuity of care when family mobility is a factor and possible decreased costs for the service provider, consumer, and community. In addition, telepractice based delivery of an established parent training program provided socialization of learning when piloted with families of preschool children with ASD. Parents reported increased confidence in helping their children to communicate and improvements in their children's social interaction, which drives improved social inclusion for both the parents and the children involved.

5. Increasing Access to Oral Health Services

Oral health is an integral part of general health. Among older people, poor oral health increases the risk of frailty (Kay and Locker 1998), and may precipitate or exacerbate other medical conditions, such as pneumonia, cerebrovascular, and cardiovascular disease, diabetes, and nutritional deficiencies (Australian Health Ministers' Advisory Council 2001; Loesche et al. 1998; Genco and Williams 2010). Good oral health and quality of life can be improved through health promotion, self-management of disease prevention, both augmented by sophisticated surgical and restorative techniques. Thus, access to oral health services is critical. However, rural and regional areas often have an "inequitable geographic distribution" of oral health workers. For example, in Australia there are critical shortages in oral medicine, pediatric dentistry, and maxillo-facial surgery (Australian Health Ministers' Advisory Council 2004; Wenger et al. 1996). These are important challenges to the wider society.

Although rural communities might have poor access to dentists and traditional dental care, the possibilities opened up by ICTs can help close the gap with peers in urban settings. These include increased use of video and multimedia content with high resolution and reliability, which makes possible connections via a computer or mobile smart device for distant oral health education and assessment, known as teledentistry. Teledentistry can supplement traditional methods of oral diagnosis and referral for patients who otherwise might not receive care and health promotion. It also empowers and upskills local community oral health-care providers in rural areas who are able to access advice from specialists in major metro teaching hospitals and practices.

Poor oral health is often associated with social isolation, particularly for the elderly (Goins et al. 2005). Additionally, many aged persons with poor oral health are also those that are on the wrong side of the digital divide thanks to their relative lack of ICTs skills and access compared to others in the community as discussed earlier in the chapter. New and emerging e-health systems for teledentistry have the potential to improve access to oral health-care information and services for high-risk populations living in underserved communities, such as older adults living in rural areas or in residential aged care facilities (RACFs). Teledentistry can provide alternative entry points into the oral health-care system and help close the gap in the provision of sustainable oral health-care services to these populations, thereby addressing both the digital and physical divide. The teledentistry model provides oral health promotion, oral health care, and dental treatment that empowers patients. It also enables better targeting, increases efficiency, and supports the delivery of culturally appropriate and responsive oral health services for older populations and migrants with English as a second language. The model incorporates alternative service-delivery systems to address the identified demand for dental care in such non-traditional settings.

This section describes the outcomes of two interconnected teledentistry projects at the Melbourne Networked Society Institute. As shown in Figure 7.3, the projects are built upon an innovative technology platform comprising an intraoral camera—a toothbrush-like object with a miniature digital camera and LED lighting at its tip and a networked computer. The first project examined the feasibility of teledentistry for telediagnosis in residential aged care and the second project assessed telediagnosis for pediatric dental trauma.

5.1 Telediagnosis in Residential Aged Care

This field trial assessed the feasibility of a teledentistry model for teleconsultation and telediagnosis in RACFs by comparing the remote assessments with those performed via traditional face-to-face oral examinations (Mariño et al. 2014). The set-up is depicted in Figure 7.4. Trained teledentistry assistants operated an intraoral camera. The model was supported by ICT training and an instructional kit for the intraoral camera operators. A total of 3

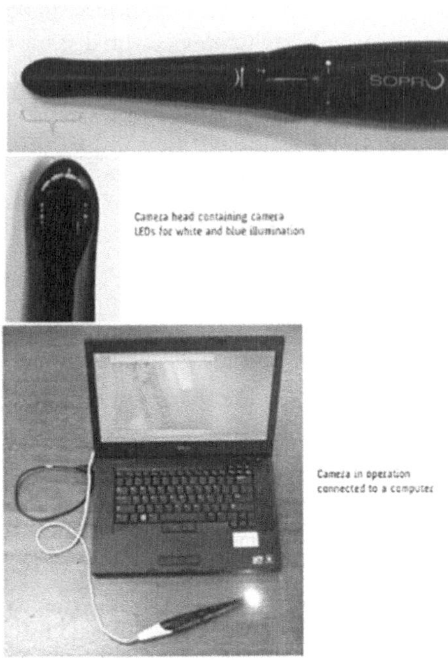

Figure 7.3 Intraoral Camera.

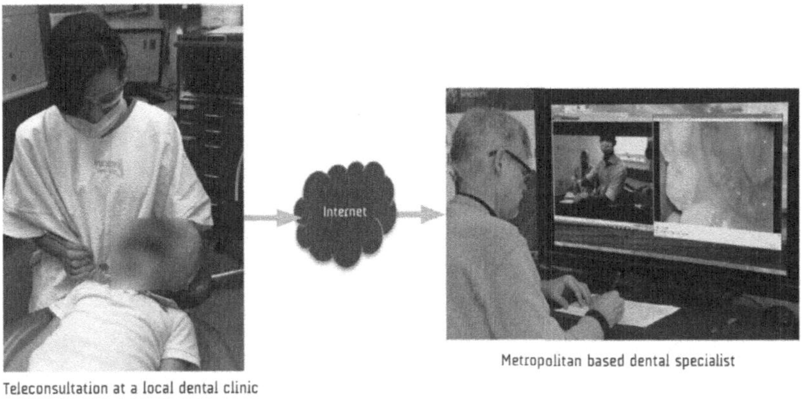

Figure 7.4 Teledentistry in Progress.

RACFs, 5 teledentistry assistants (registered nurses), and 50 RACF residents participated in this assessment.

Results indicated that the teledentistry approach for oral health screening is feasible and reliable as an alternative to traditional oral health examinations. Residents expressed high levels of satisfaction with the teledentistry service. Nurses indicated that the training provided (including a hard copy and online manual) was a helpful and easy-to-understand resource.

Thus, a telehealth assistant who seeks advice from a dentist at a distant site can perform screening examinations. This reduces the need for a dentist to travel to health-care facilities, and facilitates rapid responses (Kopycka-Kedzierawski and Billings 2006). In addition, by using this approach, the RACF avoids the disruption and difficulty of arranging travel for the patients, many of whom suffer from dementia and other cognitive disorders and would otherwise be stressed by the experience. Translation of this service-delivery model into clinical practice would extend the provision of oral health care to remote and difficult-to-serve locations, and improve access to additional patient populations, as well as lessen the impact of the shortage of oral health professionals.

5.2 Teledentistry for Rural Pediatric Consultations

The project targeted children and adolescents living in regional and remote locations and tested a prototype teledentistry system in which regional community health centers are linked either to the Melbourne Dental School or the Royal Children's Hospital in Melbourne. Three general dental practitioners working in community dental clinics in the Australian state of Victoria were trained to manipulate an intraoral camera and to use existing ICT infrastructure to communicate with a dental specialist in Melbourne. Live video images were transferred for remote assessment and triage. Patients/parents of patients participating in the study then completed a questionnaire to assess their experiences of the various aspects of the program.

Forty-two remote assessments/consultations were conducted in two specific specialist service areas: 26 cleft-lip and palette patients and 16 in orthodontics. For most patients (57.1%), the major outcome of teleconsultation was the avoidance of trips to the city for initial assessment and/or follow-ups. Accordingly, parents (65.6%) also commented that the most valuable element was avoiding disruption, difficulties, and cost of travel to the city to visit a dental specialist.

This field trial provided initial evidence on how teledentistry might improve access to specialist care, identification of broader community benefits such as the level of convenience for both the family and the dental specialist, and evidence of the levels of acceptance of the virtual examination by patients and GDP's. This approach would reduce waiting lists, a major problem in Australia. For patients, it reduced the cost of treatment and the time and expense of travel to the city to be seen by the specialist.

In concluding this section, it can be suggested that the two field trials provided innovative solutions to the problem of closing the service-delivery gap in the provision of sustainable oral health care to underserviced populations (e.g., nursing homes and rural areas). Extension of these models could provide alternative entry points into the oral health-care system, for example by incorporating a mobile dental unit to locally treat the identified demand.

The last project to be presented goes to the heart of the digital divide and social exclusion by directly addressing the needs of typically elderly patients in the home with very low ICT skills and a poor history of access to technology. It employs easy-to-use technology with automated functionality so complex menus do not have to be navigated and things "just turn themselves on."

6. Supporting Rehabilitation Programs for COPD Patients

Rehabilitation programs that deliver exercise training and disease-specific education are a core component of health care for people with chronic diseases. Such programs typically require patients to attend a health center at least twice weekly for eight weeks to exercise in a group setting with expert supervision and monitoring of important physiological signals such as oxygen levels and heart rate. In patients with Chronic Obstructive Pulmonary Disease (COPD), rehabilitation is one of the most effective treatments on offer; it improves exercise capacity, decreases breathlessness and enhances health-related quality of life (McCarthy et al. 2015). People with COPD who undertake rehabilitation spend less time in hospital, which is a positive outcome for patients, health systems and society (Griffiths et al. 2001). Despite robust evidence on these benefits, rehabilitation is delivered to less than 10% of those in Australia with COPD (Australian Institute of Health and Welfare 2013). Major barriers to uptake of this important treatment relates to the inability to travel to the rehabilitation center and poor physical mobility (Keating, Lee, and Holland 2011). Even when the rehabilitation center is close by, people with COPD report that they cannot walk the required distances to access the program (e.g., from car park to rehabilitation gym), and cited a lack of affordable transport and parking options at health centers (Keating, Lee, and Holland 2011). For this group of older people who experience distressing breathlessness and significant disability, these challenges frequently outweigh the potential benefits of attendance.

Delivery of ICT-enabled rehabilitation services directly into the home of people with COPD is a way to improve their health outcomes. These services provide peer support and community connections, enhancing social inclusion. However, there are challenges to implementation of novel healthcare solutions that use the Internet. COPD is more prevalent in older people and those from lower socio-economic groups (Pauwels 2001). These factors may influence the uptake of and familiarity with new technologies. Recent studies from around the globe suggest that 40%–50% of people with COPD are without Internet access (Delgado 2015; Duplaga 2013). People with COPD who do not have Internet access are significantly older than those who access the Internet. They also have significantly lower educational levels, lower income and are more likely to have impaired physical mobility (Duplaga 2013; Martinez 2014). Compared to patients with other respiratory diseases, people with COPD use the Internet to source health information less frequently (Duplaga 2013). Similarly, knowledge and experience of ICT-enabled health services are lower in respiratory patients who are older and have a lower educational level (Hofstede et al. 2014), features typical of many patients with COPD. It has also been reported that many people with COPD are reluctant to use the Internet to increase connectedness with their health-care team, being more likely to accept asynchronous ICT-enabled health interactions, such as booking appointments and accessing laboratory results, when compared to synchronous health interactions, such as online contact with a health-care professional (Duplaga 2013).

A telerehabilitation system for COPD was developed in 2012 by a team of researchers from the Melbourne Networked Society Institute, LaTrobe University, and the Institute for Breathing and Sleep at the Austin Hospital, Melbourne. The team developed the system in an interdisciplinary manner. The resultant system allows delivery of programs directly into the homes of people with COPD, using simple videoconferencing and physiological monitoring (Holland et al. 2013). The system uses off the shelf technology that is accessible to clinicians in rehabilitation programs, comprising a standard exercise bike, tablet computer, free videoconferencing software, and pulse oximeter for non-invasive measurement of blood oxygen levels and heart rate. This increases its accessibility. Patients are able to exercise under the supervision of a physiotherapist but can also see and interact with at least one other group participant. The telerehabilitation system was configured such that participants were only required to switch on the tablet to participate, with no requirement to interact with menus or open programs. This increases the availability and accessibility of services to counteract digital-divide barriers.

A trial of the system was conducted with users delivering telerehabilitation to eight participants with COPD severity ranging from mild to severe, with ages ranging from 56 to 83 years. Only 20% of participants were regular Internet users. The system was feasible and acceptable to participants, who attended 76% of their planned telerehabilitation sessions (twice weekly over 8 weeks). System Usability Scores (Brooke 1996), which measures users' experience of technology, were very good at 81%. The clinical outcomes of the program (improvements in exercise capacity, symptoms and quality of life) were of similar magnitude to those seen in traditional center-based rehabilitation programs for COPD (Holland et al. 2013).

Following this trial, further improvements to the telerehabilitation system have been made to increase the size of the telerehabilitation group from two to five patients and a physiotherapist, to provide the social benefits of the traditional rehabilitation program. The web platform delivers real-time measurement of oxygen levels and heart rate using a Bluetooth pulse oximeter and iPad, ensuring exercise within safe physiological limits. A menu-free interface for the oximeter further reduces barriers to participation.

Initial experience with telerehabilitation for COPD indicates that it is safe, feasible, and clinically useful. This system involved direct supervision and regular contact with a physiotherapist and other patients using videoconferencing; this was highly acceptable to our patients, despite previous concerns in the literature regarding the suitability of synchronous ICT-enabled services in COPD (Duplaga 2013). Access to the COPD system can improve the health and well-being of participants. The fact that the system involves a collaborative exercise class enables social interactions for those who often face difficulties in socializing. Social access and companionship is just as important as the exercise class. The system therefore increases social inclusion by connecting individuals across the community and minimizing barriers to participation for people with COPD across a wide range of ages and disease severity. It provides access to an important treatment that was previously beyond their reach. Wider rollout will require attention to new funding and commissioning models and good infrastructure, including quality data networks in health-care settings. There is significant opportunity to achieve health economic benefits, which should be measured by using models that include personal costs and benefits to patients with COPD, as well as the societal costs of health care. The benefits naturally include the ability to engage with peer support networks in alternative, ICT-enabled ways that enhance social inclusion.

7. Conclusion

ICTs are a valuable solution to addressing the health needs of socially excluded people. As this chapter demonstrates, the breadth and range of the solutions to the problems of social exclusion often share many common characteristics, such as allowing people to remain at home without disadvantaging them in their rights to access health services, linking of people in underserved areas with central clinicians or experts to address the shortfall in society's ability to provide them with adequate health coverage, allowing peers to communicate and share experiences, which not only improves their health literacy and outcomes but also reduces their experiences and likelihood of social isolation as thanks to e-health services they can know and feel that there are others out there in the community experiencing the same problems, and share a common bond.

A key finding for the health industry, government, and policy makers at all levels is that when done well these solutions can not only improve social inclusion but also save time, money, and allow health-care professionals to address the needs of more people and/or reach dispersed ones more easily and effectively. This is particularly important where there are professional skills shortages. Another important lesson for them is that ICTs can deliver both health care and health information in innovative ways that can connect people in similar circumstances—i.e., diabetes health literacy, chronic disease management, and improving parental interventions for autistic children. This socializes the experience, making it much more relevant and inclusive, as well as increasing the likelihood of participation and retention, particularly in rehabilitation scenarios: high dropout rates are one of the largest costs in rehabilitation not only in financial terms but also in poor long-term health outcomes.

As we have seen, e-health innovations can also offer new opportunities for leveling society's playing field in terms of services on offer—for example, local dentists learning new techniques from metro-based specialists that they can then implement in their own practices, or speech pathologists reaching out to parents of autistic children in rural and remote communities who have no local practitioner to provide them with early intervention or education services.

Just as importantly, health education and professional bodies should also recognize their duty to provide courses aimed at identifying and reducing the impacts of inequity. Governments must also recognize that alternative approaches provided by e-health, such as those described in this chapter, can improve health-care culture and positively impact on the attitudes and expectations of both providers and consumers, and should fund them accordingly. However, we must not be in thrall to technology but be careful to address the issues of the digital divide. Again, as this chapter has amply demonstrated, not all of society's members have the knowledge, access, and skills to use ICTs. New e-health services must be designed to be as intuitive and non-threatening as possible. As we have seen through the diverse examples presented in this chapter, this can mean many things: from using familiar interfaces, such as a TV and remote, having highly automated interfaces and menu-free technology choices, to simple things, such as large fonts and clear, simple webpage designs for those with poor eyesight or migrants with limited English skills.

References

American Psychiatric Association. *Diagnostic and Statistical Manual of Mental Disorders 5 2013*. Washington, DC: American Psychiatric Association.

Arberg, Peter, Torbjörn Cagenius, Olle V. Tidblad, Mats Ullerstig, and Phil Winterbottom. "Network Infrastructure for IPTV." *Ericsson Review* 84 (2007): 3.

Aresti-Bartolome, Nuria, and Begonya Garcia-Zapirain. "Technologies as Support Tools for Persons with Autistic Spectrum Disorder: A Systematic Review." *International Journal of Environmental Research and Public Health* 11, no. 8 (2014): 7767–802. doi: 10.3390/ijerph110807767.

Australian Health Ministers' Advisory Council. *Oral Health of Australians: National Planning for Oral Health Improvement*. South Australia: Department of Human Services, 2001.

Australian Health Ministers' Advisory Council. *Healthy Mouths Healthy Lives. Australia's National Oral Health Plan 2004–2013*. South Australia: Department of Health, 2004.

Australian Bureau of Statistics. "Household Use of Information Technology." Australia, 2012–13. Accessed October 4, 2014. www.abs.gov.au/ausstats/abs@.nsf/Lookup/8146.0Chapter12012-13.

Australian Bureau of Statistics. "Autism in Australia, 2009". Accessed August 21, 2017. http://www.abs.gov.au/AUSSTATS/abs@.nsf/Previousproducts/4428.0Main%20 Features42009

Australian Institute of Health and Welfare. *Monitoring Pulmonary Rehabilitation and Long-Term Oxygen Therapy for People with Chronic Obstructive Pulmonary Disease (COPD) in Australia: A Discussion Paper*. Canberra: Australian Institute of Health and Welfare, 2013.

Berkman, Nancy D., Stacey L. Sheridan, Katrina E. Donahue, David J. Halpern, and Karen Crotty. "Low Health Literacy and Health Outcomes: An Updated Systematic Review." *Annals of Internal Medicine* 155, no. 2 (2011): 97–107.

Blanco-Fernández, Yolanda, Martín López-Nores, Alberto Gil-Solla, Manuel Ramos-Cabrer, and José J. Pazos-Arias. "User-Generated Contents and Reasoning-Based Personalization: Ingredients for a Novel Model of Mobile TV." *Expert Systems with Applications* 38, no. 5 (2011): 5289–98. doi: 10.1016/j.eswa.2010.10.029.
Blobel, B. "Architectural Approach to eHealth for Enabling Paradigm Changes in Health." *Methods of Information in Medicine* 49, no. 2 (2010): 123–34. doi: 10.3414?ME9308.

Boisvert, Michelle, Russell Lang, Mary Andrianopoulos, and Mary Lynn Boscardin. "Telepractice in the Assessment and Treatment of Individuals with Autism Spectrum Disorders: A Systematic Review." *Developmental Neurorehabilitation* 13, no. 6 (2010): 423–32.

Boulos, M. N. Kamel, Fiona E. Harvey, Abdul V. Roudsari, Riccardo Bellazzi, M. Elena Hernando, Tibor Deutsch, Derek G. Cramp, and Ewart R. Carson. "A Proposed Semantic Framework for Diabetes Education Content Management, Customisation and Delivery Within the M2DM Project."*Computer Methods and Programs in Biomedicine* 83, no. 3 (2006): 188–97.

Braddock, Barbara, and Kimberly Twyman. "Access to Treatment for Toddlers with Autism Spectrum Disorders." *Clinical Pediatrics* 53, no. 3 (2014): 225–9 doi: 10.1177/0009922814521284.

Brooke, John. "SUS: A 'Quick and Dirty' Usability Scale." In *Usability Evaluation in Industry*, edited by Patrick W. Jordan, Bruce Thomas, Bernard A. Weerdmeester, and Ian L. McClelland, 189–94. London: Taylor and Francis, 1996.

Buckley, Bob. "Autism/ASD Prevalence in Australia . . . Up to 2012—APAC'13 Presentation." 2014. Accessed September 10, 2014. http://a4.org.au/a4/node/695.

Carlon, Sarah, Mark Carter, and Jennifer Stephenson. "A Review of Declared Factors Identified by Parents of Children with Autism Spectrum Disorders (ASD) in Making Intervention Decisions." *Research in Autism Spectrum Disorders* 7, no. 2 (2013): 369–81. doi: 10.1016/j.rasd.2012.10.009.

Centres for Disease Control and Prevention. "Prevalence of Autism Spectrum Disorder Among Children Aged 8 Years-Autism and Developmental Disabilities Monitoring Network, 11 Sites, United States, 2010." *Morbidity and Mortality Weekly Report. Surveillance Summaries* (Washington, DC: 2002) 63, no. 2 (2014): 1.

Chang, Na, Mhd Irvan, and Takao Terano. "A TV Program Recommender Framework." *Procedia Computer Science* 22 (2013): 561–70. doi:10.1014/j.procs.2013.09.136.Davies, Martin, and Devlin, Marcia. *Interdisciplinary Higher Education: Implications for Teaching and Learning*. Centre for the Study of Higher Education, The University of Melbourne (2007).

Delgado, Cionéia K., Mariana R. Gazzotti, Ilka L. Santoro, Andrea K. Carvalho, José R. Jardim, and Oliver A. Nascimento. "Internet Use for Health-Care Information by Subjects with COPD." *Respiratory Care* 60, no. 9 (2015): 1276–81.

Duplaga, Mariusz. "The Acceptance of e-Health Solutions Among Patients with Chronic Respiratory Conditions." *Telemedicine and e-Health* 19, no. 9 (2013): 683–91.

Genco, Robert J., and Ray C. Williams. *Periodontal Disease and Overall Health: A Clinician's Guide*. Yardley and Pennsylvania, PA: Professional Audience Communications Inc., 2010.

Goins, Turner R., Kimberly Williams, Mary Carter, Melinda Spencer, and Tatiana Solovieva. "Perceived Barriers to Health Care Access Among Rural Older Adults: A Qualitative Study." *The Journal of Rural Health* 21 (2005): 206–13.doi: 10.1111/j.1748-0361.2005.tb00084.x

Gray, Kathleen Mary, Ken Clarke, Basil Alzougool, Carolyn Hines, Gil Tidhar, and Feodor Frukhtman. "Internet Protocol Television for Personalized Home-Based Health Information: Design-Based Research on a Diabetes Education System." *JMIR Research Protocols* 3, no. 1 (2014): e13.

Griffiths, Timothy Leonard, C. J. Phillips, S. Davies, Michael Leslie Burr, and I. A. Campbell. "Cost Effectiveness of an Outpatient Multidisciplinary Pulmonary Rehabilitation Programme." *Thorax* 56, no. 10 (2001): 779–84.

Hofstede, J., J. de Bie, B. van Wijngaarden, and M. Heijmans. "Knowledge, Use and Attitude Toward eHealth Among Patients with Chronic Lung Diseases." *International Journal of Medical Informatics* 83, no. 12 (2014): 967–74.

Holland, Anne E., Catherine J. Hill, Peter Rochford, Julio Fiore, David J. Berlowitz, and Christine F. McDonald. "Telerehabilitation for People with Chronic Obstructive Pulmonary Disease: Feasibility of a Simple, Real Time Model of Supervised Exercise Training." *Journal of Telemedicine and Telecare* 19, no. 4 (2013): 222–6. House, James S., "Social Isolation Kills, But How and Why?" *Psychosomatic Medicine* 63, no. 2 (2001): 273–4.

House, James S., Karl R. Landis, and Debra Umberson. "Social Relationships and Health." *Science* 241, no. 4865 (1988): 540–5.

Houston, K. Todd. *Telepractice in Speech-Language Pathology*. San Diego, CA: Plural Publishing, 2013.

Huber, Ludwig. "Disciplinary Cultures and Social Reproduction." *European Journal of Education* 25, no. 3 (1990): 241–61.Kay, Elizabeth, and David Locker. "A Systematic Review of the Effectiveness of Health Promotion Aimed at Improving Oral Health." *Community Dental Health* 15, no. 3 (1998): 132–44.

Keating, Andrew, Annemarie Lee, and Anne E. Holland. "What Prevents People with Chronic Obstructive Pulmonary Disease from Attending Pulmonary Rehabilitation? A Systematic Review." *Chronic Respiratory Disease* 8, no. 2 (2011): 89–99.

Kopycka-Kedzierawski, Dorota T., and Ronald J. Billings. "Teledentistry in Inner-City Child-Care Centres." *Journal of Telemedicine and Telecare* 12, no. 4 (2006): 176–81.

Kumar, Sajeesh, and Ellen R. Cohn. *Telerehabilitation*. London: Springer-Verlag, 2013. doi: 10.1007/978-1-4471-4198-3.

LivBetter. "Website." Accessed June 20, 2016. http://external.livbetter.net.au/

Loesche, Walter J., Anthony Schork, Margaret S. Terpenning, Yin-Miao Chen, B. Liza Dominguez, and Natalie Grossman. "Assessing the Relationship Between Dental Disease and Coronary Heart Disease in Elderly US Veterans." *The Journal of the American Dental Association* 129, no. 3 (1998): 301–11.

Mackintosh, Virginia H., Robin P. Goin-Kochel, and Barbara J. Myers. " 'What Do You Like/Dislike About the Treatments You're Currently Using?' A Qualitative

Study of Parents of Children with Autism Spectrum Disorders." *Focus on Autism and Other Developmental Disabilities* 27, no. 1 (2012): 51–60.

Mariño, R., P. Marwaha, R. Collman, M. Hopcraft, D. Manton, M. McCullough, and I. Blackberry. "Field Testing of Remote Teledentistry Technology." In *The Sixth International Conference on eHealth, Telemedicine, and Social Medicine (eTELEMED 2014) IARIA*, 2014, 23–8.

Marmot, Michael, Sharon Friel, Ruth Bell, Tanja Houweling, and Sebastian Taylor. "Closing the Gap in a Generation: Health Equity Through Action on the Social Determinants of Health." *The Lancet* 372 (2008): 1661–9.

Martinez, Carlos H., Beth L. St Jean, Craig A. Plauschinat, Barbara Rogers, Julen Beresford, Fernando J. Martinez, Caroline R. Richardson, and MeiLan K. Han. "Internet Access and Use by COPD Patients in the National Emphysema/COPD Association Survey." *BMC Pulmonary Medicine* 14, no. 1 (2014): 1–11. doi: 10.1186/1471-2466-14-66.

McCarthy, Bernard, Dympna Casey, Declan Devane, Kathy Murphy, Edel Murphy, and Yves Lacasse. "Pulmonary Rehabilitation for Chronic Obstructive Pulmonary Disease." *Cochrane Database of Systematic Reviews*, 2 (2015). doi: 10.1002/14651858.CD003793.pub3.Meredith, Grant, Sally Firmin, and Lindy McAllister. "Digital Possibilities and Ethical Considerations: Speech-Language Pathologists and the Web." *Journal of Clinical Practice in Speech-Language Pathology* 15 (2013): 43–6.

National Autism Center. *Findings and Conclusions: National Standards Project, Phase 2*. Randolph, MA: National Autism Center, 2015.

Norman, Cameron D., and Harvey A. Skinner. "eHealth Literacy: Essential Skills for Consumer Health in a Networked World." *Journal of Medical Internet Research* 8, no. 2 (2006): e9. doi: 10.2196/jmir.8.2.e9.

Norris, Pippa. *Digital Divide: Civic Engagement, Information Poverty, and the Internet*. Cambridge: Cambridge University Press, 2001. ISBN: 0 521 80751 4.

Nutbeam, Don. "Health Literacy as a Public Health Goal: A Challenge for Contemporary Health Education and Communication Strategies into the 21st Century." *Health Promotion International* 15, no. 3 (2000): 259–67. doi: 10.1093/heapro/15.3.259.

Open IPTV Forum. "Terminology." 2013. Accessed June 20, 2016. www.webcitation.org/6LyYzPdxq.

Pauwels, Romain A., A. Sonia Buist, Peter M. A. Calverley, Christine R. Jenkins, and Suzanne S. Hurd. "Global Strategy for the Diagnosis, Management, and Prevention of Chronic Obstructive Pulmonary Disease."*American Journal of Respiratory and Critical Care Medicine* 163 (2001): 1256–76.

Payakachat, Nalin, J. Mick Tilford, Erica Kovacs, and Karen Kuhlthau. "Autism Spectrum Disorders: A Review of Measures for Clinical, Health Services and Cost—Effectiveness Applications." *Expert Review of Pharmacoeconomics & Outcomes Research* 12, no. 4 (2012): 485–503. doi: 10.1586/erp.12.29.

Prior, Margot, and Jacquline Roberts, J. *Early Intervention for Children with Autism Spectrum Disorders: 'Guidelines for Good Practice.'* Australia: Department of Families, Housing, Community Services and Indigenous Affairs, 2012.

Ruggero, Leanne, Patricia McCabe, Kirrie J. Ballard, and Natalie Munro. "Paediatric Speech-Language Pathology Service Delivery: An Exploratory Survey of Australian Parents." *International Journal of Speech-Language Pathology* 14, no. 4 (2012): 338–50. doi: 10.3109/17549507.2011.650213.

Selwyn, Neil. "Digital Division or Digital Decision? A Study of Non-Users and Low-Users of Computers." *Poetics* 34 (2006): 273–92.

Shucksmith, Mark. *Social Exclusion in Rural Areas: A Review of Recent Research.* Aberdeen, Scotland: Arkleton Centre for Rural Development Research, 2003.

Stahmer, Aubyn C., Laura Schreibman, and Allison B. Cunningham. "Toward a Technology of Treatment Individualization for Young Children with Autism Spectrum Disorders." *Brain Research* 1380 (2011): 229–39. doi: 10.1016/j.brainres.2010.09.043.

Trembath, David, and Giacomo Vivanti. "Problematic But Predictive: Individual Differences in Children with Autism Spectrum Disorders." *International Journal of Speech-Language Pathology* 16, no. 1 (2014): 57–60.

Tucker, Rod, and Adam Lodders. "Reshaping the e-Health Landscape: How Interdisciplinary Research Drives Innovation in Broadband Applications." *Telecommunications Journal of Australia* 61, no. 3 (2011): 44.1–44.7.

The University of Melbourne. 2016. Accessed October 23, 2016. http://networkedsociety.unimelb.edu.au/

US Department of Health and Human Services, and Office of Disease Prevention and Health Promotion. *Quick Guide to Health Literacy: Health Literacy Basics.* Washington, DC: US Government Printing Office, 2000.

Van Dijk, Jan A. G. M. "Digital Divide Research, Achievements and Shortcomings." *Poetics* 34 (2006): 221–35.

Vsee. 2016. Accessed October 23, 2016. http://vsee.com

Webb, Sue. "Can ICT Reduce Social Exclusion? The Case of an Adults' English Language Learning Programme." *British Educational Research Journal* 32, no. 3 (2006): 481–507. doi: 10.1080/01411920600635478.

Wenger, G. Clare, Richard Davies, Said Shahtahmasebi, and Anne Scott. "Social Isolation and Loneliness in Old Age: Review and Model Refinement."Ageing *and Society* 16, no. 3 (1996): 333–58.

Zhang, Zhengyou. "Microsoft Kinect Sensor and Its Effect." *IEEE MultiMedia* 19, no. 2 (2012): 4–10. doi: 10.1109/MMUL.2012.24.

Zimmet, Paul Z. "The Growing Pandemic of Type 2 Diabetes: A Crucial Need for Prevention and Improved Detection." *Medicographia* 33, no. 1 (2011): 15–21.

Zoom. 2016. Accessed October 23, 2016. http://zoom.us.

8 Challenging the Cost of Higher Education With the Assistance of Digital Tools

Case Studies of Protest Activity in Canada and the United States

Victoria Carty

1. Introduction

Over the past several years, there has been an explosion of protest activity among young people around the globe demanding radical changes in the existing economic and political systems as they embrace a new vision of the future. They are in many ways embodying what Pilger (2011) refers to as the "theater of the impossible"—for example, challenging the neoliberal policies that have created the most recent global economic crisis and are jeopardizing their livelihoods. Taking the future into their own hands, youth, and especially across parts of Europe and Latin America, have organized themselves against measures that attack the social welfare state with a key focus on funding for higher education. Following suit, in several locales across North America, youth have also begun to engage in contentious politics with similar fervor.

One of the resources that has helped to fuel the latest rounds of protest is digital media. Media sources have always been critical to political discontent and to the ability to organize and carry out protest politics. Today, with the availability of new communication apparatuses provided by digital technology at activists' disposal, social movement actors operate in an innovative political terrain that can increasingly foster inclusion through both electronic and face-to-face avenues of discussion, debate, critical thinking, and contentious politics. More specifically, in this new communication landscape they can efficiently, quickly, and cheaply share grievances, disseminate information through peer-to-peer networks and therefore bypass state- or corporate-controlled and owned media. They are also finding new ways to collectively make demands on authorities and create social ties and a sense of collective identity in cyberspace, which in turn assists them in organizing and participating in street protest activity.

This chapter analyzes two student-led social movements that challenge the cost of higher education as part and parcel of the neoliberal agenda and accompanying austerity measures. Both focus on how students are engaging in popular resistance to increasing student debt and proposed tuition hikes that include a broader critique of the economic, social, and political conditions promoted by neoliberal ideology at universities across Canada and the United States. These case studies provide instructive examples of how young activists are making strategic decisions and carrying out acts of political dissent that combine digital forms of information sharing and organizing as a mode of e-activism with traditional forms of protest such as direct action—a hybrid mix of online and offline strategizing and mass mobilization.

The motivation of this research is to understand how new media can be used as a tool to assist with organizing and carrying out forms of contentious politics (though not serving as a *substitute* for on-the-ground protest activism). The analyses provided in this chapter contribute to the literature on how ICTs can propel social movement activity. The findings support the work of Giroux (2012) and other scholars who recognize the new media ecology as a source that fuels collective behavior as it creates a wide range of public spheres that can accelerate political discussion and debate, and is later manifested in concrete forms of protest in the streets.

The general outline of the chapter is as follows. I begin with an overview of social movement and media studies theories that apply to the case studies. I then discuss data collection and methodology as rooted in an explorative approach. The case studies then follow; an analysis of the social movement activity in Canada and in the United States, emphasizing the crucial role than new technology played in their organizing efforts and affording a sense of inclusiveness. The chapter ends with some concluding remarks and suggestions for future research.

2. Theorizing Social Movements in the Context of the Digital Age: A Literature Review

Resource mobilization theory, cultural theories of social movements, and media studies theories are all useful in guiding our understanding of the contemporary outbreaks of contentious politics among young people in parts of the United States and Canada. Each point to how new technology is affecting mobilization efforts, tactics, and strategies that social movement actors employ, how activists are forging new forms of collective identity through "weak" social ties, and how digital tools allow them to influence public opinion by challenging authorities and holding them accountable for their response to forms of civil disobedience and dissent. They also help us understand how relationships created in the virtual world are solidified, sustained, and strengthened through face-to-face connections and networks on the ground in physical space.

Resource mobilization theory has traditionally argued that social movements rely heavily on well-established social movement organizations (SMOs) to further their agenda (Tarrow 1998; McAdam 1982; Tilly 1978; Gamson 1975; Olson 1965). In addition to SMOs, other resources include knowledge, expertise, money, media attention, time, labor, allies, and support from political elites. Another source of focus among resource mobilization scholars is the role of leaders that serve as decision makers, provide sources of inspiration, charisma and organizational strategies, frame the major grievances and demands of the movement, create a sense of collective identity and draw in new recruits, and are overall critical to mobilizing resources (Platt and Lilley 1994).

Cultural-oriented theories of social movements, on the other hand, focus more on the meaning of collective action and subjective interests as social movement actors are often immersed in commitments to others, traditions, and broader ethical or moral sentiments, and there is an immaterial quality that motivates individuals which is rooted in collective identity. Melucci (1996) defines collective identity as an interactive, shared process that links individuals or groups to a social movement through sustained interaction. Key to forging collective identity and articulating shared meanings is the way organizers "frame" their issues to resonate with potential recruits by linking participants' grievances to mainstream beliefs and values to build solidarity (Benford 1993; Snow et al. 1986; Ryan and Gamson 2006).

Theorists of social movements and collective behavior have long noted that social networks, relational ties, and friendships are an invaluable resource to social movement organizers and participants, and especially for high-risk protest movement actions (Diani 1995; Verba and Brady 1995). They have also documented that an invitation through a personal (preexisting) tie is one of the strongest predictors of individuals' engagement in activism and helps to build networks of support and a sense of collective identity (Gould 1991). Expanding on these findings, more contemporary scholars acknowledge that the introduction of digital technology has led to new forms of relationship and network building. New media platforms are providing an innovative communication field that gives activists additional resources and one that helps them to create and sustain different types of connective capabilities and ways to share information, recruit supporters to their cause, and mobilize. In this way, more recent theories of social movements blend nicely with some of the work in media studies and demonstrate how new ICTs expand the potential of these networks to develop and mutate exponentially, and especially through weak social ties which enhances a sense of inclusion.

Boulainne's (2009) findings, for example, illustrate that the dissemination of information, peer-to-peer through electronic mediums increases the likelihood of participation in protest activity, and what Jenkins (2006) calls the "spillover effect." Jurris (2014) further develops this line of theorizing and refers to these links between new media and activism as "aggregation," highlighting the importance of community building, or inclusion, through horizontal flows of information in both virtual and physical spaces. Social media, he argues, contributes to the logic of aggregation by facilitating the means through which people from diverse backgrounds can be brought together in physical space, and build and sustain solidarity. Furthermore, Juris also acknowledges the impact that independent reporting and media outlets[1] can have on activism as this can threaten the authorities' ability to control situations that unfold during outbreaks of mass direct action and enables social movement actors to control the narrative of reporting.

In other words, these new types of media potentially offer a mode of communication that is resistant to regulation by authorities. This reduces elites' capacity to repress the distribution of political communication and enhances a new type of civic engagement at the grassroots level (Bennett and Iyengar 2008). Citizens also have a new source of leverage in political struggles because they can document and circulate instances of excessive force that authorities might use against protesters, and therefore hold them accountable for their actions, which can play a key role in the recruitment of new members to the cause. All of the aforementioned theories are applicable to the student protests, as we will see during the ensuing analysis.

3. Data Collection and Methodology

This research is explorative in nature by examining how new media assist in the organizing and carrying out of mobilization efforts. I chose the cases of Canada and the United States because the student uprising in these two countries represent fairly recent social movements (the Canadian activity between early February 2012 and late September 2012, and the case in the United States between late 2011 until the spring of 2012) that were relatively unexpected—at least in popular opinion—but shared a common theme of demanding access to education at the university level by keeping costs reasonable. In each country, the students and their supporters also offered a strong critique of economic and political policies that disregard their economic distress while catering to corporate interests. Making this link refines our understanding of the evolution of social movements, while grievances and demands can expand as the movement grows and recruits new members.

The data are drawn from a variety of sources collected over a one-year span (between 2013–2014), including scholarly materials (books and journal articles cited in journals, such as *Social Forces*, *Journal of Communications*, and *Information, Communication and Technology*, among others), mainstream and alternative press (including sources, such as *The Guardian*, *The Canadian Press*, *The Montreal Gazette, and The Los Angeles Times*), and organizational sources located on the Internet (see the reference section for specifics). Most of the data collection consisted of news sources located in the LexisNexis archive and Google searches, which include the following keywords: Maple Spring, Law 78, CLASSE, UC California student debt protests, One Trillion Day Protest, and Occupy Student Debt.

4. The Mobilization Against Student Debt in Canada

The first large-scale protests in Canada began in 2012 when students engaged in disruptive activities in the province of Quebec. On February 13, they went on a general strike, which closed most of the colleges and universities throughout the province (Soltry 2012). The original grievance was a planned tuition hike by the government and the already increasing student debt. In 2012, the average debt for Canadian students was $28,000.00 and students in general had accumulated more than $15 billion in debt from student loans (Andreou 2013). The student activists demanded that the educational system remains within the domain of a public good and rejected what they perceived as its transition into a commercial entity. They viewed the proposed tuition increase as symbolic of the neoliberal values that were taking over Canadian society as a whole, and more specifically the encroachment of these into the university system.

Neoliberalism refers to an economic and political model that advocates privatization, commercial values, deregulation of markets and trade, and an increasingly important role of financial institutions in society accompanied by a smaller role of the government's ability to steer economic operations—i.e., market fundamentalism (Duggan 2003). In synch with an emphasis on privatization is a push to disinvest in social programs through the promotion of austerity measures. These ultimately deprive governments of revenue for social programs such as access to adequate education, health care and housing, employment opportunities, and other public goods.

Democratic student associations were at the core of the direct action tactics, one of the most key groups being Coalition large de l'association pour une solidarite syndical estudante (CLASSE), which was established in 2001 and represents 18 postsecondary institutions (Curran 2012). It was this student association that initiated the strike and was soon joined by other student groups including the federation etudiante collegiale du Quebec (FFCQ), Federation etudiante universitaire du Quebec (fEUQ) and Table de concertational etudiante du Quebec.

Through outreach efforts, much of it in the virtual world, CLASSE was able to forge ties between the numerous student associations as well as other community and civic organizations to collectively demand a permanent tuition freeze. Using direct action tactics, they held sit-ins in front of government offices and entrances to universities, blocked main highway entrances, carried out marches and demonstrations on campuses and surrounding neighborhoods, and most importantly, in terms of disruption, called for the general strike. The strike included not only students but also faculty and staff as well, and by the end of April, half of Quebec's students in higher education had joined the general strike (Curran 2012). This new movement was labeled Maple Spring and was the longest and largest student strike in North American history. On May 14, 2012, Line Beachawmp announced her resignation as Minister of Education because she "lost confidence in the student leaders' will to end this conflict" (Marquis 2012).

5. The Evolution of the Social Movement

Over several months, Maple Spring morphed into a movement that sought sweeping change in both the economic and political spheres. This occurred as students reached out to the broader Canadian population by incrementally amplifying their demands and framing their issues in ways that raised awareness about economic hardships, which all Canadian citizens were facing under the neoliberal model. These included austerity measures that mandated cuts in social services, factory closures, union busting, and an increase in retirement age (Heath 2012). In doing so, they tapped into other preexisting grievances and bridged the various issues under the rubric of economic injustice, which helped to further nurture a sense of collective identity among much of the public, not just across the student population. Thus, cultural theories of social movements, which emphasize the importance of framing issues in a way that resonates with potential recruits by associating participants' grievances to mainstream beliefs and values, is critical to understand how students were able to build new alliances and a sense of solidarity across diverse groups of Canadian citizens.

During the protest marches, students and their allies also began wearing what became the ubiquitous red square (Leier 2012). In a show of camaraderie protesters and civilians stitched it to their clothing and backpacks, stickers were taped on buildings and street poles, and some residents hung sheets from their windows or balconies donning the symbol (Marquis 2012). The square symbolized the rebellion by making the implicit statement that students (as well as many of their fellow citizens) were "squarely in the red"—namely, in debt. Activists later transformed the red square into a symbol of a lack of political freedom in addition to economic hardship; both perceived as a repercussion of neoliberal ideology. The tactics the activists employed included street theater and nightly marches—one of which culminated in an historic number of over 200,000 participants on May 22 (Dolphin 2012). In many of these marches participants employed the "casserole" tactic (banging pots and pans). By late May, the "Casseroles" (inspired by the cacerolazos of Chile in 1971) began as a nightly event throughout many Montreal residential neighborhoods. This is a creative form of protest and was instigated by Law 78 (described next) in that during the Casseroles, many Canadian citizens did not march on the streets, but in an act of solidarity resisting Law 78, they stood on their balconies banging pots and pans, thus circumventing the ban on unannounced street protests. This showed the creativity of the spirit of the protests, and courageous and persistent forms of defiance that participants in the struggle used to find ways around the repression of citizens' civil rights.

The passing of Law 78, an emergency law that was enacted on May 18 and ended the academic year, barred the right to assembly and university workers' right to strike of more than 50 individuals, and illegal protests were punishable by fines ranging from $5,000 to $125,000 for individuals and unions (Maharawal and Gluck 2012). This law proved to be a turning point in the struggle and worked to the students' advantage when thousands of students and their allies marched to resist the government's court injunction in an attempt to end the strike by fining anyone trying to block entrances to universities.

6. The Response From the Authorities

The Canadian demonstrators were also subject to repressive measures undertaken by the government and the police (though there were also some, albeit limited and isolated, instances of retaliation against the police among protesters). Police engaged in the beating of protesters, used pepper spray and tear gas against them, and (as mentioned) universities utilized court injunctions to try to end the strikes, thus infringing on their civil and constitutional rights (Marshall 2012). Despite the arrests and police harassment, the protests grew, and in defiance of the passing of Law 78 approximately 500,000 citizens marched in Montreal on May 25, the largest act of civil disobedience in Canadian history (Leier 2012). Nearly 1,000 were arrested for deviating from the prearranged route the police had dictated, as marchers became emboldened to challenge the permission that the police began demanding for staying on a particular route.

Therefore, the law, in addition to police violence, only fueled the cause of protesting, as public opinion shifted in favor of the students, leading to more recruits. For example, although many Canadians initially supported the tuition hikes, it was the repressive attacks by the police and the implementation of law 78 that drew in new allies, as many Canadians began to view the government as infringing on basic political rights of freedom of expression and assembly (Marshall 2012). Ultimately, what happened on the streets was a transition from a student-based movement to a mass popular movement demanding both accountability among the elites and a sustainable public sector. Once again, cultural theories of social movements are useful in understanding the surging support for the youths' cause because it was primarily through an emotional appeal emanating from the police brutality and disregard for constitutional rights of the students that stimulated the anger and energy to keep the campaign going.

7. The Role of Social Media in the Planning Stages and Reporting of Events

In addition to the on-the-ground organizing and alliance building, students also used the Internet and social media to solidify and expand the struggle. For instance, one of the tools activists used to gain supporters and influence public opinion was the wide circulation of images through a variety of digital media platforms illustrating the students' commitment to nonviolence (for the most part) despite the repressive police response (Banerjee 2012). After being exposed to the images through alternative news feeds that Mojos (mobile journalists) sent out, providing evidence of police brutality through photos and videos, the public increasingly recognized student activists as a valid constituency challenging university authorities and government policies. Aware of the impact that the reality of the situation was having on public opinion, the government, in fact, tried to stifle the mainstream press by asking television cameramen to stop shooting film of the episodes taking place on the streets (Banerjee 2012).

This exemplifies the crucial role of Mojos as a source of independent reporting and their ability to secure their own media outlets during direct action mobilizations, which is what Jurris (2014) points to. After the total number of arrests throughout Canada reached over 2,500 in June, the CLASSE website, cleverly and facetiously entitled "Somebody arrests me," had over 500,000 students posting photos of themselves breaking the law by engaging in acts of civil disobedience (Lindsay 2013). This risk-taking action via cyberspace promoted a sense of inclusiveness among students as, similar to on-the-street activity, putting oneself in a potentially threatening situation tends to increase a sentiment of collective identity among participants.

Additionally, student activists also used new media tools to spread the word and increase awareness of their struggle leading up to the outbreak of protest on the streets. For example, on the "Fight Tuition Increase" website students emphasized the fact that the financial sacrifice, which the citizens of Quebec were asked to abide by was not shared fairly, but hit the middle and lower classes disproportionately hard (Heath 2012). The site elaborates,

> In 2007, the Canadian government transferred more than 700 million in additional federal funding to the Quebec government. The Charest government could have invested this money in education, which would far have exceeded the $325 million currently collected by tuition fee increases. Instead, the richest individuals and corporations got a tax cut and students got left with nothing.
>
> (Leier 2012)

This website served as a recruitment tool for the upcoming general strike and other planned protest activities, and facilitated the building of collective identity *prior* to the protest activity that would surge for several months in public spaces. It is also illustrative of how new sources of collective identity are formed in the virtual world because it is often through electronic forms of connectivity that individuals recognize they have shared grievances that can then result in what Melucci (1996) called the creation of intermediate public spaces, and which Jenkins (2006) conceptualized as the "spillover effect," through which offline discussions translate into concrete forms of contentious politics. Similarly, Verba and Brady (1995) and Boulainne (2009) note the importance of weak social ties initiated in cyberspace through peer-to-peer sharing of information because these connections increase the likelihood of participating in grassroots mobilizing efforts on the ground. The use of digital technology further highlights Jurris's (2014) notion of the dynamics of aggregation both online and in physical, concrete spaces through direct action.

Digital feeds of the events in Canada as they unfolded gained not only national but also international support for the student struggle. Participants in the Occupy Wall Street movement in New York City, for instance, marched in solidarity with the citizens of Quebec three times and on October 15, in anticipation of the strike, supporters of the Canadian movement demonstrated in 900 cities across the globe in solidarity (cbsnews 2012). Once again, Juris's work is useful in that he contends that the connections between new media and concrete forms of activism, the dynamic that he calls aggregation, is a significant part of community building and information sharing. Often it is through social media, he suggests, that aggregation among people from diverse backgrounds takes place so as to organize and practice forms of contentious politics in physical spaces, resulting in both online and offline inclusion.

Suzanne Bilodeau, a member of a feminist group of mothers who supported the student strike, explained the critical role of social media in establishing collective identity and how they influenced public opinion. She says,

> Thanks to social media, we were shown the hidden dimension of Maple Spring . . . and especially the gap between "official" media news sources and what was really happening on the streets . . . I think this may have been the big difference between the 2012 demonstrations compared to those of previous years.
>
> Fortin 2013: 25)

This comment, in addition to the progression of the events during Maple Spring, highlights the significance of this new resource that activists have as part of their repertoire, and compliments and expands on the work of resource mobilization theorists.

The outcome of the mobilization was relatively successful. The majority of students decided to support the candidate from the Leftist Party, Pauline Marias, in the fall 2012 elections who ran on a platform of halting the tuition increase. She did this through a ministerial decree once elected, on September 20, and the student unions who had been on strike then voted to return to class (Lindsay 2013). Pauline Marias also eliminated Law 78 once in office.

8. Occupy Student Debt in the United States

The United States has perhaps felt the repercussions of the neoliberalization of higher education more severely than Canada, as its system is much more complex, with a mix of private and public colleges and universities. Until fairly recently most student aid in the United States was supplied in the form of grants, and the federal government worked together with states to ensure that universities were able to offer nearly free education for in-state residents (Duggan 2003). In the 1970s, however, the government began to decrease spending for public universities as part of the overall agenda to dismantle or privatize public services, and financial institutions began to replace the role of the government in lending financial support. The situation worsened for students when the global 2008 financial crisis sparked a new wave of social spending cuts across North America, further justifying decreases in higher education funding. At the same time, interest rates on student loans began to increase significantly and banks became less accountable to the government for their lending practices (Duggan 2003). In fact, student loans have replaced credit cards as the highest source of debt in what is being labeled the "debt-for-diploma" system (Andreou 2013). Currently, one in six student debts in the United States are in default, which is profitable for banks because students cannot default on their loans or declare bankruptcy, something which individuals can do for other loans.

In the United States, similar to their Canadian counterparts, students organized a proactive effort to stop planned tuition hikes. The average student debt is currently $37,000 for U.S. college students (cbsnews.com 2016), and this can be attributed to the following factors—all related to neoliberal policies. First, over the past 30 years, the average price of attending college has more than doubled at four-year schools (Honan 2012). Second, between 2007 and 2012, 48 states reduced funding for higher education. And third, federal funding (in the form of grants) for higher education has been drastically cut over the past several years and banks are playing an increasing role as loan providers to fill the void (Malcolm and McMinn 2013).

In addition to shared grievances, another parallel to the protest activity in Quebec is that the students in the United States used online tools to organize, recruit, and strategize during the early phases of the movement. The resistance among students in the United States began online when organizers of the Occupy Student Debt (a spinoff group of the Occupy Wall Street) encouraged their fellow students to sign a pledge on the website http:occypystudentdebt.com/owsdebtday to stop paying off their loans. The site contained a "Pledge of Refusal" which asked students to refuse to make payments toward their debt until reforms, such as free public education, were met. This site spread awareness about the growing social movement, helped students to form a virtual sense of solidarity and collective identity, and allowed them to organize themselves in ways that augmented their feelings of solidarity and inclusion.

For example, the students used digital technology to express their commitment to the cause by writing stories on Tumblr and on the "We Are the 99" website, while also posting blogs that documented their financial situation (Vasi and Suh 2013). Others posted samples of tweets and Facebook posts on the website http://younginvicibles.org, which others could use when contacting their congressional representatives. Additionally, Rolling

Jubilee, a student "strike debt" group, provided a number of social media links on its website (http://rollingjubille.org), such as the Strike Debt Facebook Page, @StrikeDebt; @StudentDebtPledge (Twitter), Why Strike Debt? (Tumblr), and Strike Debt provided videos (YouTube).

Sites, such as www.facebook.com/OccupyWallst, had hundreds of thousands of fans, and Twitter hashtags #occupywallst and #ows were sending out dozens of tweets a minute. Activists also relied heavily on Meetup (http:www.occupytogether.org) to plan and coordinate events of civil disobedience in thousands of cities across the United States. The #DreamsNotDebt hashtag allowed people to submit questions, sign petitions, participate in the student debt blog or send out an e-mail blast to other students and their circle of friends.

As Jurris (2014) attests, these forms of aggregation and inclusion that are created and thrive online are a valuable tool for social movement activists because they allow for individuals and groups to unite despite coming from different geographical locations, demographics, and cultural backgrounds. New media allow for virtual vessels through which individuals can connect and mobilize, and often relational ties and friendships are at the root of this, especially when planning for events that may be a high risk, as Verba and Brady (1995) point our attention to. Gould (1991) and Boulainne (2009) further highlight that personal preexisting ties, many of which are fostered through peer-to-peer electronic mediums, enhance the likelihood of one becoming part of a social movement. Similarly, Tufekci (2012) demonstrates that the new media ecology, which relies on a virtual infrastructure, facilitates networking building through decentralized, egalitarian, and pluralistic organizational modes, which are an important new resource for organizing and planning protest activity.

The coordination of activities and the formation of collective identity through online means also demonstrate the relationship between information sharing and protest activity in local communities—the "spillover" effect that Jenkins (2006) refers to—and the new ways of creating solidarity that digital media platforms enable. Student organizer Yates McKee, for instance, describes how students were able to solidify a sense of collective identity early on in the Occupy Student Debt struggle, originally online, which then allowed for a deeper understanding of the structural reasons responsible for their personal troubles, and which ultimately resulted in street protests. He explains,

> In two hours, several dozen people from a wide range of backgrounds and generations delivered emotionally-charged, first-person testimonials about the experience of debt servitude to Wall Street and its intermediary institutions . . . almost all of the speakers remarked that this was their very first time speaking publicly about their status as debtors. To speak as a debtor and to address others as debtors was an empowering process in its own right: the simple act of speaking built community

> and solidarity based on the shared experience of breaking with the debt shame—the insidious sense that to be indebted is an individual moral failure rather than an enforced condition of life under contemporary capitalism.
>
> (McKee 2012)

This use of digital media once again underscores the impact of the dynamics of aggregation allowed for by online tools. McKee's statement is indicative of how students and other U.S. citizens transformed the topic of personal debt into a political trajectory, funneling individual troubles into a critique of the overall neoliberal agenda. Debt as a common thread helped groups and individuals forge new alliances through the framing of issues, which then served as a recruitment tool by giving legitimacy and worthiness to the activists' grievances. This is a crucial aspect of social movement organization that Benford (1993) and Snow et al. (1986), as well as other cultural-oriented theorists of social movements, accentuate in their research.

9. Direct Action Tactics

Some of the most controversial on-the-ground events took place at the University of California and state school systems. Once again, the main grievance was the rising of tuition costs—between 2001 and 2011 tuition rose by 335% in the UC system (Wegemer 2014). In November 2011, at the University of California Berkeley, approximately 100 students pitched tents outside one of the main administrative buildings after thousands had gathered earlier in the day to protest against planned tuition hikes (Berton and Jones 2011). The police dismantled the tents and in the process beat some of the students. Video images of the abusive police conduct went viral and over one million individuals viewed the YouTube video.

This not only bolstered the efforts of activists on the Berkeley campus but also sparked forms of resistance on other UC campuses (Medina 2012). The reprisals by officials and the police violence against those engaging in acts of civil disobedience is another dynamic that the two cases of student struggle opposing increases in the cost of higher education have in common. Both reprisals and violence, however, worked in the students' favor in terms of recruiting new members, gaining sympathy, holding authorities accountable, and ultimately enriching their cause.

For example, in an incident at the University of California Davis that same month, police officers pepper sprayed and clubbed peaceful protesters, and arrested almost a dozen students who nonviolently protested against the potential increasing costs. Students taped the events and distributed the images onto the Internet, YouTube, Twitter and Facebook, and this then became a major news story in the mainstream press (Heestand 2012). After an investigation into the incident, UC Davis agreed to pay damages of

$30,000.00 to each of the 21 participants who were pepper sprayed (Gordon and Megerian 2013).

These events helped to spread interest in the cause and energized recruitment efforts. The unprovoked abuse by authorities at both Berkeley and Davis also gave the protesters leverage because the beatings earned them sympathy among the public and as a consequence increased their standing in relation to the target of their dissent. These incidents demonstrate that flagrant police violations can no longer be swept under the rug, as they could in the past, because activists can record episodes of excessive force and upload them onto a variety of media outlets and social networking sites in real time.

The mobilizations that began in California soon spread nationwide and for the next several months activists engaged in public forms of resistance. For example, students organized a nationwide "One Trillion Day" protest on April 25th of 2012 in reference to the fact that student debt had cumulatively reached the $1 trillion mark in the United States (Goodale 2012). Many set fire to debt-related documents and some held signs reading "Education in America: Don't Bank on It" (Honan 2012). During graduation ceremonies across the country, students tossed inflatable debt balls, wore plastic chains, and wrote the amount of their debt on graduation caps instead of the more celebratory messages that students typically express upon graduation.

In June of 2012, about 100 students marched around the City Universities of New York (CUNY) and Washington Square Park, and then staged a debtor's die in. In a clever play on words, activists held signs that read "You Are Not a Loan." In November of the same year, students and faculty dressed in caps and gowns at Liberty Plaza in New York City, handing out debt bills instead of diplomas (Honan 2012). These symbolic displays assisted in framing the issues at the root of student grievances, employing an injustice frame of banks preying on students. These emotions, including anger and frustration with the system, once again support social movement theories that focus on cultural reasons for protest activity, and the importance of fostering a sense of collective identity through collective action.

As the campaign evolved and spread, the message transitioned into a much more extensive set of demands and critiques of the neoliberal agenda, in many ways replicating the evolution of Maple Spring into a broader critique of neoliberalism. Under the "for-profit" university system, the students challenged the debt-for-diploma model as well as the legitimacy of the current economic, social, and political system. In terms of outcomes, the University of California students were successful, as there have not been any tuition increases in the UC system over the past three years (Gordon 2013). The students protesting against the CUNY system were less successful in the short term because administrators voted to implement tuition hikes despite the resistance among students. In the broader scheme, however, they were able to raise awareness about their dire economic situation through their participation in contentious tactics on the street and the use of digital tools for the dissemination of their message of disapproval with the system.

10. Conclusion

The arsenal that activists have in their repertoire of contention is a key component of any social movement. What the two case studies discussed in this chapter illustrate is that access to the latest technology plays an important role in contemporary mobilization efforts to influence public sentiment, achieve legitimacy, and recruit new supporters. More specifically, they depict how online tools can play a significant role in community building by developing weak social ties that can serve as an organizing outlet to identify and alert citizens about high levels of discontentment and shared grievances, while allowing activists to discuss demands, distribute information regarding upcoming protest efforts, and record and circulate images of their activity, including police abuse. In both Canada and the United States, the sharing of personal stories online and the follow-up forms of collective behavior in public spaces emboldened individuals and their supporters to further challenge authorities and the structure of higher education. Cyberspace was a key resource that activists used to organize and facilitate the emergence of their protest endeavors. This innovative media terrain helped citizens craft new, albeit virtual, political openings and forms of collective identity, solidarity, and inclusion, which facilitated participation in political discussion that were not available in the past.

Both of these case studies also illuminate how citizen journalists can influence public opinion by disseminating live coverage of events that can help with recruitment efforts by gaining sympathy for the cause and bringing in new allies. Thus, the newly emerging digital media platforms not only allow for more accurate and grassroots coverage of events but also provide a new source of political energy and communicative action.

However, there are certain limitations in this analysis, which can perhaps point to areas for social scientists to explore further. In this chapter, I cover specific aspects of the movement against increase of higher education fees in two countries only. Future studies could examine the role of new media in facilitating expressions of resistance to the neoliberal agenda concerning higher education in other countries as well. Currently, there are two countries in which students are very active in protesting the soaring cost of education: Chile and the \United Kingdom (which now has the highest rate of student debt among First World countries). Studies of protest activities in these countries, or others, would provide a broader comparison of social movement activity among youth. Finally, because both of my case studies are located in urban centers in First World countries, the digital divide, which would be a very prevalent variable to include in research on less developed countries, is devoid in the studies of youth movements in the United States and Canada.

Note

1 Some refer to these as Mojos—namely, mobile journalists who use their smartphones and other digital devices to capture live footage in real time as events transpire.

References

Andreou, Alex. "Occupy Wall Street's Debt Buying Strikes at the Heart of Capitalism." 2013. Accessed May 14, 2014. www.theguardian.com/commentsisfree/2013/nov/13/occupy-wall-st-debt-buying-hear

Banerjee, Sidhartha. *Quebec Student Protests: 2,5000 Arrests and Counting*. The Canadian Press, 2012. Accessed January 1, 2013. www.huffingtonpost.ca/2012/05/25/quebec-student-protests-arrests_n_1544938.html?ref=canada.

Benford, Robert. "Frame Disputes Within the Nuclear Disarmament Movement." *Social Forces* 71 (1993): 677–702.

Bennet, Daniel, and Shanto Iyengar. "A New Era of Minimal Effects? The Changing Foundations of Political Communication." *Journal of Communication* 58 (2008): 707–31.

Berton, Justin, and Carolyn Jones. "Protesters Guard UC Berkeley's New Occupy Camp." 2011. Accessed May 22, 2012. www.sfgate.com/bayarea/article/Protesters-guard-UC-Berkeley-s-new-Occupy-camp-2289240.php.

Boulainne, Shelley. "Does Internet Use Affect Engagement: A Meta-Analysis of Research." *Political Communication* 26 (2009): 193–211.

Cbsnews.com. 2012. Accessed May 20, 2016. http://www.cbsnews.com/news/congrats-class-of-2016-youre-the-most-indebted-yet.

Curran, Peggy. "Anatomy of a Crisis After 100 Days of Protest." *Montreal Gazette*, May 2012.

Diani, Mario. *Green Networks*. Edinburgh: Edinburgh University Press, 1995.

Dolphin, Myles. "Massive Montreal Rally Marks 100 Days of Student Protests." 2012. Accessed January 1, 2013. www.theglobeandmail.com/news/national/massivemontreal-rally-marks-100-days-of-student-protests/article4198301

Duggan, Lisa. *Twilight of Inequality? Neoliberalism, Cultural Politics, and the Attack on Democracy*. Boston, MA: Beacon Press, 2003.

Fortin, Claude. "The Maple Spring as the Flash of the Fifth Estate in Quebec or How the Millennials Appropriated Interactive Digital Technologies to Rise Up and Politically Engage." *Culture, Politics and Technology* 6 (2013): 23–52.Gamson, William. *The Strategy of Social Protest*. Homewood, II:. Dorsey Press, 1975.

Giroux, A. Henry. *Disposable Youth, Racialized Narratives and the Culture of Cruelty*. London and New York: Routledge, 2012.

Goodale, Gloria. "Student Loans: As Debts hit $1 Trillion Mark, Protesters Plan Occupy-Type Events." 2012. Accessed November 3, 2013. www.csmonitor.com/USA/Education/2012/0425/Student-loans-As-debts-hit-1-trillion-mark-protesters-plan-Occupy-type-events.

Gordon, Larry. "UC's $5 Million Plan to Aid 'Dreamer; Students Gets Mixed Reception." 2013. Accessed November 1, 2013. http://www.latimes.com/local/education/la-me-ln-uc-immigrants-20131031,0,1444146.story#axzz2lnpirxRv.

Gould, Roger. "Collective Action and Network Structure." *American Sociological Review* 58 (1991): 182–96.

Heath, Terence. "The Secrt Joy: Six Lessons from Quebec's Maple Spring." Truthout. June 8, 2012.

Heestand, Mela. "UC Davis Students and Faculty Face Prison Time for Peaceful Protest Against Bank." *AlterNet*, April 27, 2012. Accessed April 29, 2012. www.alternet.org/story/155185/uc_davis_students_and_faculty_face_11_years_in_prison_for_peaceful_protest_against_bank_

Honan, Edith. "College Students Protest Debt on 'Trillion Dollar Day.'" 2012. Accessed May 31, 2012. www.reuters.com/article/2012/04/26/us-usa-colleges-debt-idUSBRE83O1JL20120426.

Jenkins, Henry. *Convergence Culture*. New York: New York University Press, 2006.

Jurris, Jeffrey. *Conceptualizing Culture in Social Movement Research*. Edited by Britta Baumgarten, Priska Daphi, and Peter Ullrich, 227–46. Houndmills, Basingstoke, UK: Palgrave Macmillan, 2014.

Leier, Elizabeth. "Quebec's Student Strike Turning into a Citizens' Revolt." 2012. Accessed May 25, 2012. http://truth-out.org/news/item/9372-quebecs-student-strike-turning-into-a-citizens-revolt.

Lindsay, Michael. "Quebec's Student Protests: Were Really was the Dispute:" CBS. September 15, 2013.

Maharawal, Marissa, and Zoltan Gluck. "How Students Are Painting Montreal Red." 2012. Accessed December 12, 2014. http://truth-out.org/news/item/9415-how-students-are-painting-montreal-red

Malcolm, Hadley, and Sean McMinn. "Sagging State Tuition Jacks Up College Tuition." 2013. Accessed November 4, 2013. www.usatoday.com/story/money/personalfinance/2013/09/02/state-funding-declines-raise-tuition/2707837

Marquis, Eric. "Quebec Government Escalates Campaign to Break Student Strike." 2012. Accessed March 1, 2016. http://wsws.org/articles/2012/apr2012/queb-al14.shtl.

Marshall, Andrew. "From the Chilean Winter to the Maple Spring Solidarity: The Student Movements in Chile and Quebec." 2012. Accessed May 26, 2012. http://truth-out.org/news/item/9402-from-the-chilean-winter-to-the-maple-spring-solidarity-and-the-student-movements-in-chile-and-quebec.

McAdam, Doug. *Political Process and the Development of Black Insurgency 1930–1970*. Chicago, IL: University of Chicago Press, 1982.

McKee, Yates. "With September 17 Anniversary on the Horizon Debtors Embrace a Connective Thread for OWS." 2012. Accessed January 14, 2013. http://wagingnonviolence.org/feature/with-september-17-anniversary-on-the-horizon-debt-emerges-as-connective-thread-for-ows.

Medina, Jennifer. "Campus Task Force Criticizes Pepper Spraying of Protesters." 2012. Accessed April 13, 2012. www.nytimes.com/2012/04/12/us/task-force-criticizes-pepper-spraying-of-protesters-at-uc-davis.html.

Melucci, Alberto. *Challenging Codes of Collective Action in the Information Age*. Cambridge, MA: Cambridge University Press, 1996.Olson, Mancur. *The Logic of Collective Action: Public Goods and the Theory of Groups*. Cambridge, MA: Harvard University Press, 1965.

Pilger, John. "The Revolt in Egypt Is Coming Home." 2011. Accessed January 10, 2014. www.truth-out.org/the-revolt-egypt-is-coming-home67624.

Platt, Gerald M., and Stephen J. Lilley. "Multiple Images of a Leader: Constructing Martin Luther King Jr.'s Leadership." In *Self, Collective Behavior and Society: Essays Honoring the Contributions of Ralph H. Turner*, edited by Gerald Platt and Chad Gordon, 55–74. Greenwich, CT: JAI Press, 1994.Ryan, Charlotte and William Gamson. "The Art of Reframing Political Debates." *Contexts* 5, no. 1 (2006): 13–18.

Snow, David A., Burke Rochford, Steven K. Worden, and Robert D. Benford. "Frame Alignment Processes, Micromobilization, and Movement Participation." *American Sociological Review* 51 (1986): 464–81.

Soltry, Ingar. "Canada's 'Maple Spring': From Quebec Student Strike to the Movement Against Neoliberalism." 2012. Accessed January 2014. www.socialist project.ca/bullet/752.php.Tarrow, Sidney. *Power in Movement: Social Movements, Collective Action and Politics*. Cambridge, MA: Cambridge University Press, 1998.

Tilly, Charles. *From Mobilization to Revolution*. Reading, MA: Addison-Wesley Press, 1978.

Tufekci, Zeynep. "New Media and the People-Powered Uprisings." 2012. Accessed November 4, 2014. www.technologyreview.com/view/425280/new-media-and-the-people-powered-uprisings.

Vasi, Ion Bogdan, and Chan Suk Suh. "Predicting the Spread of Protests with Internet and Social Media Activity." *Unpublished Paper*, Columbia University, New York, 2013.

Verba, S., K. L. Scholzman, and H. Brady. *Voice and Equality: Civic Voluntarism in American Politics*. Cambridge, MA: Harvard University Press, 1995.

Wegemer, Christopher. "Rent or Tuition? The Growing Student Dilemma." 2014. Accessed November 8, 2014. www.truth-out.org/news/item/25258-rent-or-tuition-the-growing-student-dilemma?tsk=adminpreview.

9 Telework Impact on Productivity and Well-Being

An Australian Study

Rachelle Bosua, Sherah Kurnia, Marianne Gloet, and Antonette Mendoza

1. Introduction

The proliferation of collaboration and networking tools such as HipChat, Yammer, Quip, Smartsheet, Salesforce Community Cloud, mobile devices, and smartphones create multiple opportunities to work in many locations away from the traditional office. We define *telework* or *telecommuting* as a flexible work arrangement that allows people to work from any location other than the traditional office on either a temporary or regular basis (Di Martino and Wirth 1990; Maruyama, Hopkinson and James 2009).

Telework offers many potential benefits to individuals, organizations, and the wider community. It provides opportunities for enhancing social inclusion by enabling individuals with specific constraints to participate in the workforce. In addition, a number of studies indicate that telework provides temporal and spatial flexibility, improves productivity, and enhances work-life balance while it also reduces stress levels, organizational, operational and infrastructure costs, and traffic congestion (Baruch 2000; Hilbrecht et al. 2008; Maruyama and Tietze 2012; Troup and Rose 2012).

Telework significantly disrupts the workplace in the form of new work environments and work patterns that necessitate the design of new work locations or spaces that foster telework (Bayrak 2012; Maruyama and Tietze 2012). Examples include the rise in activity-based work and hot-desking and the design of smart work hubs (Humphry 2014; Jones 2013; Shieh and Searle 2013). Examples of telework success stories include IBM, Sun and Cisco Systems who each currently has more than 50% of its workforce working at least one or more weekdays away from the traditional office (Meister and Willyerd 2010; Ruth 2009).

Telework can take on many forms depending on roles and responsibilities of workers. Not all work is conducive to a mobile work arrangements, for example service delivery whereby the physical presence of workers is required (e.g., technicians or field engineers), or production systems that require the presence of workers to observe and oversee production processes. Forecasts indicate that telework arrangements are rising globally, for example a forecast predicted a 43% increase in the U.S. mobile workforce by 2016 (Shadler 2009), while similar trends have been predicted for Australia. Recent predictions indicate that 12% of Australia's government workforce will telework by 2020 (McCrindle 2013; Insights (AIM) 2012)

Despite many potential benefits of telework, few organizations offer telework arrangements due to challenges of supervising teleworkers and measuring worker productivity (O' Sullivan 2013). Currently, there is a lack of in-depth studies exploring the actual impact of telework and how benefits can be achieved. Specifically, existing studies of the impact of telework on productivity and well-being show conflicting results (Hilbrecht et al. 2008; Maruyama and Tietze 2012; Troup and Rose 2012), while some studies indicates that telework may lead to social isolation of workers (Pyoria 2011; Weisberg and Porell 2011).

Considering gaps in telework research regarding productivity and well-being, this study explores the impact of telework on team productivity and individual well-being of Australian employees by posing the following research questions:

- *How does telework enhance productivity and well-being of employees?*
- *How can organizations ensure improvement in productivity and well-being through telework?*

We present our work as response to these questions in this chapter. The next section highlights background telework literature, followed by a description of the research methodology, key findings, discussion, and summary of the key themes arising from the research, and directions for further research.

2. Telework: Definition and Key Concepts

The notion of "work" and where work is conducted have changed dramatically over the last few decades. Introduced in the early 1990s, telework or telecommuting has since become prevalent as a result of developments in collaboration and networking technologies and the rise of mobile devices. Telework is described as any form of work conducted away from the physical workplace allowing workers to work either part time or full time, using ICTs to communicate and collaborate (Bentleigh et al. 2016; Bosua et al. 2013; Di Martino and Wirth 1990). There are many forms and variations of telework, based on the type of arrangements and locations where work is conducted.

While telework was originally conducted predominantly from a home office, workers are nowadays more mobile and work virtually from anywhere. Examples include *activity-based* work (working in different areas of a workplace determined by work *activities* one engages in—e.g., a meeting room, a social breakout space or alike), hot-desking, satellite office

Table 9.1 Different Types of Telework Arrangements

Type of telework arrangement	*Description*
Day extender	Informal or formal agreements involving the shifting of start and ending work hours to avoid peak traffic hours and facilitate commuting to and from the workplace (Bekkeheien et al. 1999; Shin and Kinsella 2003).
Hybrid telework	Informal or formal agreements whereby teleworkers spend one to maximum three work days out of the office and the remainder of the week in the office (Gray, Hodson, Gordon 1993; Standen 1997).
Full-time telework or virtual work	More formal, permanent agreements whereby workers spend all their time working either from home, a satellite office or any other location (equivalent to a virtual team; Humphry 2014; Pyoria 2011; Whitehousem Diamond, and Lafferty 2002).

(working in an office located away from the main office), or smart work hubs (a shared office space that offers flexible work practice support in an alternative work location closer to home). The latter work environment is becoming more popular, particularly with start-ups and small business initiatives that cannot afford expensive office space in large cities (Shieh and Searle 2013). In addition, the proliferation of mobile and handheld devices allows telework from many different locations, for example vehicles, coffee shops, open collaborative spaces, and meeting rooms. In this chapter, "telework" refers to a working arrangement that extends beyond the normal home base, and includes working from anywhere.

Telework arrangements manifest in different forms based on the location and the type of arrangement made (Table 9.1). Arrangements vary from permanent to temporary or ad hoc, and can be formal or informal (Maruyama and Tietze 2012; Troup and Rose 2012). Permanent agreements offer long-term regular occurrences of flexible work, while temporary/ad hoc arrangements are usually made on a needs basis (e.g., illness or caring of family members).

3. Benefits and Limitations of Telework

Literature reports many telework benefits for organizations, managers, and workers. Organizational benefits include real estate cost savings and fringe benefit costs, such as electricity savings and parking costs (Cascio 2000; Harker and MacDonnell 2012; Kelly and Moen 2007; O'Sullivan 2013). In addition, some organizations report on the ability to retain excellent

Workers, increased worker productivity and efficiency, and reductions in worker absenteeism (Bailey and Kurland 2002; Gajendran and Harrison 2007; Silva 2007; Somers 2007). Worker benefits include better control and management of working time, juggling work and family life, and reductions in commuting and travel expenses (Hilbrecht et al. 2008; O'Sullivan 2013; Troup and Rose 2012). Some studies report job satisfaction and productivity as positive aspects associated with telework (Golden and Veiga 2005; Hilbrecht et al. 2008; Maruyama and Tietze 2012).

Apart from benefits, there are limitations associated with telework. Management limitations include difficulties to manage and measure the productivity impact of telework, while social exclusion is also a concern (Pyoria 2011; Weisberg and Porell 2011; Baruch 2000; Felstead, Jewson, and Walters 2003; Pyoria 2011; Valsecchi 2006). Previous studies indicate that telework does not always work effectively, can be in conflict with the workplace and management style (Bayrak 2012; Hoang et al. 2008; Maruyama and Tietze 2012; Peters and Heusinkveld 2010). For employees, some studies report that social exclusion and limited interaction with colleagues may affect promotion and career progression (Pyoria 2011; Weisberg and Porell 2011). Other authors (Hilbrecht et al. 2008; Maruyama and Tietze 2012; Troup and Rose 2012) claim that long telework hours add to workers' stress levels and frustration while previous studies highlight that lack of management-teleworker support inhibits productivity and overall job satisfaction (Maruyama and Tietze 2012; Myers and Hearn 2000; Peter and Heusinkveld 2010). Table 9.2 lists key studies on different perspectives of teleworker benefits and limitations.

3.1 Telework and Productivity

Literature on telework productivity is limited, mostly self-reported with a few authors highlighting the complexity of measuring productivity for teleworkers (Baker, Avery, and Crawford 2007; Blok et al. 2010; De Menezes and Kelliher 2011; Neufeld and Fang 2005). Productivity claims on telework are unanimous and state that telework could boost both organizational and individual productivity (De Menezes and Kelliher 2011; Pyöria 2011). However, claims about higher productivity are often made without careful consideration of how managers perceive individual and team productivity. In addition, there is a lack of appropriate measurement tools to measure and monitor productivity effectively (O'Sullivan 2013).

Westfall (2004) proposes four factors that need to be incorporated in a productivity measurement equation: amount of work, intensity of work, efficiency of work and adjustments (i.e., additional organizational costs required to telework). This framework formed a basis for our study with teleworkers and managers regarding their views on productivity. Since an accurate measurement of telework productivity is difficult, other factors that may positively impact an individual's productivity need to be considered. Examples include frequency of social interactions with managers and other team members and task completion (Baker, Avery, and Crawford 2007; De Menezes & Kelliher 2011; Neufeld and Fang 2005), and managers' views of individual and team productivity.

Table 9.2 Organizational and Employee Benefits and Limitations of Telework

Telework Benefits and Limitations	*Key References*
Organizational and management benefits	
• Real estate and energy cost savings • Retaining and accessing talent • Continuity of operations in disasters • Improved employee productivity and efficiency	Cascio 2000; Kelly and Moen 2007; Bailey and Kurland 2002; Gajendran and Harrison 2007; Jordan 2006; Silva 2007; Harker and MacDonnell 2012; Philpott 2007; Somers 2007
Teleworker benefits	
• Work-life-family balance • Job satisfaction • Commuting and transport cost savings • Increased flexibility and control	Hilbrecht et al. 2008; Golden and Veiga 2005; Troup and Rose 2012; Maruyama and Tietze 2012;
Organizational limitations	
• Social exclusion of workers • Productivity • Workplace culture and management support • Difficulty in managing teleworkers)	Golden, Veiga, and Dino 2008; Morganson et al. 2010; Harker and MacDonnell 2012; Gani and Toleman 2006; Hoang et al. 2008; Huws, Jagger, and O'Regan 1999; Peters and Heusinkveld 2010; Pyoria 2011; Felstead, Jewson, and Walters 2003; Valsecchi 2006; Baruch 2000
Teleworker limitations	
• Employee isolation/social exclusion • Concerns about promotion • Anxiety and stress about work • Longer working hours • Lack or management support	Pyoria 2011; Weisberg and Porell 2011; Illegems and Verbeke 2004; Baruch 2000; Maruyama and Tietze 2012; Troup and Rose 2012; Hilbrecht et al. 2008; Myers and Hearn 2000; Peter and Heusinkveld 2010

3.2 Telework and Well-Being

Studies report that telework contributes to work-life-family balance and job satisfaction (see Table 9.2). In addition, prior studies on well-being argue that the level and combination of certain job-related characteristics (e.g., difficulty and stress levels, level of autonomy, task demands and workplace social support) affect individuals' well-being (Bentleigh et al. 2016; Jeurissen and Nyklicek 2001; Warr 1990). These findings suggest that individual well-being positively influences attitudes and perceptions toward work and is consistent with improved job satisfaction, morale, flexibility, and family/work-life balance (Baker, Avery, and Crawford 2007; Hartig, Kylin, and Johansson 2007).

Former Australian telework studies focus on the following aspects of telework: work-family interaction and balance, well-being and productivity (Standen, Daniels, and Lamond 1999; Whitehouse, Diamond, and Lafferty 2002), trends and incidence of telework (Lindorff 2000; Myers and Hearn 2000), adoption and management experiences (Daniels, Lamond, and Standen 2001; Lafferty and Whitehouse 2000; Whitehouse et al. 2002), impact of workplace culture on telework (Gani and Toleman 2006), policy impacts (Alizadeh 2013; Van den Broek and Keating 2011), telework arrangements and work-family outcomes (Troup and Rose 2012), and rural telework (Simpson et al. 2003).

Despite numerous telework studies, there is a noticeable gap in research exploring the impact of telework on both productivity and well-being from a management and employee perspective, especially in Australia. In addition, no studies have investigated how telework improves employee productivity and well-being, the key elements for effective telework, and the challenges managers face to facilitate effective telework. Such insights are critical to develop appropriate and effective strategies for telework in Australia, therefore our focus on employee well-being and productivity from both an individual employee and manager perspectives.

4. Research Methodology

Due to the exploratory nature of the study and our interest in a contemporary phenomenon, we chose a multiple case study research approach. This approach enabled the collection of rich data on productivity and well-being of individual workers and their managers in real-life settings (Yin 2013).

The study was conducted in 2012 to 2013 and involved data collection through interviews conducted with six medium to large Australian organizations, which focused on hearing manager and teleworker perspectives on telework productivity and well-being. We also collected data through daily data logs from three different teams over four consecutive workdays of a week chosen by each team. Data logs also revealed additional insights into how telework impacted social inclusion, productivity, and well-being while the two data sets allowed for better data triangulation.

4.1 Overview of the Case Organization

Six Australian organizations from education and government institutions and the private sector (see Table 9.3) participated in our study. The inclusion of diverse sectors was expected to enhance the generalizability of the study findings. We interviewed 25 hybrid teleworkers across six organizations. We selected *hybrid teleworkers* to hear both "in-the-office" and "out-of-office" experiences. Participants were based in various locations, including Greater Melbourne, Sydney, Newcastle, Brisbane, Perth, Adelaide, and Dubbo, and worked at least two days from home each week. Notably, the

Table 9.3 Case Study Participant and Data Collection Details

Organization and size	*Description*	*Number of Interviewees*	*Interviewees*	*Type of data collected*
1. EduCo (Large)	Tertiary Research Institute	1 manager	Director	Face-to-face interview
2. EducoIT (Large)	Technical and further education	1 manager; 2 workers	Infrastructure and IT manager, Senior Systems Administrator and Network Administrator	Face-to-face interviews
3. NetworkCo (Large)	Network and collab-orative solutions	5 managers; 6 workers	Regional Sales and Project managers (including one non-teleworker), consulting engineer, and regional sales team members	1) Video-call and face-to-face interviews 2) One week of data logs sharing telework experiences
4. GovernCo (Medium)	City council	3 managers; 1 worker	Project managers and education officer	Face-to-face interviews
5. InfraStrucCo (Medium)	Network infrastructure solutions	2 managers; 1 senior worker	Strategic solutions director, national solutions architect, manager, and solutions architect	1) Video-call interviews 2) One week of data logs sharing telework experiences
6. TestCo (Large)	Investment Banking and Superannuation services	1 manager; 6 workers	Application Testing manager and six Testers	1) Face-to-face interviews 2) One week of data logs sharing telework experiences

EduCo participant spoke as a team manager and a teleworker, which supported both perspectives.

The participating organizations provided varying levels of IT support for telework as outlined in Table 9.4). Each organization's different IT tools in use were considered collectively and categorized into three levels of telework support: high-, medium- and low level (Bosua et al. 2013).

NetworkCo and InfrastrucCo provided high-level IT support since they used multiple tools and devices to support telework. Media-rich tools such as videoconferencing, awareness, and collaborative tools were provided by each organization, which allowed seamless interactions between teleworkers, office workers, and managers. Both organizations deployed audio, videoconferencing, online presence tools (e.g., WebX and Skype), instant messaging/chat tools (e.g., Jabber, MSN), smart mobile and handheld devices (iPhones, iPads), and high-speed Internet connection. They also used shared electronic calendar tools (iCal), multiple screens for visual display of work, virtual desktops, multiple web-based tools, and cloud services, including web-based e-mail, Microsoft Exchange, and SharePoint online.

TestCo and EducoIT offered medium-level IT support to facilitate telework. Their networking and collaboration tools provided employees only access to remote files and servers, and audio/video communication through teleconferencing tools. Such support facilitated limited online collaboration between team members, involving limited use of handheld devices. Dial-up conferencing tools, medium-speed Internet connection, and some use of collaborative web-based tools were commonly identified in both organizations.

Lastly, GovernCo and EduCo provided low-level IT support for telework since there were limited tools and devices offered by the organizations to

Table 9.4 Case Organizations and the Types of IT Support to Facilitate Telework

Case Organization	*Type of IT Support*	*Description*
NetworkCo InfraStrucCo	High-level IT support	Multiple communications, networking (e.g., high connectivity videoconferencing) tools and devices are used. In most cases, a suite of tools was provided by the organization to fully support telework.
TestCo EducoIT	Medium-level IT support	Organizational support limited to "Bring Your Own Devices," with limited networking tools provided by the organization (e.g., voice conferencing tools).
GovernCo EduCo	Low-level IT support	Limited to no organizational tool and device support for telework. Mostly mobile phone connectivity with access to an e-mail server.

support telework. Both organizations had low-speed Internet connection to access files on servers, relied strongly on web-based e-mail systems and provided limited to no mobile phones and devices to their employees to collaborate virtually.

4.2 Data Collection and Analysis

Data collection involved two phases: phase 1 comprised a 30 to 45 minutes face-to-face/ video-call interview with all our participants (13 managers and 16 teleworkers). Interview questions focused on teleworker productivity, teamwork, and well-being. Phase 2 required participating teleworkers from three teams to log their daily experiences of work on four consecutive days in one week of their choice (i.e., two telework days and two non-telework days). To support this second phase of data collection, a website was developed to facilitate data logging activities across teams and workdays (see Figure 9.1). The daily data logs enabled the researchers to get real-world insights into individuals' work patterns and their perceptions of, and feelings associated with working away from the office.

Employees from three organizational teams NetworkCo (six participants), TestCo (nine participants), and EducoIT (two participants) participated in data logging activities. Recruiting participation in data logging activities was challenging and required significant time commitment from the participants. Data logs captured travel expenses; cost savings associated with working away from the office—e.g., Internet and utility costs; actual hours worked each day; daily feelings about productivity and well-being and attitudes toward work; number and type of daily tasks planned and completed; and number and types of work-related interruptions. Due to the rich data provided, the number of phase 2 participants was adequate to provide meaningful observations and to corroborate findings with the interviews.

Participants' feelings about their daily tasks (morale, control over their work, job satisfaction, intensity of the work and pressure on telework days and non-telework days) were logged using a 7-point Likert scale. Figure 1 presents one of the screens from the daily data log website. We analyzed captured data logs and conducted a follow-up interview with each of the participating data log team's managers. The aim was to correlate the initial stage 1 manager interviews and data logs with the second-round manager interviews. At the end of phase 2, data logs were verified with team managers to assess collective tasks planned and completed by team members to verify individual worker and overall team productivity.

All interview and logged data was transcribed and thematically analyzed using qualitative data analysis techniques in the form of open, close and selective coding (Neuman 2014; Yin 2013). The focus was on identifying key themes related to employee productivity and well-being, from both management and worker perspectives.

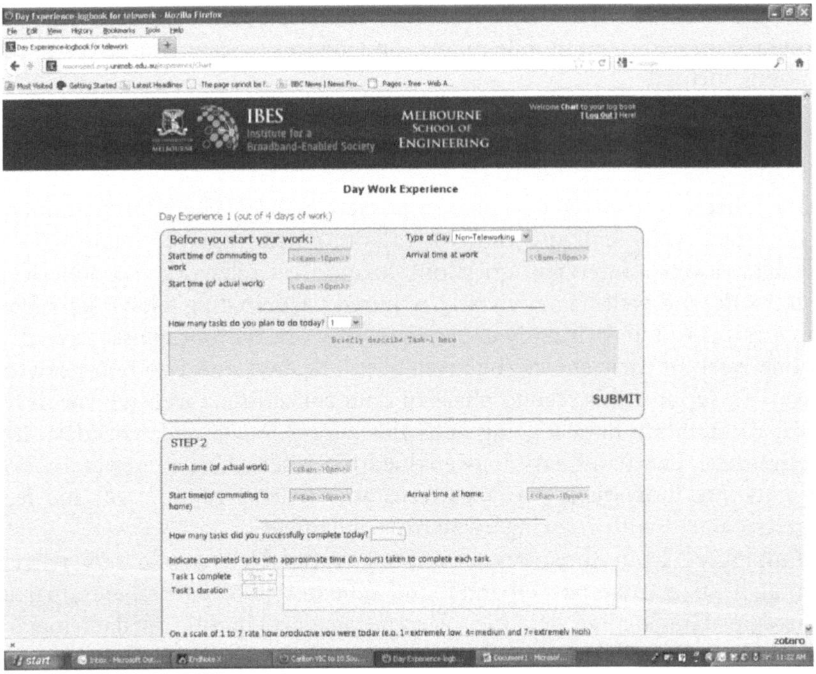

Figure 9.1 Example of a Data Log Website Screen.

5. Findings

We discuss our findings based on the key themes that emerged from the data analysis. In total, we identified six key themes that are particularly relevant as presented next.

5.1 Theme 1: Telework Improves Productivity Through Fewer Work-Related Interruptions

The findings demonstrate that telework contributes to individual and team productivity. This was particularly evident as two participants from the high-level IT support case organizations commented,

> I am a lot more productive when I telework, I can remove myself from distractions, focus and disappear from people quite easily. When I'm in the office, I can turn down my phone, put "do not disturb" on the door but people can see I'm there, still knock on my door and interrupt me.
> (Strategic Solutions Director, InfraStrucCo)

> Telework gives you space in a different environment, whether at home, the coffee shop, or in the office—wherever you can find some space to

> make sure you have a plan of attack for the day, week, month, year and making sure you're tracking to it.
>
> (Project Manager, NetworkCo)

Even participants from the medium to low-level IT support organizations reported higher productivity as two project managers commented,

> I'm more productive when I work from home. In my job, I get interrupted all the time—sometimes it's important, sometimes not. When I'm here [office], people come to me all the time. It's very rare that I set myself five things to do that will get done [at work].
>
> (Senior Systems Administrator, EducoIT)

> I'm much better off working from home . . . than in my open plan office. I work much better in a secluded environment [where] I can concentrate deeply. I can't do that at work.
>
> (Project Manager, GovernCo)

Data logs also confirmed higher productivity on telework days as opposed to non-telework days as two TestCo teleworkers commented,

> I was very productive today as I completed my tasks with few interruptions . . . working from home meant less interruptions from colleagues, and I was able to get though a large amount of processing with no disruptions.

Data log analyses indicated that more planned tasks were completed on telework days and TestCo team members rated their productivity higher on telework days compared to non-telework days. Their team manager follow-up interview indicated the manager perceived her teams as productive in the particular data log week. Her positive response indicated that she was "*very pleased*" with her team's performance on telework days. Other team managers whose teams participated in the data logs (InfraStrucCo and NetworkCo) responded similarly. All team managers indicated overall satisfaction with individual and team teleworker productivity on telework days.

5.2 Theme 2: Telework Enhances Well-Being Through Flexibility, Work-Life and Family Balance, and Less Stress

Participants reported a positive relationship between telework and well-being adding that flexibility and "head-space" contributed to well-being, which in turn generated a feeling of being more productive,

> The positives are less stress, no commuting on telework days, feeling comfortable when I'm working. I experience less anxiety on telework days, and have better work-life balance.
>
> (Project Manager, NetworkCo)

The EduCo manager confirmed the link between job satisfaction and well-being,

> I feel more refreshed when at home, almost like it's a weekend sometimes because I enjoy my work. It's not something I feel I have to do. Normally by Thursday if I don't telework, I feel my energy ebbing away, then thank God it's Friday. I would be tired by the end of the week and need the weekend to recover. Having Wednesday as a telework day, I have much more energy remaining for the end of the week. Maybe that's because I get a chance to do stocktaking, clear the decks and my e-mails when I telework mid-week.

A NetworkCo project manager mentioned flexibility, and work-life family balance,

> I get flexibility from telework—I can see more of my kids, take my daughter to school and pick her up—things that regular office workers wouldn't be able to do.

In addition, day-experience data logs indicated high well-being rating (5 to 7 on the 7-point Likert scale) for telework days, which data log confirmed,

> Good balance between home/work, being able to assist with the family and still get a number of tasks completed (NetworkCo), and I'm more than happy and stress free.
>
> (TestCo)

Less enthusiastic data log experiences were reported on non-telework days,

> I feel I have achieved an average amount of work [on my telework day], so I will be behind again [tomorrow on the non-telework day]—[feeling] a little pressured.
>
> (NetworkCo)

One manager linked telework and his sense of well-being to his work and family-life balance, which decreased stress,

> In terms of well-being, I find that a certain element of stress is completely gone from my life. When I go and watch Ollie [the son] play tennis, I'm actually enjoying it.
>
> (Manager, NetworkCo)

5.3 Theme 3: IT Support Affects Telework Productivity

Across all cases, productivity was closely linked to the availability and use of adequate enabling and supportive IT tools and applications for telework (i.e., collaboration and networking tools, easy access to content, virtual workspaces, and supportive network infrastructures).

Participants from high-level IT support organizations indicated seamless work from anywhere, and felt more productive than those with medium or low levels of IT support,

> We're at the higher end of the scale and are provided with a "virtual office environment." I have a router in my home, so my home office effectively becomes an extension of [NetworkCo's] environment. There is no difference in the environment between office and home, so technology is the enabler.
>
> (Regional Sales Manager, NetworkCo)

A participant from a medium-level IT support organization indicated,

> I think that better technology could improve it [telework]. I know of some companies that use Skype or other technologies that might be useful. You might sometimes e-mail someone with a question and you're waiting on a response and find it's taking a bit of time.
>
> (Tester, TestCo)

Teleworkers with limited networking and collaborative tools, devices, and low-speed Internet access flagged a negative impact on productivity,

> The problem is logging into [organizational] systems, it's slow . . . I take a USB home the night before teleworking with all the big files on them so I don't constantly download from our server during the day. I try not to having to do remote login because it takes forever to download files.
>
> (Manager, EduCo)

5.4 *Theme 4: A Supportive Organizational Telework Culture Contributes to Productivity*

Findings across the cases emphasize the need for an understanding telework culture that embraces work both in and out of the office as one manager commented,

> With the right culture and with the right employee type you get great productivity gains. That will aid the employees' life and their work.
>
> (Manager, NetworkCo)

TestCo's supportive telework culture of working one to two weekdays away from the office required all workers to inform their managers when they intended to telework the next week. This supportive culture positively impacted employees' overall job satisfaction and productivity since expectations were clear to all employees.

In addition, a supportive organizational telework culture also contributed to job satisfaction,

> My manager tells us work hard [be productive], do what you need to do, if you need to see the doctor or want to spend a day or two [working] at home, as long as you're accessible and deliver the results [be productive] linked to your KPIs.
>
> (Manager, NetworkCo)

5.5 Theme 5: Managers' Trust in Teleworkers Influences Productivity

Managers trusted their workers to deliver what was expected when they teleworked—i.e., being productive in terms of the timely completion of assigned tasks. Teleworkers in turn felt their managers fully trusted their ability to complete assigned tasks within the required time, which positively impacted their productivity as a TestCo manager indicated,

> I found it difficult to identify what [teleworkers] were doing—it all came down to trust, could we trust them to be productive and work without being in the office? It came down to getting to know the team and we soon knew who could be trusted [in terms of completing the work] and who couldn't. Those suspects [who] weren't doing the right thing were monitored closely.
>
> (Test Manager, TestCo)

NetworkCo, a global leader in telework, was highly selective in recruiting employees and easily terminated employees that were unproductive. NetworkCo required employees to work in the office for a period of time before they could telework. A six-month probation period served as an opportunity for managers to sense individual employee productivity levels and build trust,

> When you start the job you are given the trust, and it's yours to lose if you don't do the right thing. And then you are judged by results, so if I deliver the results then I have the trust of my managers.
>
> (Project Manager, NetworkCo)

5.6 Theme 6: Hybrid Telework as the Most Feasible Telework Mode

Discussions with teleworkers on preferred and most feasible telework arrangements indicated strong support for a maximum of one to three days away from the office. These days were particularly noted in data logs as 'more productive workdays' than the days spent in the office, while full teleworking were considered too isolated,

> If I had to telework 100% of the time I'd probably hate it because you would be too isolated. I have friends that I work with and you are

stimulated by coming into the office seeing people and hearing about different things that are going on, seeing customers or partners and so on . . . working from home every day, I would become isolated pretty quickly.

(Teleworker, NetworkCo)

Some participants had experience of working away from the office for periods of time (NetworkCo and InfraStrucCo), but most participants agreed that hybrid telework was more effective in keeping up with social aspects of the work environment as one participant confirmed,

I would miss out on the gossip and socialization if I don't come to work in the week.

(Teleworker, InfrastrucCo)

6. Discussion

We have explored the impact of telework on individual and team productivity and well-being by asking the following research questions:

How does telework enhance productivity and well-being of employees?
How can organizations ensure improvement in productivity and well-being through telework?

Our analyses offer a number of important observations that complement the findings of prior studies on the impact of telework on productivity and well-being.

Figure 9.2 summarizes the key findings related to how telework may enhance productivity and well-being of employees and the key elements that affect the improvement in productivity and well-being through telework. We explain Figure 9.2 by discussing the answers to the two research questions.

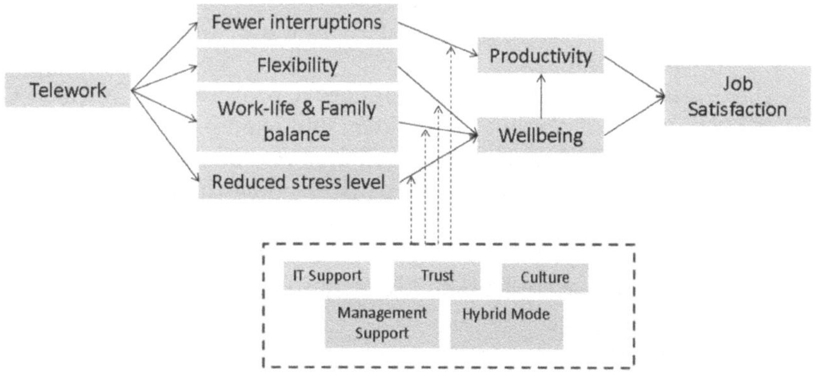

Figure 9.2 Telework Benefits and Telework Impact on Productivity and Well-Being.

In response to the *first research question*, our findings indicate that telework impacts positively on individual and team productivity and well-being. Our study demonstrates that telework enables workers to be more productive since there are *fewer work-related interruptions* (e.g., colleagues asking questions, informal discussions, and phone calls) when working away from the offices. For example, TestCo testers have highly cognitive tasks they need to perform as part of testing critical parts of software components. Telework days give them the opportunity to work without interruptions on these tests. A GovernCo participant involved in the highly cognitive task of policy development shared the same sentiments that other studies report (Baruch 2002; Hilbrecht et al. 2008; O'Sullivan 2013), indicating that telework enables more efficient work with fewer distractions.

Findings across the six organizations also indicate strong support for a better *work-life and family balance* that engenders *job satisfaction* and a *positive attitude* toward work. All participants unanimously agreed that telework allowed *greater flexibility* and *a sense of autonomy* over work, which *lowered stress levels* and allowed for *more productive* work. Working parents of NetworkCo valued the opportunity to attend some afternoon sporting events of their children, while they could catch-up on their office work during evenings. This allowed workers to better *balance family* and *work life*, while the ability to have a presence at home contributed to a more happy family life. These important observations of how telework enhances employee productivity and well-being are captured in Figure 9.2. While our findings are consistent with previous studies that identify the positive impact of telework on well-being (Hilbrecht et al. 2008; Maruyama and Tietze 2012; Troup and Rose 2012), this study further shows that improvement in well-being increases the productivity of employees.

In response to the second research question, our study identified that five critical elements, when employed improve employee productivity and well-being through teleworking: IT support, trust, culture, a hybrid telework mode, and management support as shown in Figure 9.2.

1) Making *appropriate IT support* tools available to facilitate telework: IT is a key enabler for telework and appropriate IT tools allowing teleworkers to communicate, collaborate, and access content from anywhere is essential. High-level IT supportive tools (e.g., in NetworkCo and InfrastrucCo) can boost individual and team productivity, while a mid-level IT support environment (e.g., TestCo and EduCoIT) can significantly facilitate productive telework. Findings indicate that careful consideration needs to be given to the choice of supportive IT tools and infrastructures for telework.
2) *Trust* enhances telework productivity and well-being as confirmed by Peters and Den Dulk (2003). NetworkCo, InfraStrucCo, and TestCo participants emphasized the importance of trust in telework relationships between managers and workers. Our study indicates that some

managers were more willing than others to trust employee productivity in a telework environment (NetworkCo and InfraStrucCo), probably as a result of probationary work periods, which allowed managers to better understand individual workers' productivity levels. We found that managers who teleworked better understood the context of telework, were less concerned about trust, and focused more on getting the required outputs.

3) An *organizational culture* conducive to telework, reflects a supportive environment in which telework is accepted and valued. Across all our cases employees and managers upheld a strong telework culture. A telework "culture manifests in the different types of agreements between managers and teleworkers such as friendly notifications to managers and team members when teleworkers were not in the office" (e.g., TestCo). Cultural elements support work-life and family balance, which many of the employees require to promote their sense of well-being. This finding strongly corresponds to other studies that report the positive impact of an organizational telework culture (e.g., Daniels, Lamond, and Standen 2001).
4) A *hybrid telework mode* clearly reenergizes workers, enabling workers flexible work from different work locations (often more tranquil home-settings). Our participants unanimously agreed their preference for hybrid telework, as they were dependent upon social interactions and networking with colleagues when in the office. Workers felt that personal interactions with colleagues enhance individual learning experiences, personal growth, and the generation of new ideas.
5) Appropriate *management support* is necessary to ensure frequent and effective communication, and clear agreements on targets and outcomes that could be monitored regularly. Thus, management support forms a sound basis for managing teleworkers and can assist in developing an underlying platform of trust. Adequate management support can be achieved through more formal policies and agreements that identify worker and management expectations.

7. Conclusion

Considering the positive outcomes of telework from a well-being and productivity perspective, we conclude that telework will increase as more organizations adopt this new form of work and accompanying technologies that enable telework. In addition, we speculate that this trend may impact management perceptions about work mobility and productivity. Hence, telework might open up access to new innovative forms of work that might include individuals who are currently excluded from the workforce. Examples include mothers returning to the workforce, workers with physical disabilities who find access to the physical workplace difficult, and workers who are not fully functional in busy office environments.

Our study contributes to broader telework studies indicating that enhanced productivity and well-being are both outcomes of hybrid telework arrangements that lead to job satisfaction. Workers can be more productive when they are hybrid teleworkers, provided the right combination of the following elements are present: ICT support, management support, and trust, a culturally supportive environment and access to hybrid telework arrangements. A limitation in our study was the small number of participants. Future research should consider detailed attributes and strategies required to foster and manage work mobility, while larger and more comprehensive studies are required to identify how productivity can be attained, and the supportive cultural elements required for teleworker productivity, well-being, and job satisfaction.

Acknowledgments

The authors would like to thank Melbourne Networked Society Institute at the University of Melbourne who funded this project.

References

Alizadeh, T. "Planning Implications of Telework: A Policy Analysis of the Sydney Metropolitan Strategy." *Australian Planner* 50, no. 4 (2013): 4–315.

Bailey, D. E., and N. B. Kurland. "A Review of Telework Research: Findings, New Directions, and Lessons for the Study of Modern Work." *Journal of Organizational Behavior* 23 (2002): 383–400.

Baker, E., G. C. Avery, and J. Crawford. "Satisfaction and Perceived Productivity When Professionals Work from Home." *Research and Practice in Human Resource Management* 15, no. 1 (2007): 37–62.

Baruch, Y. "Teleworking: Benefits and Pitfalls as Perceived by Professionals and Managers." *New Technology, Work and Employment* 15, no. 1 (2000): 34–49.

Bayrak, T. "IT Support for Telecommuting Workforce." *Telematics and Informatics* 29 (2012): 286–93.

Bekkeheien, M., Ø. Håland, R. Klovening, and R. Stokholm. "Energy Demand Patterns Towards 2050." In *Energy: The Next Fifty Years, Organization for Economic Co-Operation and Development*. Paris: OECD Publication Services, 1999.

Bentleigh, T. A., S. T. T. Teo, L. McLeod, F. Tan, R. Bosua, and M. Gloet. "The Role of Organizational Support in Teleworker Wellbeing: A Socio-Technical Systems Approach." *Applied Ergonomics* 52 (2016): 207–15.

Blok, M. M., L. Groenesteijn, R. Schelvis, and P. Vink. "New Ways of Working: Does Flexibility in Time and Location of Work Change Work Behaviour and Affect Business Outcomes." *Work* 41 (2010): 5075–80.

Bosua, R., M. Gloet, S. Kurnia, A. Mendoza, and J. Yong. "Telework, Productivity and Wellbeing: An Australian Perspective." *Telecommunications Journal of Australia* 63, no. 1 (2013).

Cascio, W. F. "Managing a Virtual Workplace." *The Academy of Management Executive* 14, no. 3 (2000): 81–90.

Daniels, K., D. Lamond, and P. Standen. "Teleworking: Frameworks for Organizational Research." *Journal of Management Studies* 28, no. 8 (2001): 1151–85.

De Menezes, L. M., and C. Kelliher. "Flexible Working and Performance: A Systematic Review of the Evidence for a Business Case." *International Journal of Management Reviews* 13 (2011): 452–74.

Di Martino, V., and L. Wirth. "Telework: A New Way of Working and Living." *International Labour Review* 129, no. 5 (1990): 529–54.

Felstead, A., N. Jewson, and S. Walters. "Managerial Control of Employees Working at Home." *British Journal of Industrial Relations* 41, no. 2 (2003): 241–64.

Gajendran, R. S., and D. A. Harrison. "The Good, the Bad, and the Unknown About Telecommuting: Meta-Analysis of Psychological Mediators and Individual Consequences." *Journal of Applied Psychology* 92 (2007): 1524–41.

Gani, Z., and M. Toleman. "Success Factors and Barriers to Telework Adoption in e-Business in Australia and Singapore: The Influence of Culture and Organizational Culture." *Journal of Theoretical and Applied Electronic Commerce Research* 1, no. 3 (2006): 81–92.

Golden, T. D., and J. F. Veiga. "The Impact of Extent of Telecommuting on Job Satisfaction: Resolving Inconsistent Findings." *Journal of Management* 31, no. 2 (2005): 301–18.

Golden, T. D., J. F. Veiga, and R. N. Dino. "The Impact of Professional Isolation on Teleworker Job Performance and Turnover Intentions: Does Time Spent Teleworking, Interacting Face-to-Face, or Having Access to Communication-Enhancing Technology Matter?" *Journal of Applied Psychology* 93, no. 6 (2008): 1412.

Gray, M., N. Hodson, and G. Gordon. *Teleworking Explained*. Chichester: Wiley, 1993.

Harker, M. B., and R. MacDonnell. "Is Telework Effective for Organizations? A Meta-Analysis of Empirical Research on Perceptions of Telework and Organizational Outcomes." *Management Research Review* 35, no. 7 (2012): 602–16.

Hartig, T., C. Kylin, and G. Johansson. "The Telework Trade-Off: Stress Mitigation vs. Constrained Restoration." *Work and Stress* 15, no. 3 (2007): 254–64.

Hilbrecht, M., S. M. Shaw, L. C. Johnson, and J. Andrey. "I'm Home for the Kids': Contradictory Implications for Work—Life Balance of Teleworking Mothers." *Gender, Work and Organization* 15, no. 5 (2008): 454–76.

Hoang, A. T., R. C. Nickerson, P. Beckman, and J. Eng. "Telecommuting and Corporate Culture: Implications for the Mobile Enterprise." *Information Knowledge Systems Management* 7 (2008): 77–97.

Humphry, J. "Visualising the Future of Work: Myth, Media and Mobilities." *Media, Culture & Society* 36, no. 3 (2014): 351–66.

Huws, U., N. Jagger, and S. O'Regan. *Teleworking and Globalisation*. London: London Institute for Employment Studies, 1999.

Illegems, V., and A. Verbeke. "Telework: What Does it Mean for Management?" *Long Range Planning* 37, no. 4 (2004): 319–34.

Insights (AIM). "Managing in a Flexible Work Environment", *Australia Institute of Management (AIM) White Paper*, 2013.

Jeurissen, T., and I. Nyklicek. "Testing the Vitamin Model of Job Stress in Dutch Health Care Workers." *Work and Stress* 15, no. 3 (2001): 254–64.

Jones, A. M. *The Fifth Age of Work: How Companies Can Redesign Work to Become More Innovative in a Cloud Economy*. Portland: Night Owls Press LLC, 2013.

Jordan, B. "Telework's Growing Popularity: And It Makes Good Homeland Security Sense, Too." *Homeland Defense Journal* 4, no. 6 (2006): 16–22.

Kelly, E.L. and P. Moen. "Rethinking the Clockwork of Work: Why Schedule Control May Pay Off at Work and at Home." *Advances in Developing Human Resources* 9, no. 4 (2007): 487–506.

Lafferty, G., and G. Whitehouse. "Telework in Australia: Findings from the National Survey in Selected Industries." *Australian Bulletin of Labour* 26, no. 3 (2000): 236–52.

Lindorff, M. "Home-Based Telework and Telecommuting in Australia: More Myth than Modern Work Form." *Asia Pacific Journal of Human Resources* 38, no. 3 (2000): 1–11.

Maruyama, T., P. G. Hopkinson, and P. W. James. "A Multivariate Analysis of Work–Life Balance Outcomes from a Large-Scale Telework Programme." *New Technology, Work and Employment* 24, no. 2 (2009): 76-88.

Maruyama, T., and S. Tietze. "From Anxiety to Assurance: Concerns and Outcomes of Telework." *Personnel Review* 41, no. 4 (2012): 450–69.

McCrindle. "The McCrindle Blog: Teleworking in Australia: Latest Trends and Perceptions." 2013. Accessed August 12, 2015. http://mccrindle.com.au/the-mccrindle-blog/teleworking-in-australia-work-remote

Meister, J. C., and K. Willyerd. *The 2020 Workplace*. New York: HarperBusiness, 2010.

Morganson, V. J., D. A. Major, K. L. Oborn, J. M. Verive, and M. P. Heelan. "Comparing Telework Locations and Traditional Work Arrangements: Differences in Work-Life Balance Support, Job Satisfaction, and Inclusion." *Journal of Managerial Psychology* 25, no. 6 (2010): 578–95.

Myers, N., and G. N. Hearn. "Communication and Control: Case Studies in Australian Telecommuting." *Australian Journal of Communication* 27, no. 2 (2000): 39–64.

Neufeld, D. J., and Y. Fang. "Individual, Social and Situational Determinants of Telecommuter Productivity." *Information & Management* 42 (2005): 1037–49.

Neuman, W. L. *Social Research Methods Qualitative and Quantitative Approaches*. Boston, MA: Pearson, 2014.

O'Sullivan, G. "Literature Review: Teleworking in Human Services." Southern Area Consortium of Human Services, School of Social Work, San Diego, CA, 2013, 1–42.

Peters, P., and L. Den Dulk. "Cross Cultural Differences in Managers' Support for Home-Based Telework a Theoretical Elaboration." *International Journal of Cross Cultural Management* 3, no. 3 (2003): 329–46.

Peters, P., and S. Heusinkveld. "Institutional Explanations for Managers' Attitudes Towards Telehomeworking." *Human Relations* 63, no. 1 (2010): 107–35.

Philpott, D. "Telework and Continuity of Operations (COOP)." *Homeland Defence Journal* (Special Report). 2007. Accessed August 22, 2017. http://www.cisco.com/c/dam/en_us/solutions/industries/docs/gov/teleworkWebExRpt072107.pdf

Pyoria, P. "Managing Telework: Risks, Fears and Rules." *Management Research Review* 34, no. 4 (2011): 386–99.

Ruth, S. "Green It More than a Three Percent Solution?" *Internet Computing IEEE* 13, no. 4 (2009): 74–8.

Shadler, T. "US Telecommuting Forecast 2009 to 2016—a Digital Home Report." [Internet]. 2009. Accessed November 8, 2012. www.forrester.com/US+Telecommuting+Forecast+2009+To+2016/fulltext/-/E-RES46635?objectid=RES46635

Shin, B. and D. C Kinsella. "Increasing Business Value of Communications Infrastructure: The Case of Internet-Based Virtual Private Networks." In *Creating Business Value with Information Technology: Challenges and Solutions*, edited by N. Shin. Hersley, London: IRM Press, 2003.

Shieh, A., and G. Searle. "Telework and Spatial Trends in Australian Cities: A Critical Review." In *SOAC 2013: 6th State of Australian Cities Conference*, State of Australian Cities Research Network, 2013, 1–8.

Silva, C. "From 60 to 0: Denver Group Halts Turnover at Area Businesses Through Telework." *Employee Benefit News* 21, no 8 (2007): p.44.

Simpson, L., L. Daws, B. Pini, and L. Wood. "Rural Telework: Case Studies from the Australaian Outback." *New Technology, Work and Employment* 18, no. 2 (2003): 115–26.

Somers, S. "Survey and Assessment of Planning for Operational Continuity in Public Works." *Public Works Management & Policy* 12, no. 2 (2007): 451–65.

Standen, P. "Home, Work and Management in the Information Age." *Journal of the Australian and New Zealand Academy of Management* 3 (1997): 1–14.

Standen, P., K. Daniels, and D. Lamond. "The Home as Workplace: Work-Family Interaction and Psychological Well-Being in Telework." *Journal of Occupational Health Psychology* 4 (1999): 368–81.

Troup, C. and J. Rose. "Working From Home: Do Formal or Informal Telework Arrangements Provide Better Work–Family Outcomes?" *Community, Work & Family* 15, no. 4, (2012): 471–86.

Valsecchi, R. "Visible Moves and Invisible Bodies: The Case of Teleworking in an Italian Call Centre." *New Technology, Work and Employment* 21, no. 2 (2006): 123–38.

Van den Broek, D., and E. Keating. "Rights to a Process for the Masses or Select Privileges for the Few? Telework Policy and Labour Market Inequality in Australia." *Policy Studies* 32, no. 1 (2011): 21–33.

Warr, P. B. "Decision Latitude, Job Demands, and Employee Well-Being." *Work and Stress* 4, no. 4 (1990): 285–94.

Weisberg, A., and M. Porell. "Moving Telework from Compliance to Competitiveness." *The Public Manager* 40, no. 1 (2011): 12–4.

Westfall, R. D. "Does Telecommuting Really Increase Productivity?" *Communications of the ACM* 47, no. 8 (2004): 93–6.

Whitehouse, G., C. Diamond, and G. Lafferty. "Assessing the Benefits of Telework: Australian Case Study Evidence." *New Zealand Journal of Industrial Relations* 2, no. 3 (2002): 257–68.

Yin, R. K. *Case Study Research Design and Method*. Thousand Oaks, CA: Sage Publications, 2013.

10 Supporting Regional Food Supply Chains With an E-Commerce Application

Sherah Kurnia, Md Mahbubur Rahim, Serenity Hill, Kirsten Larsen, Patrice Braun, Danny Samson, and Prakash Singh

1. Introduction

In recent years, sustainability has become an important issue for businesses and the broader society (Porter and Kramer 2011). An increasing number of organizations have endeavored to improve their business operations within their organization and supply chains to enable them to achieve economic, environmental, and social benefits, which is known as the Triple Bottom Line (TBL) (Elkington 1997). Various sustainability initiatives introduced are based on the sustainable development principle (Brundtland 1987) that encourages the fulfillment of the needs of the current generation without compromising the ability of the future generations to meet their needs.

Specifically, within the food supply chain context, food waste is identified as a significant supply chain issue due to growing supermarket domination in many countries (Blay-Palmer et al. 2013). Large supermarket chains around the globe typically apply high quality control, which results in the rejection of imperfect produce supplied by local farmers. In addition, because of the restricted agreement on the supply imposed by large players within the industry, excess fruit and vegetables exists within supply chains and farmers have limited means of distributing these products (Blay-Palmer et al. 2013). Such a situation threatens the sustainability of regional supply chains that impact on the economic, environmental, and social condition of the local communities in regional areas.

In response to the current food supply chain issues, the Food Hub concept has emerged as a way to improve local food supply systems. Food Hubs represent an alternative model to the mainstream food supply and are defined as organizations that aggregate, distribute, and market food products primarily sourced from local and regional producers though simple, and more sustainable supply chains (Fischer et al. 2013). They facilitate a closer connection between producers and consumers (Matson and Thayer 2013). The existing literature has identified several existing and potential benefits Food Hubs provide to producers, consumers, and local economies (Flaccavento 2016; Rose 2015; Fischer et al. 2013).

However, Food Hubs have experienced several key challenges that threaten their viability over a long term. For example, finding appropriate value chain partners for distributing products and developing mechanisms for value chain decision making, transparency, and trust are difficult (Stevenson et al. 2011). In practice, many Food Hubs lack coordinated marketing mechanisms that raise brand awareness to local customers, and many are unable to establish effective strategies for product differentiation, branding, and regional identity (Rose and Larsen 2013). Furthermore, Food Hubs have not been well recognized by government and mainstream players within the food industry (Rose and Larsen 2013).

With the advancement of the broadband Internet technology, ICT has the potential to support Food Hubs operations and overcome some of their challenges. Specifically, the application of electronic markets that was rapidly introduced in the late 1990s appears to be relevant to address the issues within local food supply chains. The value proposition of an electronic market is to reduce search costs for buyers and sellers through the use of a hub, facilitate product evaluation through transparency of information, and help discover the right price of products (Bakos 1998; Alt and Klein 2011). One recent innovative applications of e-market to support regional food supply chains is Open Food Network (OFN)[1] OFN has been established in the state of Victoria, Australia to connect various players of regional food supply chains in Australia and support the operations of Food Hubs. However, since such an application of an e-market is relatively new, there exists limited research conducted to assess the effectiveness of such an ICT-enabled innovation in supporting the sustainability and resilience of Food Hubs and the regional food systems. Furthermore, OFN was just launched in Australia in mid-2015. Hence, little is known about the actual use and the impact of OFN on the stakeholders. Without an understanding of the effectiveness of OFN, it is difficult to devise appropriate strategies to enhance the features and encourage wider adoption within regional food supply chain parties.

This study is part of a large research project that aims to understand how ICT can help implement sustainability practices within organizations and supply chains. We systematically assess how the OFN is used by the early adopters who are regional food supply chain players and the impacts on their operations. The overall research question and sub-questions addressed in this study are:

- *How Does OFN Support Regional Food Supply Chains in Australia?*
- Sub-question 1: How is OFN used by the adopting supply chain players?
- Sub-question 2: What are the benefits and challenges experienced by the OFN adopters?

These research questions are addressed through conducting a focus group with nine active OFN enterprise users to investigate the use and impact of OFN. Through this investigation, we enhance the current understanding of how OFN, as an example of simple e-commerce application, can benefit the supply chain participants and potentially enable social inclusion of disadvantaged local farmers to establish a more sustainable and fair trading environment. There were only limited research participants involved since the OFN is relative new in Victoria. Despite the small sample size, we have identified several issues and challenges faced by the participants in using OFN that will be valuable for the OFN technology provider to further improve the OFN features and devise appropriate strategies that may encourage wider adoption. Understanding obtained from this study may benefit future design of ICT-enabled initiatives that address sustainability in different contexts.

2 Literature Review

2.1 Food Hubs: Benefits and Challenges

The growing interest on a healthier diet to avoid controllable diseases, and concern over the environmental impact of supply chains has given rise to the popularity of supporting the concept of "localization." Localization of food supply chains simply means that food should be consumed as close to the point of origin as possible (Seyfang 2006; Barham et al. 2012). The concept of localization stems from the consumers' interest in understanding where and how their product is grown to ensure their food has high nutritional value without damaging the environment (Seyfang 2006). The concept had led to the emergence of "Food Hubs" in regional areas in the United States, United Kingdom, and Australia.

Food Hubs could be viewed as an intermediary that uses an innovative business model to connect small and medium-sized producers and local consumers (Woods et al. 2013). Existing research indicates that Food Hubs may offer several environmental, social, and economic benefits to the food industry and communities. For example, Food Hubs support environment friendly production such as organic items that do not use harmful ingredients such as pesticides and chemicals (Flaccavento 2009, 2016; Stevenson et al. 2011). They provide nutritious products in their local community by shortening the delivery timeline (Rose and Larsen 2013). They also facilitate civic agriculture by helping community members understand the origin of their food and its supply system (Lyson 2005). They help improve the health of low-income communities, school children, and other institutions (e.g., universities, prisons and hospitals) by providing them with greater access to fresh food (Conner, Campbell-Arvai, and Hamm 2008; Erlbaum, McManus, and Nowak 2011). Furthermore, studies indicate that Food Hubs ensure equitable income for farmers and food system workers, fair prices for consumers (Flaccavento 2009; Matson and Thayer 2013), and create more jobs in the local economy to support the development and operations of Food Hubs and regional food distribution systems (Rose and

Larsen 2013). Finally, Food Hubs potentially minimize fuel consumption and carbon emissions by using optimization technology for route scheduling and truck optimization (Rose and Larsen 2013).

Despite the growing popularity of Food Hubs, their operations encounter many challenges. Pricing is one of the key challenges. Food Hubs management needs to ensure fair trading for the producers and consumers (Fischer et al. 2013). As the volume of transactions increase, many have found it challenging to manage the rise in suppliers, buyers, and operational costs associated with the growth and many have identified balancing supply and demand as a challenge (Fischer et al. 2013). Meeting capital requirements (e.g., new infrastructure cost, start-up costs, and distribution costs) represent a challenge (Clancy and Ruhf 2010; Melone et al. 2010; Fischer et al. 2013; Flaccavento 2016). Finding appropriate value chain partners and developing mechanisms for value chain decision making, transparency, and trust, determining effective strategies for product differentiation, branding and regional identity, and determining appropriate strategies for product pricing based on understanding true cost structures are other challenges faced (Martinez et al. 2010; Tropp and Barham 2008).

In Australia, Rose and Larsen (2013) and Rose (2015) identify specific barriers and obstacles faced for the development of local food economies as a whole in the Southern Melbourne region. They include difficulty in securing capital for business expansion and growth, lack of recognition of the new sector from government/mainstream players, and inability to represent themselves and advocate their needs for large food manufacturers who can lobby to influence regulations. Furthermore, there is lack of integrated food and agricultural policy, regulations around food safety affecting innovation, lack of coordination among policy makers, lack of highly skilled labor, and lack of coordinated marketing mechanisms to raise brand awareness.

2.2 Roles of ICT to Support Sustainability Initiatives

Currently, scant literature exists linking ICT with Food Hubs, although some IT-related challenges in the sector have been identified. Previous studies have indicated the roles that ICT can play in supporting organizations to practice sustainability initiatives within the supply chains (Dao, Langella, and Carbo 2011; Porter and Kramer 2011). These roles include automating, informating, and transforming and providing infrastructure to support organizational activities in such a way that sustainability goals can be met (Dao, Langella, and Carbo 2011; Kurnia, Rahim, and Gloet 2012). Kurnia, Rahim, and Gloet (2012), in particular, synthesize several key practices along the three dimensions of sustainability and conceptualize how each practice is supported by different roles played by ICT based on Dao et al.'s (2011) classification. They find informating and infrastructure provision to be the two major roles in supporting various sustainability practices. They also highlight the significance of ICT's role in transforming business processes.

However, limited studies exist with empirical evidence for explaining how different roles of ICT support organizations to improve the TBL performance (Elliot 2011; Kurnia, Rahim, and Gloet 2012).

Furthermore, ICT has been useful to support information sharing and facilitate collaboration among different organizations, opening the potential to address one of the challenges related to collaborating and goal alignment among different parties involved in Food Hubs operations (Dewet and Jones 2001). However, Food Hubs often have limited ICT skills, lack of understanding of ICT requirements to support business operations, lack of technical assistance related to web and data management, organizational management, product development, and food safety knowledge and compliance (Day-Farnsworth et al. 2009; Flaccavento 2016). A survey study involving Food Hubs in 2013 indicated that technology is one of their top three challenges (Fischer et al. 2013). There is some evidence that information technology developments in food supply chains that connect to traceability, efficiency in distribution, quality systems, market information, and product development, are also adapted to shorter, localized food chains (Barham et al. 2012; Matteson and Hunt 2012), although there is little research on specific ICT interventions and impact. Therefore, further studies are required to better understand how ICT can help support Food Hubs and enable them to effectively run their operations, and how this is specifically linked to environmental, social, and economic benefits.

2.3 Electronic Marketplace

A number of the capacity limitations and challenges faced by Food Hubs and identified in the existing literature may be overcome by introducing an online intermediary—namely, an electronic marketplace (EM)—to connect the producers and consumers and assist with the aggregation and distribution of fresh product. EM can be broadly defined as a virtual place where buyers and sellers meet to exchange goods/services (Segev, Gebauer, and Faeber 1999). There are many benefits incorporating EMs in a business model. From a buyer's perspective, EMs have the capacity to reduce search costs by making product and pricing information accessible, which raises competition among suppliers, resulting in lower prices (Bakos 1998; Soh, Markus, and Goh 2006). EMs also allow sellers to compete for a wider range of customers, who could have been unreachable, with lower customer acquisition and transaction costs (Mahadevan 2003). Additionally, EMs can enable sellers to adopt price discrimination strategy efficiently to charge different prices to different buyers (Bakos 1998).

From the perspective of buyer/seller orientation, OFN falls into the category of a "neutral" EM involving a third party equally favoring buyers and sellers (Grieger 2003). These EMs are called "market makers" and their typical business model is to facilitate interactions between buyers and sellers, charging a fee for this service (Benjamin and Wigand 1995). Being

IT-based and unrestrained by brick and mortar resources, market makers have a strong potential to upscale (Kaplan and Sawhney 2000). However, they must recruit a large volume of buyers before they can attract sellers, or vice versa, placing them in a "chicken-and-egg" predicament (Grieger 2003).

Additionally, the interests of buyers are in direct conflict with those of sellers in regards to the level of pricing transparency (Soh, Markus, and Goh 2006). While buyers are attracted to price transparency, this may discourage sellers to participate in EM. Flexible pricing mechanisms (e.g., forward and backward auctions), and pricing transparency features (e.g., direct comparisons alter the price discovery process) raise competition (Bakos 1998). Neutral EMs like OFN must make a strategic decision that satisfies both parties. Soh, Markus, and Goh (2006) asserted that EMs must compensate buyers/sellers with other sources of value if the pricing transparency policy goes against their interests. In a discussion of the failings of past EMs, research pointed to inadequate revenue as a common downfall of market maker EMs (Grieger 2003; Kaplan and Sawhney 2000). Establishing a pricing model that appeals to customers and generates adequate revenue to sustain operations has proven difficult.

3 Open Food Network

The core vision of OFN is to develop a diverse, transparent, and decentralized food system. It provides food enterprises such as Food Hubs and other organizations involved in regional food distributions with an online marketing platform and tools for aiding their operational and administration activities. OFN enables Food Hubs to take advantage of e-commerce opportunities, without needing a large amount of capital to invest in the software. For customers, the OFN website is a collection of online stores, organized as a directory. Customers can browse profiles and stores to learn about where their food is coming from and place and pay for orders.

OFN is being used by various enterprises as their shopfront and as a tool for managing back-end activities. Current adopters of OFN include farmers selling directly to customers, not-for-profit food coops, and Food Hubs with multiple suppliers, buying group customers, and commercial customers. These early adopters provide a test to whether OFN can facilitate financially viable, alternative business models. Functionalities of OFN include reporting tools, inventory management, and capacity to integrate accounts with trading partners (such as suppliers and buying group customers). This makes cooperative relationships and trading partnerships easier to manage, and lowers the administration burden for hubs to collaborate, therefore reducing overall costs in the supply chain. Ultimately, OFN's goal is to create a website with functionality that allows enterprises to establish an online trading presence in a relatively short amount of time and at little cost.

4 Research Method

This study aims to understand the role of ICT focusing on an EC application in supporting regional food supply chains in Australia. In particular, we focus on assessing the effectiveness of OFN in supporting Food Hubs and other regional food supply chain parties, and address the challenges they face. Focus group was used in this study to obtain in-depth, qualitative information from the early adopters of OFN. This method is beneficial for this type of exploratory study that investigates how ICT supports regional food supply chains as perceived by several companies (known as adopters) that adopted OFN for use. The use of focus group interviews in examining adopters' views on e-commerce and its consequences is not new, while the use of single focus group-based interviewees is also acknowledged as an important tool for qualitative data collection (Agar and MacDonald 1995).

OFN was launched in June 2015 and there are currently still a limited number of adopters. The focus group involved nine users of OFN as outlined in Table 10.1. All participants are enterprises that act as "hubs." A professional moderator, assisted by a team of researchers, led the focus group session. The moderator ensured that no coercive influence or burden was placed on the participants to share their views; rather an amicable and neutral style was maintained to encourage the participants to share their honest and insightful views on the challenges and positive outcomes of adopting OFN. Hence, as pointed out by Agar and MacDonald (1995), directive styles of moderation were deliberately avoided. The focus group session was held in June 2015, lasted for over two hours and was recorded, while subsequently it was transcribed for analysis. The focus group interview session included three main sections. After a short introduction, the degree of OFN adoption by the participating companies (as reflected by the users), their perceived benefits, and the burning issues that affected the adoption of those companies were discussed. All the key ideas were coded

Table 10.1 Overview of Focus Group Participants

Organization	*Type*	*Year in Operation*	*Number of Participants*	*Participant Number*
A	Food Hub	5 years	2	Participant 1 Participant 2
B	Food Hub	In planning	1	Participant 3
C	Food Hub	1 year	1	Participant 4
D	Farm-Based Hub	3 years	1	Participant 5
E	School-Based Hub	Pilot	2	Participant 6 Participant 7
F	Food Hub	2 years	1	Participant 8
G	Food Hub	3 years	1	Participant 9

under relevant thematic headings (Miles and Huberman 1994) to explore OFN use, benefits, and challenges as perceived by OFN adopters.

5 Study Findings

5.1 Use of OFN

The focus group findings show that OFN is being incorporated into a diverse range of food enterprise models. First, OFN has been used to *facilitate more efficient and customer friendly taking and processing orders*. This use is specifically relevant for Food Hubs that hold complex operations. Specifically, Food Hub enterprises typically aggregate products from nearby producers to be distributed to multiple buying groups and institutional customers. They accept unique customer orders, in contrast to fixed mixed vegetable boxes, representing a relatively high degree of operational complexity. Product availability information is taken from farmers weekly. This information is then displayed on OFN accordingly. Each buying group has a unique shopfront created within the hub's order cycle, allowing for customized shopfronts according to the preferences of each buying group.

With the use of OFN, individual customers place orders through their respective group shopfronts and upon closure of the order cycle, the central Food Hub views the order totals aggregated across all buying groups and places orders with farmers accordingly. Product is then brought to a central packing location either by delivery from the farmer, or collection by the hub. Orders are packed and then picked up and delivered to the buying group pick up locations by a member of each buying group.

For a number of focus group participants who have been performing direct marketing of their meat products, OFN has been used as a marketing tool to reach regional and metropolitan customers. This enables the participating enterprises to obtain advance customer orders. Taking orders in advance (through OFN) of slaughtering and packing gives an accurate indication of demand, reducing the risk of over or under supplying. Customers are also given flexibility to buy a "mixed box," with the option of adding "extras." This can secure purchases of the less popular cuts of meat while offering the option to order additional products.

OFN has also been used as part of a project of the Rural City Council *to increase accessibility of fresh fruits and vegetables to the community*, with the goal of reducing obesity. Specifically, OFN has been used to sell fresh produce from a local wholesaler to buying groups established in local schools.

5.2 Benefits of OFN to Enterprise Users

Several benefits of OFN have been achieved by the early adopters participating in our study, which are discussed next. Benefits of EMs that have been identified in different contexts within the literature appear to be relevant.

5.2.1 Expanding Marketing Channels

The ability to market online is valuable to enterprises as it gives them an independent means of taking their product to market and allows customers to shop conveniently. Adopting OFN's e-commerce functionality gives enterprises a substantial advantage over traditional marketing channels (e.g., farm gate stalls, markets) because it helps enterprises connect with the rising number of Australian consumers who prefer the convenience of e-commerce over the traditional marketing channels. This sentiment is echoed in the following excerpts by Participant 9,

> That rapidity with which you can scale up from having no online presence to being a fully fledged online store. People have all kinds of expectations of the ways that they want to engage with a food business these days. If you don't have an online presence, it's fairly difficult.

5.2.2 Ease of Use

The ease-of-product trial and use is very important to the emerging Food Hub movement. Minimizing the learning and transactional costs is critical as there exist many hurdles preventing customers from engaging with innovation. OFN has ensured that its shopfront and checkout processes are familiar to customers and easy to use, which benefits the current and potential adopters. For example, Participant 5 noted, "*My customers say it is really easy to use, they like it, it's really visually appealing.*"

5.2.3 Administrative Efficiency

Administrative activities involved in operating a Food Hub include establishing product availability, accepting orders from customers, and managing accounts payable/receivable. OFN's reporting functionality allows users to easily view and interpret data relating to their trading activities. Several focus group participants indicated that the usability OFN's reporting functionality is excellent. It gathers data about their products, customers, and orders, and allows them to display and interpret this data in different ways. OFN has made the task of managing orders less demanding: "*I've got so many hours back every week in my life. I don't need to respond with individual e-mails to every order*" (Participant 5). For multi-party trading arrangements, this is particularly valuable, as the tasks of manually tracking orders and accounts is a burden on these—often under resourced—organizations.

5.2.4 Operational Efficiency

The operations of local Food Hub businesses are complex. This is due to the seasonality and perishability of products; the inconsistency and unpredictability of supply, with the involvement of multiple suppliers; the multiple incoming and outgoing logistics; and the tracking of payments from customers to suppliers. Achieving efficiency in operations as well as automating and standardizing processes are particularly important for Food Hubs that are low on human resources or rely on contributions from under skilled and often transient volunteers.

The OFN software enables Food Hubs to efficiently manage these operational complexities, making them more manageable, and bringing feasibility to their business models. Participants 1 and 2 were acutely aware of the challenges they faced in juggling their operations. "*We had a desperate need for some sort of software to help us organize ourselves*" (Participant 1). They were also impressed by the "phenomenal flexibility" of the system, as it allowed them to change the way they articulated their problem and their model and still find a way to adapt the system to find a solution. Another hub manager explained the simplicity of incorporating a new supplier into the system. "*It can happen very quickly . . . all of a sudden they're on the order cycle and we're beginning to move their stuff*" (Participant 4). The flexibility of OFN is highly valued by users, as it can accommodate shifts in their operations and manage the elements of uncertainty which are inherent in their businesses.

OFN's order cycle model encourages enterprises to accept and fulfill orders in a periodic, routine manner. This allows enterprises to aggregate demand on a weekly, fortnightly, or longer basis, leading to more efficient operations and logistics. The order cycle function is being used by Participant 5, who has two complementary order cycles, which are used to coordinate butchering and packing activities. As mentioned in the limitations section, shopping in this cyclical manner involves a learning curve for customers.

5.2.5 Visibility of and Access to New Customers

Functioning as a network of interconnected enterprises, rather than in isolation, OFN brings multiple related benefits to local food producers: 1) reach target customer segments (conscious consumers in their locale), 2) connect with and form symbiotic trading relationships with other enterprises who share aligned interests, 3) strengthen their image or brand through association with the OFN's values (such as the value "stronger together than alone").

Multiple focus group participants mentioned about attracted new customers from their involvement and exposure on OFN: "*Then the first week, wham I got ten new customers*" said Participant 5, and Participant 4 said, "*We've had at least one or two producers and some customers coming along because of some kind of contact with or awareness of OFN.*"

Enterprises using OFN are able to share the OFN's brand equity and strong brand values,

> I think a lot of people out there, a lot of consumers, want that information. And it's an added value thing . . . So that is the big selling point

(of OFN) for both people setting it up for their own organization and people coming in and buying from it.

(Participant 3)

5.2.6 Raising the Profile of Food Hubs and Alternative Food Distribution Models

OFN is giving momentum to Australia's local food movement—a movement that is still in its infancy and in need of promotion and substantiation. Among the public, government, and mainstream players, there is a lack of recognition of Food Hubs and their potential role in regional food systems. By raising the profile and credibility of these novel models, OFN is improving the receptivity of government, suppliers, and consumers to the innovation, which makes for a more amenable environment for emerging Food Hubs. A local council employee who had piloted a school-based Food Hub explained, "*OFN enabled us to have conversations that we haven't had before*" (Participant 5).

5.2.7 Reducing Transaction Costs

The administrative demands of coordinating multi-party trading can be prohibitive for many organizations. OFN has functions that facilitate coordination and ease the exchange between parties, making the formation of complex, multi-party trading arrangements feasible. One hub operator explained how the OFN technology had led them to consider a greater diversity of food distribution models: "*It's (expanded) our notion of how local food comes in to a central point and how it's distributed and goes out.*" "*Just using the word hub, we're able to see a much bigger role, for our group, than just food in and food out to families via veggie bag systems*" (Participant 1). Participants 1 and 2 have since explored supplying through wholesale channels, bulk-buying groups, and the local community house, who serves low-income families, amplifying their impact.

5.3 Issues with OFN to Enterprise Users

As OFN is new, several issues are encountered by the early adopters, which are discussed next. Recognizing and identifying these issues are critical for further enhancement of the OFN features and devising appropriate strategies for encouraging wider adoption among regional food supply chain players.

5.3.1 Subscription and Advanced Customer Accounts

Presently, OFN cannot attach detailed information to a customer's account, restricting the ways that enterprises can use OFN for managing customer relationships. The development of more detailed and integrated customer accounts would allow for several added features directly requested by participants, including: managing subscription/membership within the OFN system (Participant 9), creating "member only" order cycles (Participant 5), and creating automatic "standing orders" placed by subscribers (Participant 5).

It is common for Food Hubs and Community Supported Agriculture enterprises (CSAs) to operate under a membership or subscription model. Currently, OFN does not offer functionality for managing subscription or creating "member only" shops, which was raised as a major concern among the study participants: "*We run on a subscription model. So we have customers subscribe for a 12 week season, and so not having a subscription management system as part of it, it's the biggest limitation for us*" (Participant 9). Hence, an additional external subscription database is required and manual data handling is involved to deal with non-member customers.

An extension of this shortcoming is the inability to create a recurring order for customers who have subscribed or paid upfront for a repetitive order (e.g., a weekly vegetable box delivery). Thus, a request was placed by an enterprise (Participant 9) for "standing orders for particular products, and having repeating orders" to prevent enterprises from needing to manually place orders for their committed subscribers each week. These limitations show that there is space for OFN to create customer accounts that contain membership status and recurring order information, and can be used to further streamline manual administrative tasks.

5.3.2 Restricted Checkout (Requiring a Minimum Order or Membership)

Participants discussed several scenarios where they would like to restrict customers from placing an order. This includes when costumers are not members, have not purchased one essential item, have not reached a minimum spending amount, or are located beyond a set distance from the hub. At the time of writing, the OFN cannot restrict checkout based on any of the aforementioned scenarios. The type of desired restriction varied among study participants, depending on the enterprise. One producer wanted all customers to purchase the mixed box in order to checkout, with the option of adding additional extras: "*I would prefer it's a minimum order as a box, rather than a dollar amount*" (Participant 5). For another hub, a set spending amount was more important, as this would make it possible for them to manage margins: "*That's something that I foresee could be an issue, for people like us. Feasibility of fixed costs and everything, so you need to have a certain tipping point*" (Participant 3).

5.3.3 Lack of Integration with Other Software

OFN does not integrate with any other software. One participant declared that they wanted to be able to integrate between their labeling scales, accounting package, and their OFN account. This would allow them to perform multiple functions in the same place, saving time. In addition to the significant efficiency gains to be achieved from coordinating OFN with other software products, integration would also reduce the barriers for enterprises who are migrating their existing operations to OFN and improve the ease of uptake.

5.3.4 Cross-Platform Performance and Responsiveness

Several participants raised concerns about how well OFN performed on devises such as smartphones and tablets: "*I think the biggest [issue that customers are facing] is the mobile and tablet thing*" (Participant 9). One enterprise had customers who used tablets and mobiles exclusively, as they didn't own computers. Another hub was trying to secure hospitality customers, who traditionally ordered by fax. For them, this transition would be smoother if they could order on their smartphones in the restaurant: "*Moving online was sort of blowing them away a little bit. And they were like, well how will our chefs do that? And I was like, well they've got smartphones*" (Participant 6). While OFN is designed responsively, its current speed performance impedes streamlined use from smartphones and tablets.

5.3.5 Setup Time for Existing Enterprises

Depending on the size and complexity of the enterprise, there is a significant time commitment associated with establishing an enterprise on OFN. This time is for data entry (creating profiles and products) and familiarizing oneself with the different functions and how the site works: "*Working out the nuts and bolts, and ironing out the creases to start with, that's a very, very long process*" (Participant 3). This was raised as a major challenge and a process that enterprise users wanted simplified. Specifically, Participant 5 said, "*If I could have just imported a spreadsheet, and all my products would be there, that would have been nice.*" While this user did persevere, she stated, "*Initial setup becomes quite a mental hurdle and finding the time to populate the site was difficult.*"

5.3.6 Lack of Enterprise Capacity and Need for a Forum for Knowledge Sharing

Throughout the focus group, participants raised issues that confronted their business and were not directly related to the OFN's core functions. Participants recognized that there was a huge range of skills and resources that enterprises need to be successful in this domain,

> When you get into this, you realize there's quite a big suite of skills and knowledge you need to run a local food economy and these people

> don't have that . . . and I wonder whether there is that reason why a lot start up and crash.
>
> (Participant 6)

One participant highlighted that the "*Open Food Network already is somewhat directly educating people and providing that kind of function, and it could be expanded upon, sharing learnings in different ways. I don't know. I see a lot of potential for that*" (Participant 4). An example of such a resource which proved useful for one participant was a collection of Food Hub case studies, housed at foodhubs.org.au: "*I had never worked in this area before and I read that website quite a lot before I started working there, and it really helped*" (Participant 6). A further suggestion was the provision of "*something like a development kit, for people starting up in this space, in other places*" (Participant 9). Or simply, "*Even something associated to the OFN website, whereby you could go on and visit the help section for people using it*" (Participant 8).

The participants also recognized, "*There's a huge amount of knowledge sitting in the people who work around OFN, it would be a shame not to share that and to share the learning*" (Participant 4). There was consensus that some kind of online forum for sharing knowledge and having discussions with other OFN users would be very valuable,

> We could have people learning about Food Hubs and having that dialogue about, what do you put in your constitution, what sort of structure did you set up, and also some technical stuff about OFN, like how do I do this or that.
>
> (Participant 3)

One participant said in this respect,

> I'd love to hear about how other people operate, and get little ideas from them. Just the little things, like putting recipes out. What people are doing and how you might be able to improve yours, or change it a little bit.
>
> (Participant 6)

5.3.7 Product Property and Quality Control (e.g., Free Range)

At the time of the focus group, there was no way for hubs or producers to attach a property field to products, to share any third party certification or labeling which applied to their products. This information was also not apparent to customers in the shop: "*The main thing is just putting a filter for certified organics*" (Participant 4). Such product information has become important for customers who are generally becoming more health conscious.

6 Discussion and Future Study

Based on the focus group conducted to explore how OFN, as an example of an ICT application, supports sustainability practices within regional food supply chains in Australia, we have identified some evidence of how different roles of ICT are played out by OFN to provide infrastructure to support sustainability initiatives and to automate, informate, and transform business processes within Australian regional supply chains. In line with a number of previous studies (Dao, Langella, and Carbo 2011; Kurnia, Rahim, and Gloet 2012; Elliot 2011; Dewet and Jones 2001; Jablonski, Perez-Burgos, and Gomez 2011; Barham et al. 2012), our study shows that OFN enables regional food supply chain parties to achieve economic, environmental, and social benefits by playing different roles to support various sustainability practices, but the most significant roles played are informating and infrastructure provision.

Table 10.2 demonstrates a number of key practices along the three pillars of sustainability that have been enabled by OFN through various roles played, as suggested in Kurnia, Rahim, and Gloet (2012). The first two key practices related to profit margin and cash flow (Kurnia, Rahim, and Gloet 2012) are combined into "enhancing operational efficiency" to suit the study context. The shaded rows indicate the relevance of the key practices and the ICT roles proposed in Kurnia, Rahim, and Gloet (2012) for this study.

In terms of economic sustainability, our study shows that OFN provides a useful infrastructure for regional supply chain parties to improve business administration and operational efficiency, increase the numbers of suppliers and customers by improving visibility and expanding marketing channels efficiently. These improvements are achieved through the automation and increase access to relevant information that is required to manage supply chain operations and control the quality of products. Hence, Food Hubs and other enterprises involved in the regional food supply chains can enhance sales and improve customer satisfaction.

In terms of environmental sustainability, the findings indicate that OFN enables the elimination of traditional marketing activities that lead to reductions in the use of paper and printing requirements, transportations to visit specific farms and markets, and the use of space for facilitating interactions among the players. With the timely availability of supply and demand information, enterprises involved in the logistics can optimize their routing schedule to maximize the truckloads and minimize fuel consumptions. Finally, in terms of social benefits, the findings point toward useful evidence of how OFN through its roles in providing infrastructure, automation, informating, and transformation can help generate some social benefits. For example, by playing those four roles, OFN affects the way food enterprises interact with suppliers and customers that in turn contributes to the community relations through online forums to exchange ideas and mutual

Table 10.2 Sustainability Practices and the Roles of OFN

Dimension	*Key Practice*	*ICT Roles Demonstrated Through OFN*			
		Automate	*Informate*	*Transform*	*Infrastructure*
Economy	Enhancing operational efficiency	Y	Y		Y
	Achieving consumer satisfaction		Y		Y
	Creating repeat customers		Y		Y
	Enhancing sales	Y	Y		Y
	Quality initiatives		Y		Y
	Creating competitive advantages		Y	Y	Y
Environmental	Eco-design of products	Y	Y	Y	Y
	Green purchasing	Y	Y	Y	Y
	Clean/Lean Production		Y	Y	Y
	Green Distribution		Y	Y	Y
	Reverse Logistics		Y	Y	Y
Social	Community relations		Y	Y	Y
	Employee well-being		Y	Y	Y
	Human rights		Y		Y
	Work safety/healthier community		Y		Y
	Ethical considerations		Y		Y
	Purchasing from minority-owned suppliers	Y	Y		Y
	Product safety/quality		Y	Y	Y
	Education support		Y		Y

supports. Likewise, OFN helps improve food enterprises' employee well-being through enhanced skills and greater operational efficiency and education support to the community. Through product quality and information transparency, OFN even potentially creates a healthier and more resilient community.

Besides the benefits of OFN, this study has identified several challenges that need to be addressed before we can expect a wider adoption of OFN. The challenges, however, can be addressed through continuous improvement of the OFN features and ongoing education support to the regional communities to leverage the potential of ICT and to harness on knowledge sharing and collaboration within the communities to establish sustainable local food economy. This study confirms OFN as a useful ICT tool for supporting sustainability initiatives within regional food supply chains and extends the current understanding of how ICT can support sustainability practices with organizations and supply chains. It shows how OFN can potentially enhance the inclusion of disadvantaged local producers in a more sustainable and fair trading environment. A longitudinal study is necessary to find out the longer term impacts of OFN on those early adopters and the overall sustainability practices within the same regional community. Furthermore, a comparative study with the UK and U.S. experience as OFN is being rolled out globally would be part of our future research. All this will contribute to the theory development regarding the importance and roles of ICT in supporting sustainability initiatives.

Note

1 For more information on OFN, see at: https://openfoodnetwork.org/

References

Agar, M., and J. MacDonald. "Focus Groups and Ethnography." *Human Organisation* 54 (1995): 78–86.

Alt, R., and S. Klein. "Twenty Years of Electronic Markets Research—Looking Backwards Towards the Future." *Electronic Markets* 21, no. 1 (2011): 41–51.

Bakos, Y. "The Emerging Role of Electronic Marketplaces on the Internet." *Communications of the ACM* 41, no. 8 (1998): 35–42.

Barham, J., D. Tropp, K. Enterline, J. Farbman, J. Fisk, and S. Kiraly. *Regional Food Hub Resource Guide*. Washington, DC: United States Department of Agriculture, Agricultural Marketing Service, 2012.

Benjamin, R., and R. Wigand. "Electronic Markets and Virtual Value Chains on the Information Superhighway." *Sloan Management Review* 36, no. 2 (1995): 62–72.

Blay Palmer, A., et al. "Constructing Resilient, Rransformative Communities Through Sustainable 'Food Hubs.'" *Local Environment* 18, no. 5 (2013): 521–8.

Brundtland, G. H. *Report of the World Commission on Environment and Development: Our Common Future*. New York: United Nations, 1987.

Clancy, K., and K. Ruhf. "Report on Some Regional Values Chains in the Northeast." February 2010, 1–21.

Conner, D. S., V. Campbell-Arvai, and M. W. Hamm. "Value in the Values: Pasture-Raised Livestock Products Offer Opportunities for Reconnecting Producers and Consumers." *Renewable Agriculture and Food Systems* 23, no. 1 (2008): 62–9.

Dao, V., I. Langella, and J. Carbo. "From Green to Sustainability: Information Technology and an Integrated Sustainability Framework." *The Journal of Strategic Information Systems* 20, no. 1 (March 2011): 63–79.

Day-Farnsworth, L., B. McCown, M. Miller, and A. Pfeiffer. *Scaling Up: Meeting the Demand for Local Food*, 1–40. Madison, WI: Center for Integrated Agricultural Systems, University of Wisconsin, December 2009.

Dewett, T., and G. R. Jones. "The Role of Information Technology in the Organization: A Review, Model and Assessment." *Journal of Management* 27, no. 3 (2001): 313–46.

Elkington, J. *Cannibals with Forks: The Triple Bottom Line of the 21st Century*. Oxford: Capstone, 1997.

Elliot, S. "Transdisciplinary Perspectives on Environmental Sustainability: A Resource Base and Framework for IT-Enabled Business Transformation." *MIS Quarterly* 35, no. 1 (2011): 197–236.

Erlbaum, J., K. McManus, and A. Nowak. "Colorado Local Food Hubs for Farm to School Products: An Initial Feasibility Analysis of Local Food Hubs for Colorado Producers and Schools." *Colorado: Real Food Colorado* (2011): 1–53.

Fischer, M., M. Hamm, R. Pirog, J. Fisk, J. Farbman, and S. Kiraly. *Findings of the 2013 National Food Hub Survey*. Michigan: Michigan State University Centre for Regional Food Systems & The Wallace Center at Winrock International, 2013. http://foodsystems.msu.edu/uploads/files/2013-food-hub-survey.pdf

Flaccavento, A. *Building a Healthy Economy from the Bottom Up: Harnessing Real World Experience for Transformative Change*. Lexington, KY: The University of Kentucky Press, 2016.

Flaccavento, A. "Healthy Food Systems: A Toolkit for Building Value Chains, Appalachian Sustainable Development." 2009.

Grieger, M. "Electronic Marketplaces: A Literature Review and a Call for Supply Chain Management Research." *European Journal of Operational Research* 144, no. 2 (2003): 280–94.

Jablonski, B. B. R., J. Perez-Burgos, and M. I. Gomez. "Food Value Chain Development in Central New York: CNY Bounty." *Journal of Agriculture, Food Systems, and Community Development* 1, no. 4 (2011): 129–41.

Kaplan, S., and M. Sawhney. "E-Hubs: The New B2B Marketplaces." *Harvard Business Review* (May–June 2000): 97–103.

Kurnia, S., M. M. Rahim, and M. Gloet. "Understanding the Roles of IS/IT in Sustainable Supply Chain Management." In *Pacific Asia Conference on Information Systems*, July 2012, pp. 1–12.

Lyson, T. "Civic Agriculture and Community-Solving Problems." *Culture and Agriculture* 27, no. 2 (2005): 92–8.

Mahadevan, B. "Making Sense of Emerging Market Structures in B2B Ecommerce." *California Management Review* 46, no. 1 (2003): 86–100.

Martinez, S., M. S. Hand, M. Da Pra, S. Pollack, K. Ralston, T. Smith, S. Vogel, S. Clark, L. Lohr, S. A. Low, and C. Newman. "Local Food Systems: Concepts, Impacts, and Issues." *United States Department of Agriculture: Economic Research Report No. 97*, May 2010.

Matson, J., and J. Thayer. "The Role of Food Hubs in Food Supply Chains." *Journal of Agriculture, Food Systems, and Community Development* 3, no. 4 (2013): 43–7.

Matteson, G., and A. R. Hunt. "The Emergence of Retail Agriculture: Its Outlook, Capital Needs, and Role in Supporting Young, Beginning, and Small Farmers." Local Food Strategies, LLC report to the Farm Credit Council 2012.

Melone, B., Cardenas, E., Cochran, J., Gross, J., Reinbold, J., Brenneis, L., Sierra, L., Cech, S., and Zajfen, V. "A California Network of Regional Food Hubs: A Vision Statement and Strategic Implementation." In *The Regional Food Hub Advisory Council*, September 2010.

Miles, M. B., and A. M. Huberman. *Qualitative Data Analysis*. Thousand Oaks, California: Sage, 1994.

Porter, M. E., and M. R. Kramer. "Creating Shared Value." *Harvard Business Review* 89, no. 1/2 (2011): 62–77.

Rose, N. *Fair Food: Stories from a Movement Changing the World*. Brisbane: University of Queensland Press, 2015.

Rose, N., and K. Larsen. "Economic Benefits of 'Creative Food Economies': Evidence, Case Studies and Actions for Southern Melbourne." 2013. www.foodalliance.org.au/wp-content/uploads/2014/02/Creative-Food-Economies-N-Rose-K-Larsen-3.9.13.pdf

Segev, A., J. Gebauer, and F. Faeber, "Internet-Based Electronic Markets." *Electronic Markets* 9, no. 3 (1999): 138–46.

Seyfang, G. "Ecological Citizenship and Sustainable Consumption: Examining Local Organic Food Networks." *Journal of Rural Studies* 22, no. 4 (2006): 383–95.

Soh, C., M. L. Markus, and K. H. Goh. "Electronic Marketplaces and Price Transparency: Strategy, Information Technology, and Success." *MIS Quarterly* 30, no. 3 (2006):705–23.

Stevenson, G. W., R. King, L. Lev, and M. Ostrom. "Midscale Food Value Chains: An Introduction." *Journal of Agriculture, Food Systems, and Community Development* 1, no. 4 (2011): 27–34.

Tropp, D., and J. Barham. "National Farmers Market Summit Proceedings Report." *United States Department of Agriculture: Agricultural Marketing Service*, 2008. www.ams.usda.gov/AMSv1.0/getfile?dDocName=STELPRDC5066926

Woods et al. "Local Food Systems Markets and Supply Chains." 2013. www.choicesmagazine.org/choices-magazine/theme-articles/developing-local-food-systems-in-the-south/local-food-systems-markets-and-supply-chains

Part III

Adoption, Usage, and Management Aspects Surrounding Social Inclusion and Usability of ICT-Enabled Services

11 Digital Divides, Usability, and Social Inclusion

Evidence From the Field of E-Services in the United Kingdom

Bianca C. Reisdorf and Darja Groselj

1. Introduction

ICTs and Internet-enabled devices are becoming increasingly ubiquitous in highly advanced societies across the world. In many countries, core functions of society, such as health information, education, and government services are moved online to enable easier and more efficient use. A prominent example of this development is the United Kingdom's "digital by default" strategy that aims to encourage citizens to use government services primarily online (Cabinet Office 2012). Consequently, inequalities in how people can access and use such services can have a profound impact on individuals' lives (Robinson et al. 2015). It is thus crucial to acknowledge the shrinking group of people who are completely offline in high diffusion countries and to study the implications of lack of access to ICT-enabled services on their lives. Moreover, it is equally important to investigate those who are online, but barely so—i.e., low users of the Internet (Blank and Groselj 2014; Reisdorf and Groselj 2017).

According to the Association for Computing Machinery (ACM) code of ethics, "In a fair society, all individuals would have equal opportunity to participate in, or benefit from, the use of computer resources regardless of race, sex, religion, age, disability, national origin or other such similar factors" (ACM 2016). This would require technologies to be designed so that they are usable for every citizen, regardless of the aforementioned factors. One problem is the diversity of users. According to Shneiderman (2000: 86), universal usability would mean "[a]ccommodating users with different skills, knowledge, age, gender, disabilities, disabling conditions (mobility, sunlight, noise), literacy, culture, income, and so forth." From the digital-divide literature, we know that a considerable number of citizens in highly developed countries are still offline, which supports the notion that universal usability has not been achieved thus far.

A number of questions come to mind when we consider the relationship between digital inclusion and usability of Internet-enabled services: Do Internet users engage with services offered online? Who are the users who do engage, and who is left out? How can we design services that are easy to use even for those who do not use the Internet (regularly)? Keeping these questions in mind, this chapter aims to further our understanding of the relationship between digital inclusion and usability of ICT-enabled services.

2. Theories of Digital Divides and Digital Inequalities

Here we briefly summarize theories[1] of digital divides and digital inequalities as well as recent theories on the links between social exclusion and digital exclusion (Helsper 2012a; Tsatsou 2013). This section also discusses aspects of usability and digital divides, and particularly digital skills that have been identified as one of the key aspects of meaningful engagement with technologies (Hargittai 2003; Hargittai and Shafer 2006; Van Dijk and Van Deursen 2014).

2.1 Digital Divides and Digital Inequalities

Since the late 1990s and early 2000s, digital-divide theories have moved away from a binary definition of access versus no access (Rogers 2001) and toward a more refined understanding of digital inequalities and "second-level" divides (DiMaggio et al. 2004; Hargittai 2001; Selwyn 2004). Van Dijk (2005) developed a sequential model of digital divides, which takes into consideration motivations as well as access, skills, and usage opportunities. Recent research in technologically advanced societies has focused more and more on these dimensions—such as attitudes (Reisdorf and Groselj 2017), access (Gonzales 2016), and skills (Van Deursen and Van Dijk 2015)—in addition to socio-economic factors that are related to digital divides. These additional factors—and the understanding that digital divides are not only about who is online and who is offline but also about how much and how well those who are online use and utilize the Internet—has led to an increasing number of studies in the field. The topic has also gained increasing attention from policy makers who argue that everyone is better off if everyone is online (Go ON UK n.d.). At the same time, many services are moving online. For example, many British government services have moved to being "digital by default" (Cabinet Office 2012), leaving those who are offline or those with lower quality access or lower Internet skills behind or with second tier service. This creates concern among researchers who emphasize strong links between social and digital exclusion (Helsper 2012a; Robinson et al. 2015).

2.2 The Links Between Social and Digital Exclusion

Digital-divide researchers examined quite early on the links between digital and social exclusion (Haddon 2000; Selwyn 2004). While some found that social exclusion increases digital exclusion and vice versa (Zillien and Hargittai 2009), others emphasized the "curative potential of new media" in decreasing social exclusion (Mansell 2002; in Tsatsou 2011: 319).

In recent years, a number of researchers have developed the theoretical understanding of the relationships between social and digital inequalities. Helsper (2012a) asserts that the previous attempts at linking social and digital exclusion did not go far enough. Although digital divides have been identified as a multifaceted phenomenon that includes a variety of factors related to both social and digital exclusion, such as socio-demographic factors, access, skills, and attitudes (DiMaggio et al. 2004; Van Dijk 2005), they do not explicitly take into consideration the various corresponding fields that are associated with social and digital exclusion. Helsper's (2012a) corresponding fields model argues that social exclusion in fields of economic, social, cultural, and personal resources is linked to exclusion from similar resources in the digital realm via social impact mediators (access, skills, and attitudes). Furthermore, Helsper hypothesizes a reverse path where exclusion from digital fields of economic, social, cultural, and personal resources shapes social inclusion in these fields via digital impact mediators (relevance, quality, ownership, and sustainability of online engagement). Similarly, Tsatsou (2013) argues for a need to include micro-perspectives of digital inclusion into digital-divide research, accounting for identity and literacy in addition to the traditional digital-divide factors, which are mainly related to socio-economic status. Building on the large number of previous studies on digital exclusion, Robinson et al. (2015: 578) make the case that "research on digital inequality is in its infancy and is evolving rapidly, along with its object of study." They emphasize the links between social and digital inequalities, which influence each other, rather than a one-way impact of offline factors on Internet use (or lack thereof).

Accordingly, it is important to examine aspects of digital inclusion and usability that may affect uptake of online services. Next, we link concepts of usability to concepts of digital inclusion by examining the principles of usability and comparing them to the principles of digital skills, which have been identified as one of the key factors in digital inequalities.

2.3 Usability and Digital Skills

An increasing field of study in digital-divide research is the consideration of digital skills, which are closely related to the concept of usability. With physical access theoretically available to all citizens of highly technologized countries (e.g., in public libraries or community centers), skills are increasingly mentioned as a factor that holds non-users back from going online (Helsper and Reisdorf 2017).

According to Shneiderman (2000), universal usability should enable technology design that is accessible to all individuals in a society regardless of their socio-demographic backgrounds or their literacy, culture, or any other factors. However, this is clearly not the case at the moment, as use of digital technologies is stratified by all of these factors. In an attempt to match usability and digital inclusion Aleixo, Nunes, and Isaias (2012: 221) argue that usability standards should "incorporate inclusion and accessibility guidelines." Aleixo and his colleagues developed a set of usability guidelines that aim to decrease the complexity of webpages to enable easier access, even for users with low digital literacy skills. They then matched these guidelines to a list of digital literacy skills that had been previously developed by the London Society of College National and University Librarians. As an example, the usability guideline "Avoid Cluttered Displays" would be directly connected to the digital literacy competence "Recognise information" (Aleixo, Nunes, and Isaias 2012: 237).

Although Van Dijk (2005) included the concept of digital skills into his theoretical model of digital divides early on, the measure itself was not included into many studies of digital inequalities until recently. In a number of studies and publications, Van Deursen and Van Dijk (2009, 2011; Van Dijk and Van Deursen 2014) developed a set of types of skills related to digital divides. They differentiated four types of skills: operational, formal, information, and strategic. While operational and formal skills are medium-related (i.e., operating and navigating hardware, software, browsers, etc), information and strategic skills are content-related (i.e., locating specific information and taking advantage of it) (Van Deursen and Van Dijk 2011). Van Deursen and Van Dijk (2009) found that lower education and older age are negatively related to medium-related skills while higher levels of education are positively related to content-related skills. In addition, their study showed that only 62% of all information skills tasks and 22% of all strategic skills tasks are solved successfully. This finding is important, especially in light of many governments expecting their citizens to perform important tasks, such as tax returns, online. Just because someone is on the right side of the digital divide does not mean that they will be able to perform complex online tasks. Usability, reflected in ease of use and user-friendly design, needs to be considered by any entity that aims to offer their services primarily online (Hargittai 2003).

These studies and concepts show that digital skills are strongly related to digital inclusion. Lower skills lead to lower engagement with the Internet and services offered online. Usability is hence a key factor that can have a positive impact on increasing digital inclusion.

3. Digital Divides Research: Evidence From the Field of E-Services

In recent years, digital-divide research has largely moved toward gradations of use (Blank and Groselj 2014; Reisdorf and Groselj 2017). These studies have shown that socio-demographic factors, skills, and attitudes are not only linked to who is online and who is offline but also to how much time people spend online and what kinds of activities they engage in. For example, in their study of "status-specific" Internet use Zillien and Hargittai (2009) found that users who come from advantaged socio-economic backgrounds (e.g., higher education or income) engage in more "capital-enhancing" usage than users who come from lower socio-economic backgrounds.

3.1 Digital Divides and E-Government

A number of studies have examined digital divides in relation to e-government, most commonly in the context of specific countries (Choudrie, Ghinea, and Songonuga 2013; Griffin, Foster, and Halpin 2013; Reddick and Turner 2012; Taipale 2013). While some studies focus on the implementation of government services online, showing a digital divide between large and affluent districts and small poorer districts (Hermana and Silfianti 2011), other studies focus on use of e-government services and find a strong connection between socio-economic background and the use of government services online (Venkatesh, Sykes, and Venkatraman 2014). In their study of digital divides and e-government use in the United States, Nam and Sayogo (2011) found that socio-demographic background, perceived usefulness of online government services, higher Internet usage intensity, and general trust in the government have a positive effect on e-government use. In addition to socio-economic factors and trust, Internet skills and online experience were found to have a positive impact on the likelihood of using e-government services (Bélanger and Carter 2009).

In light of these results, it is questionable whether the growing rollout of government services online—especially targeted at public welfare benefits receivers—is useful or whether it increases social exclusion by providing services for socio-economically disadvantaged populations online when those are the groups that are least likely to be online (Hall and Owens 2011). E-Government services often display poor usability, as they are not designed for low-skilled Internet users (Helbig, Gil-García, and Ferro 2009). This is partly due to the misconception that Internet (and e-government) users are one homogenous group. In our case study that follows, we will focus on users and non-users as well as different types of Internet users to account for differences within the user population.

3.2 Digital Divides and E-Health

Another important area related to digital divides and inequalities is the study of e-health—namely, whether and how people engage with ICT-enabled health services. E-health can have a direct impact on personal well-being, illness, and life expectancy, which are issues that are closely related to social inequalities (Bartley 2004; Glymour, Avendano, and Kawachi 2014).

A number of scholars have connected digital literacy to the concept of e-health literacy, often referred to as eHEALS,[3] which is a scale frequently used in researching digital health literacy. Similar to general digital literacy, e-health literacy differs significantly along socio-demographic dimensions, such as age and education, with those who are younger and more highly educated showing significantly higher e-health literacy than those who are older or less well educated (Kontos et al. 2014; Neter and Brainin 2012). As Neter and Brainin (2012: n.p.) have stated about the e-health-literate participants in their study: "They were also more active consumers of all types of information on the Internet, used more search strategies, and scrutinized information more carefully than did the less e-health-literate respondents." According to this, respondents with higher e-health literacy had more and better access to ICTs, used the Internet more frequently, and perceived their general digital skills as higher than those with lower e-health literacy.

3.3 Digital Divides and E-Learning

Arguably one of the most important and most desired aspects of the digital revolution has been its potential to bring free information and education—and thereby social inclusion—to everyone, even those who cannot afford the time or money to obtain formal education. However, digital divides and inequalities persist in this realm as much as in other online services, as e-learning requires access to computers and the Internet as well as a minimum of digital skills to navigate e-learning and sharing platforms, such as YouTube or Pinterest.[4]

Early on, researchers made a connection between digital inequalities and e-learning, showing that digital divides persist between schools in poorer and more affluent areas, leaving behind those students who happen to live in the "wrong" school district (Haythornthwaite 2007: 104): "e-learning—in its distributed, 'owner-operated' form (i.e., the learner is responsible for the equipment)—is yet another means for the elite to reap further benefit from being part of the elite (in this case in the form of education)" (ibid, 112). However, almost ten years and innumerable new gadgets and sharing platforms later, we have to reconsider whether things are still this Black and White or whether e-learning has become more informal and more widespread.

In this ever-changing context, Eynon and Helsper (2011) examined the importance of choice versus exclusion in using the Internet for both formal and informal learning activities. "For uptake of more formal learning opportunities, the most important factors were those related to exclusion [. . .] and, for informal learning, the most important factors were aspects of exclusion, socialization and choice" (ibid., 546). This shows that we need to differentiate between different types of learning as well as a number of factors that could be related to why some people use the Internet more for e-learning than others.

4. A Case Study of E-Government and Digital Divides in Britain

Building on the findings from previous research, this section presents a case study examining digital divides and use of (e-)government services in Great Britain based on quantitative secondary data from the Oxford Internet Surveys (OxIS). The case study provides an empirical illustration of key issues related to inequalities in access to and use of ICT-enabled services. It is structured around two core research questions that follow from the earlier discussion.

1. *Are Internet users more likely to use any government services—whether online or offline—as compared to Internet non-users?*
2. *What factors shape use of online services namely e-government services, among British Internet users?*

The case study provides evidence of how access to ICTs shapes engagement with e-government services, and how quality of Internet access, skills, and attitudes shape engagement.

4.1 Digital Britain: The Case of "Digital by Default"

The case of e-government is interesting in light of the British government's strategy of providing government services "digital by default," hence encouraging British citizens to use services online instead of in person or by telephone. In 2012, the British government published a Digital Efficiency Report (Cabinet Office 2012: 8), which estimated that for some government services "the average cost of a digital transaction is almost 20 times lower than the cost of a telephone transaction, about 30 times lower than the cost of postal transaction and about 50 times lower than a face-to-face transaction." Although expected benefits of the "digital by default" strategy also include increased convenience for users, optimization and related cost savings are the main reasons behind the shift to the digital provision of government services. In fact, the British government estimated that moving to the "digital by default" service provision could incur savings worth up to £1.8 billion per year (ibid, 8).

However, savings on the side of the government do not necessarily translate into savings regarding cost, time, and stress for users of these government services. NGOs, academics and activists have warned against the negative consequence of the "digital by default" provision of government services, which is likely to leave behind digitally excluded individuals (Helsper 2012b; Rust 2014). The move to e-government not only affects individuals who are completely offline but also individuals who are online, but lack confidence and skills to make meaningful use of e-government services. Thus, it is important to understand who is making use of services online and who is not, and what factors are driving differentiated engagement with e-government services. The empirical evidence presented in this section sheds light on concrete implications of digital inequalities on social inclusion in the realm of ICT-enabled services.

4.2 Methodology

The data are taken from the 2013[5] OxIS. OxIS is a cross-sectional, population-based survey that collects data on British Internet users and non-users.

Face-to-face interviews with individuals 14 years and older are conducted across Great Britain, using a random multi-stage sampling strategy. In 2013, there were 2,053 respondents in the sample, which is representative of the population of Great Britain when sampling weights are applied. The use of this dataset is especially valuable to the questions we ask here, as it is nationally representative, and the data allow us to connect trends in e-government use to socio-demographic and other relevant data on a larger scale. OxIS distinguishes between three groups of people: those who use Internet technologies (users), those who have never used the Internet (non-users), and those who used to go online in the past (ex-users). In 2013, 78% of the British population were online, 18% were non-users, and 4% were ex-users. Ex-users were excluded from further analyses, since previous research showed they are distinctly different from non-users (Dutton and Blank 2013; Helsper and Reisdorf 2013). We refrained from analyzing this group separately, as the number of cases is too small to conduct any meaningful analyses.

In this chapter, we examine use of four specific government services for both Internet users and non-users. Internet users were asked whether they (1) get information about local council services; (2) pay for a local council tax, fine, rent, or service; (3) get information about central government services, such as benefits, taxes, driving license or passport; (4) pay for a central government tax, fine, license, or service. Internet users were able to choose between four different responses: do not use a service; use a service offline; use a service online; use a service offline and online. Internet non-users were asked the same set of questions where questions 2 and 4 specified that these services are performed online.[6] Non-users were only able to respond yes/no to these questions. The slight difference in these questions and their answer-options poses a limitation for the results of our analyses. Although we recoded the questions to account for these differences, not asking about offline/online for all non-user questions limits the comparability between user and non-user data.

We conducted descriptive and logistic regression analyses (Long and Freese 2014) to determine what socio-economic and digital inclusion factors shape engagement with government services. While the descriptive statistics give a general overview of the use of government services in Britain, the logistic regression analyses allow us to explore which factors contribute positively and negatively to using these services. We examine socio-economic variables that are traditionally related to digital inequalities (Blank and Groselj 2014; Helsper and Reisdorf 2013; Van Deursen and Van Dijk 2014), such as age, gender, education, life stage, income, disability, children in the household, living alone versus with a partner, ethnicity, and rural versus urban residence.

Following theoretical work on digital inclusion (Helsper 2012a; Tsatsou 2013), we also include three digital inclusion factors: access, skills, and motivation. Quality and ubiquity of access (Helsper 2012a; DiMaggio et al.

2004) are measured as a sum of a number of locations users report to go online at (access at home, on the go, at another person's home, at work, at school, at Internet café, at a library) and devices owned (one, two, or three PCs, smartphone, tablet, e-reader, Internet-enabled TV; range 0–14; mean = 4.7; SD = 3.3). Digital skills are measured with six items asking about users' confidence to perform different things online, such as "Judge the reliability of online content" or "Upload photos to a website."[7] The skills variable is created by averaging users' scores (range: 1–5; mean = 3.4; SD = 1.1). Motivation to engage with Internet technologies was measured with a "general technology attitudes"[8] variable (range: 1–5; mean = 3.4; SD = 0.8), which represents users' average score on six items such as "Technology is making things better for people like me" and "When new technologies or gadgets are invented, it is a good idea to try them."[9]

4.3 Findings: E-Government in Britain

The analysis that follows is structured around the two research questions. The first part of this section focuses on differences in engagement with government services online and offline between Internet users and non-users. The second part focuses on the population of Internet users and examines what factors shape users' decision to perform government services online rather than offline.

4.3.1 Are British Internet Users More Likely to Use Government Services Than Non-users?

The data show that in Britain, individuals who use Internet technologies are more likely to use government services in general. The first two bars in Figure 11.1 display proportions of the populations of Internet users and non-users who said they got information about central government and about local council services either offline or online in the past 12 months. About half of Internet users and about a fifth of non-users said they got information on government services. Somewhat lower proportions of users reported paying for central government (44%) and for local council (29%) services online. Unsurprisingly, a smaller proportion of non-users say they paid for government services online: only 4% say they paid for a central government service online, whereas 12% say they paid for local tax, fine, or service online.

This means that about one in eight non-users pays for local council services online. This may seem surprising since non-users do not use Internet technologies themselves. We suspect that they use e-government services via proxy users, meaning that other people perform online activities on their behalf. Research shows that many non-users access online content and services indirectly via members of their social networks, such as family

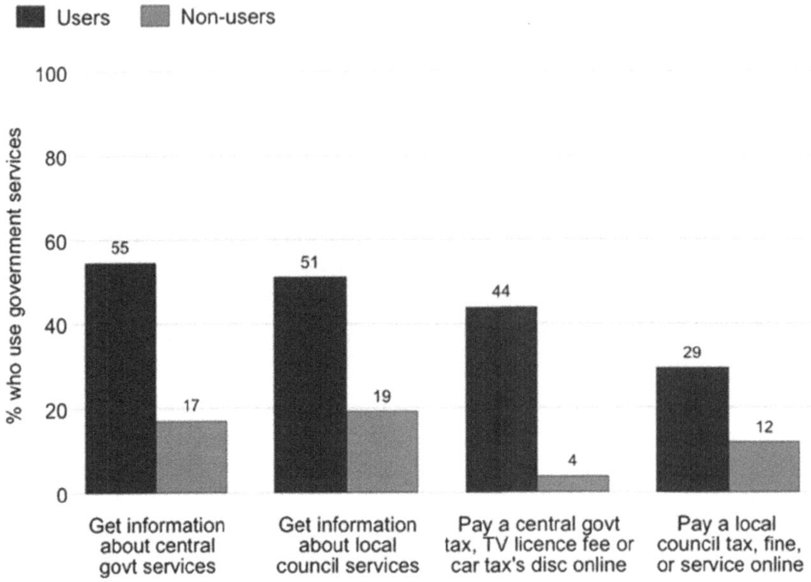

Figure 11.1 Use of Government Services by Users and Non-users.

Source: OxIS 2013: Internet users and non-users; N = 1,982. Notes: govt = government.

or friends (Eynon and Geniets 2012; Reisdorf, Axelsson, and Söderholm 2012).

To test whether use of government services is significantly related to Internet use and non-use we ran a set of logistic regressions that take individuals' socio-economic characteristics into account. The output in Table 11.1 is presented in odds ratios, which present the increased or decreased likelihood of the outcome yes versus no (using government services) on the basis of the reference category of the independent variable in question. The models have moderate explanatory power, ranging from 11% to 20%.

Age (only significant in one model, with a small effect), gender, disability, children in household, and place of residence are not related to use of government services. Education has a significant and large effect on the likelihood of engagement with all four examined services. Having higher education as opposed to no educational qualifications increases the likelihood of using government services between 2.4 to 4 times, depending on the service. Lifestage has mixed effects across the four examined activities. Compared to retired individuals, students are less likely and employed individuals are more likely to pay for local council services online; they are also more likely to get information about central government services, whereas unemployed individuals are less likely to pay for central government services online. However, all of these effects are rather small. People with higher incomes are more likely to engage in all activities apart from getting information about central government services. People who live alone are less likely to pay for

Table 11.1 Logistic Regression Coefficients (Odds Ratios): Likelihood of Using Government Services

	Get info about central government services (No)	*Pay for central government service online (No)*	*Get info about local council services (No)*	*Pay for local council service online (No)*
Age	1.00	0.99	1.01*	1.00
Gender (Female)				
Male	1.09	1.30	0.79	0.98
Education (None)				
Secondary	1.50*	2.01**	1.48*	1.13
Further	1.88**	2.35**	2.35***	1.56
Higher	3.23***	4.07***	3.21***	2.43***
Life stage (Retired)				
Students	0.94	0.43	0.54	0.33*
Employed	1.54*	1.11	1.03	1.69*
Unemployed	1.42	0.51*	1.00	0.66
Income	1.08	1.17**	1.10*	1.14**
Disability (Yes)				
No	0.91	1.23	1.32	0.92
Children in household (No)				
Yes	1.08	0.96	0.99	1.17
Living alone (Yes)				
No	1.11	1.81***	1.71***	1.34
Residence (Rural)				
Urban	1.33	1.01	1.37	1.48
Ethnicity (White)				
Non-white	1.10	1.39	1.53	1.94**
Internet use (Non-user)				
User	3.17***	8.99***	3.18***	1.28
Constant	0.07***	0.02***	0.04***	0.07***
N	1,754	1,754	1,750	1,747
McFadden's R^2	0.11	0.20	0.12	0.12

Source: OxIS 2013. Notes: Reference category in (). *p < 0.05; **p < 0.01; ***p < 0.001

central government services online and get information about local council services, and individuals who are not White are 1.9 times more likely to pay for local council services online. Controlling for all of the aforementioned factors, Internet users are about 3.2 times more likely to get information about central government and local council services online and 9 times more likely to pay for central government services online as compared to non-users. Together with higher education, being an Internet user is the strongest factor in predicting engagement with these three government services, whether they are offered online or offline. Interestingly, being an Internet user is not significantly related to the likelihood of paying for local council services online. This result is in line with the rather high number of non-users who reported paying for local services online—most likely via proxy users—which we discussed earlier.

4.3.2 What Factors Shape the Use of E-Government Services Among British Internet Users?

Internet users are more likely than non-users to engage with government services offline and online. However, being an Internet user does not necessarily translate into use of government services online. To examine whether the mode of engagement with government services—i.e., offline or online—varies among Internet users, we use Reisdorf and Groselj's (2017) typology of low (27%), regular (38%), and broad (35%) Internet users. The typology is based on time spent online and breadth of activities.

Figure 11.2 shows significant differences between low, regular, and broad users regarding use and mode of use of government services. About half of the low users say they do not perform any activities, and only 21% to 31% say they use government services online. Similarly, about half of the regular users report they do not get information about central and local government services or pay for local council services online. However, compared to low users, regular users are more likely to perform government services online rather than offline. Low and regular Internet users are most similar when it comes to paying for local council services.

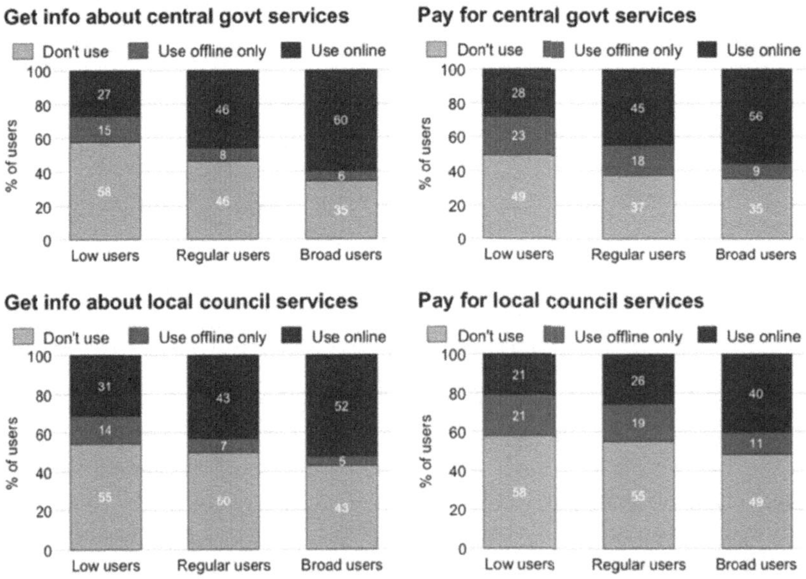

Figure 11.2 Non-use and Use of Government Services by Types of Internet Users.

Source: OxIS Internet users: 2013 N=1,611.

Generally, broad Internet users are most likely to use e-government services. Over half of the broad users report getting information about central and local council services and paying for central government services online. Figure 11.2 suggests that users who have more spend more time online are more likely to use government services, and to use them online.

We ran a set of logistic regressions to examine which specific digital inclusion factors influence the likelihood of use of e-government services

Table 11.2 Logistic Regression Coefficients (Odds Ratios): Likelihood of Using E-Government Services (vs. Offline Government Services)

	Get info about central govt services (Yes, offline)	*Pay for central govt service (Yes, offline)*	*Get info about local council services (Yes, offline)*	*Pay for local council service (Yes, offline)*
Age	1.00	0.96***	0.98	0.98
Gender (Female)				
Male	1.06	1.18	1.02	0.90
Education (None)				
Secondary	1.10	1.06	1.06	0.64
Further	1.11	0.86	1.62	0.65
Higher	1.60	1.54	3.86**	1.06
Lifestage (Retired)				
Students	0.64	0.20*	0.2	0.35
Employed	1.02	0.53	1.25	1.23
Unemployed	0.78	0.36*	0.92	0.63
Income	1.18	1.11	0.93	1.11
Disability (Yes)				
No	1.00	1.39	1.63	1.31
Children in household (No)				
Yes	1.33	0.86	1.39	1.13
Living alone (Yes)				
No	0.53*	1.31	0.90	0.95
Residence (Rural)				
Urban	0.88	0.54*	0.86	0.97
Ethnicity (White)				
Non-white	0.76	1.56	0.52	1.48
Online skills	1.45**	1.39**	1.42*	1.39**
N of devices and locations	1.08	1.15**	1.09	1.12*
General technology attitudes	1.22	1.23	1.58*	1.20
Constant	0.43	1.28	0.26	0.24
N	714	815	702	637
McFadden's R^2	0.11	0.16	0.171	0.14

Source: OxIS 2013. Notes: Reference category in (). *p < 0.05; **p < 0.01; ***p < 0.001

versus offline government services among Internet users (see Table 11.2). The models explain between 11% to 17% of variance in the use of e-government services.

Of the socio-economic factors, gender, income, disability, presence of children in household, and ethnicity have no significant effects on the use of e-government services versus offline government services. Age, being a student or unemployed (vs. retired), and living in urban areas are all negatively related to paying for central government services online. Having higher education is positively related to getting information about local council services online. Living alone has a negative effect on the likelihood of getting information about central government services online. Taken together, socio-economic factors have scattered effects on use of e-government services among Internet users with no clear pattern. Online skills and quality of access have more consistent—however fairly small—effects on use of e-government services. Skills are relevant in predicting the likelihood of using e-government services—those who are more confident in their digital skills are more likely to use all four examined services online. Users who can go online on a larger number of devices and locations are more likely to use transactional e-government services. Lastly, having more positive attitudes toward technology is positively related only to the likelihood of getting information about local council services online.

5. Discussion and Conclusion

This chapter aimed to shed light on who engages with services offered online and who is left out. Specifically, we wanted to know whether Internet users are more likely to use any government services—whether online or offline—as compared to Internet non-users and what factors shape use of e-government services among British Internet users. The literature review combined with the multivariate analyses show that apart from socio-demographic factors (see also Venkatesh, Sykes, and Venkatraman 2014), digital skills and high quality, ubiquitous Internet access are crucial for use of e-government services.

The empirical analysis of use of government services conveyed two messages: First, controlling for a range of socio-economic factors, Internet users are more likely to get information about government services in general—whether online or offline. This suggests that people who use Internet technologies are better informed about government services than people who do not use the Internet themselves. A possible explanation is that information about government services is more easily and readily accessible online making non-users less likely to obtain such information. Second, in contrast to the first point, Internet use does not have a significant effect on the likelihood of paying for local council services online. This is interesting because one would expect that Internet users are more likely to perform any online activity as compared to non-users. We speculate that non-users rely on their family and friends to pay for local council services online on their behalf. While we do not have sufficient data to explain why this is the case, one possible explanation is that some of local council services have to be performed online because they are only available online—i.e., "digital by default."

However, being online does not simply translate into engagement with e-government. In fact, low and regular Internet users are more likely not to use government services or to use them offline rather than online. This is mostly due to limited digital skills and access, which prevent a large number of Internet users from using e-government services. Regardless of a range of socio-economic factors, individuals' Internet access, and their technology attitudes, those with less confidence in their ability to perform things online opt out from use of e-government services (Bélanger and Carter 2009; Van Deursen and Van Dijk 2009). Thus, if the British government is to fully adopt the "digital by default" strategy for service provision, sufficient resources need to be secured for assisting people in using such services and developing their digital skills (Helsper 2012b). Furthermore, it is interesting that characteristics of users' Internet access are only associated with transactional e-government services and not with finding information about government services online. This suggests that using transactional online services is related to choices users can make in terms of where and on which device they can go online. Such nuances need to be acknowledged when such online services are being designed and promoted (Aleixo, Nunes, and Isaias 2012).

The results from our analyses can be transferred to other ICT-enabled services, such as e-health, which have been shown to be used mainly by Internet users who have high levels of (perceived) digital literacy (Neter and Brainin 2012). If we assume that this is the case for most online services, this has implications for the necessity for high levels of usability in the design of e-services and the acknowledgment that Internet users are not simply one homogeneous group (Aleixo, Nunes, and Isaias 2012). Providing crucial services "digital by default" is a risky undertaking that may exacerbate existing social and digital exclusion by leaving behind those who are most in need of, for example, government services, but are the least likely to access these services online (Helsper 2012b).

These results have implications for policy makers, researchers, and industry alike. Confirming previous results from similar research (Bélanger and Carter 2009; Hargittai 2003; Van Deursen and Van Dijk 2009) and building on Aleixo, Nunes, and Isaias's (2012) assessment of usability and digital skills, the three sectors have a responsibility to combine their forces to address the issues that our research uncovered. Policy makers need to rethink their strategy of going "digital by default" when a large number of citizens are not making use of these resources. They need to work closely with researchers and the industry to enable access, resources, skills, and easy-to-navigate services that match users' skills, rather than expecting a high level of skills from all Internet users and non-users to navigate the often complicated infrastructures and layouts that many e-services display to date.

While our research produced some interesting results, there are limitations resulting from the survey instrument (e.g., different questions for Internet users and non-users) and the fact that we were only able to examine data from one country. More in-depth research is needed that takes into account the users' perspectives and experiences (Aleixo, Nunes, and Isaias 2012) and that includes collaboration with interface designers who can test and tweak different versions of e-services to serve the broadest audience possible. Usability standards need to acknowledge and take into account digital skills and digital inequalities, at the same time that policy makers need to tackle digital inclusion beyond factors of access and socio-demographic factors (Reisdorf and Groselj 2017).

Notes

1 As theories and empirical work in the area of digital divides and inequalities are closely intertwined, we also give examples of empirical research in this section.

2 eHEALS is an "8-item measure of eHealth literacy developed to measure consumers' combined knowledge, comfort, and perceived skills at finding, evaluating, and applying electronic health information to health problems" (Norman and Skinner 2006: n.p.).

3 As Pinterest includes a large number of photo tutorials on diverse activities, we consider it a sharing platform that enables e-learning.

4 The last OxIS data were collected in 2013. Due to funding cuts, no data were collected in 2015. However, the 2013 data represent the most comprehensive dataset on Internet use and non-use in the United Kingdom to date. Similar datasets from the United Kingdom do not include as many factors and variables as were needed for our analyses.

5 The reason that the questions regarding the payment of council tax, fine, rent or services for non-users specified online is that these services were some of the first to be moved to "digital by default." If non-users answered yes to these questions, it does not necessarily mean that they actually use the Internet themselves, as many non-users report proxy-use—i.e., having other people perform some tasks for them online. Questions 1 and 3 did not specify online, as we assume that non-users would use other avenues to obtain information about services.

6 Other included items: Remove a virus that infected your computer; Participate in a discussion online; Make new friends online; Download and save music (MP3s). These six skills measured by the OxIS are related to Van Deursen and Van Dijk's (2011) operational, formal, information and strategic skills.

7 Van Dijk's "notion of motivational access is primarily shaped by attitudes towards technology" (Van Deursen and Van Dijk 2015: 3). Following this notion, we used attitudinal variables to measure motivation to engage with technologies. A more detailed discussion of the relationship between motivations and attitudes can be found in Reisdorf and Groselj 2017.

8 Other included items: I do not trust technologies, because they fail when you need them the most (reversed in combined variable). I get nervous using technologies, because I might break something (reversed). Often it is easier to do things without using technologies (reversed). I find it difficult to keep up with new technology

(reversed). The PCAs used to design this variable and the relevant factor loadings can be found in Reisdorf and Groselj 2017.

References

Aleixo, Carlos, Miguel Nunes, and Pedro Isaias. "Usability and Digital Inclusion: Standards and Guidelines." *International Journal of Public Administration* 35, no. 3 (2012): 221–39.

Association for Computing Machinery. *ACM Code of Ethics and Professional Conduct*. 2016. www.acm.org/about-acm/acm-code-of-ethics-and-professional-conduct

Bartley, Mel. *Health Inequality: An Introduction to Theories, Concepts and Methods*. Cambridge: Polity, 2004.

Bélanger, France, and Lemuria Carter. "The Impact of the Digital Divide on e-Government Use." *Communications of the ACM* 52, no. 4 (2009): 132–5. doi:10.1145/1498765.1498801.

Blank, Grant, and Darja Groselj. "Dimensions of Internet Use: Amount, Variety and Types." *Information, Communication & Society* 17, no. 4 (2014): 417–35. doi:10.1080/1369118X.2014.889189.

Cabinet Office. "Government Digital Strategy." 2012. www.gov.uk/government/uploads/system/uploads/attachment_data/file/296336/Government_Digital_Stratetegy_-_November_2012.pdf

Choudrie, Jyoti, Gheorghita Ghinea, and Vivian Nwamaka Songonuga. "Silver Surfers, e-Government and the Digital Divide: An Exploratory Study of UK Local Authority Websites and Older Citizens." *Interacting with Computers* 25, no. 6 (2013): 417–42.

DiMaggio, Paul, Eszter Hargittai, Coral Celeste, and Steven Shafer. "Digital Inequality: From Unequal Access to Differentiated Use." In *Social Inequality*, edited by Kathryn Neckerman, 355–400. New York: Russell Sage Foundation, 2004.

Dutton, William H., Grant Blank, and Darja Groselj. *Cultures of the Internet: The Internet in Britain*. Oxford: University of Oxford, 2013. http://oxis.oii.ox.ac.uk/reports

Eynon, Rebecca, and Anne Geniets. *On the Periphery? Understanding Low and Discontinued Internet Use Amongst Young People in Britain*. Oxford: University of Oxford, 2012. www.oii.ox.ac.uk/research/projects/?id=87

Eynon, Rebecca, and Ellen J. Helsper. "Adults Learning Online: Digital Choice and/or Digital Exclusion?" *New Media and Society* 13, no. 4 (2011): 534–51. doi:10.1177/1461444810374789.

Glymour, M. Maria, Mauricio Avendano, and Ichiro Kawachi. "Socioeconomic Status and Health." In *Social Epidemiology*, edited by Lisa F. Berkman, Ichiro Kawachi, and M. Maria Glymour, 17–62. Oxford: Oxford University Press, 2014.

Go ON UK, n.d. www.go-on.co.uk/.

Gonzales, Amy. "The Contemporary US Digital Divide: From Initial Access to Technology Maintenance." *Information, Communication & Society* 19, no. 2 (2016): 234–48. doi:10.1080/1369118X.2015.1050438.

Griffin, Dave, Amanda Foster, and Edward Halpin. "Joined-Up E-Government: An Exploratory Study of UK Local Government Progress." In *Proceedings of the Annual Conference of CAIS/Actes du congrès annuel de l'ACSI*, 2013.

Haddon, Leslie. "Social Exclusion and Information and Communication Technologies: Lessons from Studies of Single Parents and the Young Elderly." *New Media and Society* 2, no. 4 (2000): 387–406. doi: 10.1177/14614448000002004001.

Hall, Thad E., and Jennifer Owens. "The Digital Divide and e-Government Services." In *Proceedings of the 5th International Conference on Theory and Practice of Electronic Governance*, ACM, 2011, 37–44.

Hargittai, Eszter. "Second-Level Digital Divide: Mapping Differences in People's Online Skills." *arXiv preprint* cs/0109068, 2001.

Hargittai, Eszter. "Serving Citizens' Needs: Minimizing Online Hurdles to Accessing Government Information." *IT& Society* 1, no. 3 (2003): 27–41.

Hargittai, Eszter, and Steven Shafer. "Differences in Actual and Perceived Online Skills: The Role of Gender." *Social Science Quarterly* 87, no. 2 (2006): 432–48.

Haythornthwaite, Caroline. "Digital Divide and e-Learning." In *The Handbook of e-Learning Research*, edited by Richard Andrews and Caroline Haythornthwaite, 97–118. London: Sage, 2007.

Helbig, Natalie, J. Ramón Gil-García, and Enrico Ferro. "Understanding the Complexity of Electronic Government: Implications from the Digital Divide Literature." *Government Information Quarterly* 26, no. 1 (2009): 89–97. doi:10.1016/j.giq.2008.05.004.

Helsper, Ellen J. "A Corresponding Fields Model for the Links Between Social and Digital Exclusion." *Communication Theory* 22, no. 4 (2012a): 403–26. doi:10.1111/j.1468-2885.2012.01416.x.

Helsper, Ellen J. "Social Digital Series: Digital by Default—Excluded by Default?" 2012b. http://blogs.lse.ac.uk/mediapolicyproject/2012/03/21/digital-by-default-excluded-by-default/

Helsper, Ellen J., and Bianca C. Reisdorf. "A Quantitative Examination of Explanations for Reasons for Internet Nonuse." *Cyberpsychology, Behavior and Social Networking* 16, no. 2 (2013): 94–9. doi:10.1089/cyber.2012.0257.

Helsper, Ellen J., and Bianca C. Reisdorf. "The Emergence of a "Digital Underclass" in Great Britain and Sweden: Changing Reasons for Digital Exclusion." *New Media & Society* 19, no. 8 (2017): 1253–70. doi: 10.1177/1461444816634676

Hermana, Budi, and Widya Silfianti. "Evaluating e-Government Implementation by Local Government: Digital Divide in Internet Based Public Services in Indonesia." *International Journal of Business and Social Science* 2, no. 3 (2011): 156–63.

Kontos, Emily, Kelly D. Blake, Wen-Ying Sylvia Chou, and Abby Prestin. "Predictors of eHealth Usage: Insights on the Digital Divide from the Health Information National Trends Survey 2012." *Journal of Medical Internet Research* 16, no. 7 (2014): e172. doi:10.2196/jmir.3117.

Long, J. Scott, and Jeremy Freese. *Regression Models for Categorical Dependent Variables Using Stata*, 3rd ed. College Station, TX: Stata Press, 2014.

Nam, Taewoo, and Djoko Sigit Sayogo. "Who Uses e-Government?: Examining the Digital Divide in e-Government Use." In *Proceedings of the 5th International Conference on Theory and Practice of Electronic Governance*, ACM, 2011, 27–36. doi:10.1145/2072069.2072075.

Neter, Efrat, and Esther Brainin. "eHealth Literacy: Extending the Digital Divide to the Realm of Health Information." *Journal of Medical Internet Research* 14, no.1 (2012): e19. doi:10.2196/jmir.1619.

Norman, Cameron D., and Harvey A. Skinner. "eHEALS: The eHealth Literacy Scale." *Journal of Medical Internet Research* 8, no. 4 (2006): e27. doi:10.2196/jmir.8.4.e27.

Reddick, Christopher G., and Michael Turner. "Channel Choice and Public Service Delivery in Canada: Comparing e-Government to Traditional Service Delivery." *Government Information Quarterly* 29, no. 1 (2012): 1–11 doi:10.1016/j.giq.2011.03.005.

Reisdorf, Bianca C., Ann-Sofie Axelsson, and Hanna Maurin Söderholm. "Living Offline: A Qualitative Study of Internet Non-Use in Great Britain and Sweden." *Selected Papers of Internet Research*, Association of Internet Researchers, Manchaster, 2012, 1–28. http://spir.aoir.org/index.php/spir/article/view/10

Reisdorf, Bianca C., and Darja Groselj. "Internet (Non-)Use Types and Motivational Access: Implications for Digital Inequalities Research." *New Media and Society* 19, no. 8 (2017): 1157–76. doi:10.1177/1461444815621539.

Robinson, Laura, Shelia R. cotton, Hiroshi Ono, Anabel Quan-Haase, Gustavo Mesch, Wenhong Chen, Jeremy Schulz, Timothy M. Hale, and Michael J. Stern. "Digital Inequalities and Why They Matter." *Information, Communication & Society* 18, no. 5 (2015): 569–82. doi:10.1080/1369118X.2015.1012532.

Rogers, Everett M. "The Digital Divide." *Convergence* 7, no. 4 (2001): 96–111. doi:10.1177/135485650100700406

Rust, Elizabeth. "When the UK Goes 'Digital by Default,' Who Will Be Left Behind?" *The Guardian*, June 23, 2014. www.theguardian.com/technology/2014/jun/23/when-the-uk-goes-digital-by-default-who-will-be-left-behind

Selwyn, Neil. "Reconsidering Political and Popular Understandings of the Digital Divide." *New Media and Society* 6, no. 3 (2004): 341–62. doi:10.1177/1461444804042519.

Shneiderman, Ben. "Universal Usability." *Communications of the ACM* 43, no. 5 (2000): 84–91.

Taipale, Sakari. "The use of e-government services and the Internet: The role of socio-demographic, economic and geographical predictors." *Telecommunications Policy* 37, no. 4 (2013): 413–22. doi: https://doi.org/10.1016/j.telpol.2012.05.005

Tsatsou, Panayiota. "Digital Divides Revisited: What Is New About Divides and Their Research?" *Media, Culture and Society* 33, no. 2 (2011): 317–31. doi:10.1177/0163443710393865.

Tsatsou, Panayiota. "Digital Inclusion." In *Digital World: Connectivity, Creativity and Rights*, edited by Gillian Youngs. London: Routledge, 2013.

Van Dijk, Jan A. G. M. *The Deepening Divide: Inequality in the Information Society*. Thousand Oaks, CA: Sage Publications, 2005.

Van Dijk, Jan A. G. M., and Alexander J. A. M. van Deursen. *Digital Skills: Unlocking the Information Society*. New York: Palgrave Macmillan, 2014.

Van Deursen, Alexander J. A. M., and Jan A. G. M. van Dijk. "Improving Digital Skills for the Use of Online Public Information and Services." *Government Information Quarterly* 26, no. 2 (2009): 333–40. doi:10.1016/j.giq.2008.11.002

Van Deursen, Alexander J. A. M., and Jan A. G. M. van Dijk. "Internet Skills and the Digital Divide." *New Media and Society* 13, no. 6 (2011): 893–911. doi:10.1177/1461444810386774.

Van Deursen, Alexander J. A. M., and Jan A. G. M. van Dijk. "The Digital Divide Shifts to Differences in Usage." *New Media and Society* 16, no. 3 (2014): 507–26. doi:10.1177/1461444813487959.

Van Deursen, Alexander J. A. M., and Jan A. G. M. van Dijk. "Toward a Multifaceted Model of Internet Access for Understanding Digital Divides: An Empirical Investigation." *The Information Society* 31, no. 5 (2015): 379–91. doi:10.1080/01972243.2015.1069770.

Venkatesh, Viswanath, Tracy Ann Sykes, and Srinivasan Venkatraman. "Understanding e-Government Portal Use in Rural India: Role of Demographic and Personality Characteristics." *Information Systems Journal* 24, no. 3 (2014): 249–69. doi: 10.1111/isj.12008

Zillien, Nicole, and Eszter Hargittai. "Digital Distinction: Status-Specific Types of Internet Usage." *Social Science Quarterly* 90, no. 2 (2009): 274–91. doi: 10.1111/j.1540-6237.2009.00617.x

12 Mobility of Work

Usability of Digital Infrastructures and Technological Divide

Mohammad Hossein Jarrahi and Luke Williamson

1. Introduction

Knowledge workers constitute a growing class in the global workforce (Florida 2002; Liegl 2014). In fact, in the U.S.A., this class of workers has doubled in one generation, from 22% in 1960 to 43% in 2006 (Rainie and Wellman 2012). This global phenomenon issues from an economic evolution where actors engage with commercial commodities on an abstract rather than concrete plane (Erickson et al. 2014); researchers often call these changes information revolution, as economic and social transformation has partially been fueled by the rise of new ICTs (Castells 2006; Webster 2006). Moreover, the effects of the Great Recession and the expansion of the global economy have engendered modular and project-focused knowledge-work models (Barley and Kunda 2006; Schultze and Boland 2000). Organizations and individuals in knowledge-intensive contexts have enacted work arrangements such as telecommuting, home-working, shared offices, hot-desking, global virtual team working, and mobile working (Sørensen 2011).

For virtue of their relatively high levels of independence, many knowledge workers can now work apart from conventional corporate structures or on the move (Harmer and Pauleen 2012). As a subclass of knowledge workers, mobile knowledge workers are defined by work that is both knowledge intensive and mobile (Rainie and Wellman 2012). These workers must travel more than episodically, and are required regularly to navigate a set of boundaries—temporal, spatial, organizational, infrastructural, social, and cultural—in accomplishing their everyday professional responsibilities.

Primarily, mobile work arrangements can be considered a product of technological, social, and economic factors, as well as personal choice (Barley and Kunda 2006). A critical factor for knowledge-work mobility is use of technology, specifically ICTs. The autonomous and intellectual nature of knowledge work lends itself well to technological mediation (Ciolfi and de Carvalho 2014; Thomson and Jarrahi 2015), the integration of ICTs into knowledge-work practices, (Jarrahi and Sawyer 2015; Mazmanian, Orlikowski, and Yates 2013), enabling flexible work practices, and extended mobility among the knowledge workforce. Thus, the maturation of ICTs (Cousins and Robey 2005; Su and Mark 2008) has ushered a socio-technical revolution that has spawned a recognizable cadre of mobile knowledge workers, satellites orbiting spheres of both traditional and non-traditional entities of enterprise and commerce (Costas 2013). In this respect, the advancement in ICTs' features and magnitude has been one of the most critical factors that have given rise to the mobile knowledge workforce (de Carvalho, Ciolfi, and Gray 2011). It is the synergistic effects between social actors and technology that extend the ICTs' de facto effectiveness. Without the ubiquity of networked infrastructures today, mobile knowledge work would be impossible.

However, just as ICTs are enablers, they can also be disablers. When technological breakdowns occur, mobile work is disturbed. The increased dynamism of work via physical mobility introduces infrastructural challenges for mobile workers and, unlike those in conventional work settings, they cannot take technological resources for granted. This may create a "digital divide" between traditional office workers and those who may work remotely or on the move. Thus, in the context of mobile work, knowledge is employed not only in one's work but also for managing work and reacting to or avoiding technological breakdowns. While mobile knowledge workers may hope for interoperability among multiple infrastructures, they are often faced with barriers that cause breakdowns, curbing operative intent. Consequently, the adaptive strategies that these workers employ determine their own operational effectiveness.

The research question motivating this work focuses on different forms of barriers existing in digital infrastructures that undergird mobile knowledge work and specific strategies used by workers to overcome these barriers. Peering through the lens of individual mobile knowledge workers' accounts of their technology and information use, their impressions, experiences of success or failure, and active use, we can identify infrastructural barriers that mobile knowledge workers face. By building upon an empirical study of 16 mobile knowledge workers and the language of seams (Vertesi 2014), which describes physical and digital connecting links and disconnects among layered infrastructures, we attempt to identify these barriers and along with them, the strategies that mobile knowledge workers enact. The findings presented here are especially useful for raising an understanding of the work practices of workers affiliated with microenterprises that are typically resource constrained (Wolcott, Kamal, and Qureshi 2008). A common denominator of many microenterprises is the tendency to lower overhead cost by working remotely while being untethered from a fixed notion of office. Thus, a defining characteristic of most micropreneurs' work arrangements is increasing nomadicity and mobility of work. Surveys have demonstrated that self-employed workers are almost twice as likely to be mobile as paid workers—42% and 26%, respectively (PearnKandola 2007).

This study can help system designers, managers and policy makers generate a greater understanding of the challenges affiliated with ICT use by mobile knowledge workers; understanding these challenges can potentially lead to more inclusive management decisions and systems' designs that consider the contingencies of mobile work.

2. Related Work

There has been a significant amount of research on the digital divide and social inclusion, mobile work, and mobile knowledge work, particularly in the last two decades. What follows is a brief overview of the relevant literature, which explores the sociological, physical, and technical aspects of mobile knowledge work practices.

2.1 Social Inclusion and Digital Divide

Digital divide literature encompasses two traditions of thought, both socially charged and sparking political and ideological debate. Early research leaned toward hard technological determinism while later research promoted soft. While both maintain that technology exerts social and economic impact, hard technological determinism sees technology as the driving force, whereas "soft determinists" consider technology either as influencing social phenomena or symptomatic of them (Rodino-Colocino 2006).

The NTIA's initial reports (Department of Commerce 1995) on the digital divide, oriented in the hard technologically deterministic view, gauged technology as a ladder to success in which disadvantaged groups cannot compete with those possessing abilities to "access, accumulate, and assimilate information" digital information. Based on the belief that new economic growth is entrenched in an "information revolution" digitally borne, the report promoted the need for increasing Internet access to underprivileged groups. Calling for the same, Hoffman, Novak, and Schlosser (2001) correlated Internet access with economic and social advancement and argued for technological education initiatives to level the playing field.

Avoiding the binary qualifier of access, much research shows aspects of soft technological determinism. Barzilai-Nahon (2006) criticized the U.S. government's approach to the digital divide, which naively likens it with the "Mercedes divide," urging a systematic method for contextual examination; she recognizes that there exist groups with different circumstances and needs. While researchers admit the necessity for acquiring digital infrastructure (e.g., computers and peripheral devices), Internet access, and a skillset for competently using ICTs (e.g., Servon 2008; Van Dijk 1999; Warschauer 2002) these authors credit training, content creation, and social networks as largely overlooked variables in the digital divide.

Notably, Van Dijk and Hacker (2003), Warschauer (2002), and Mossberger, Tolbert, and Stansbury (2003) stress the importance of social networks and users' socio-cultural and educational needs, concluding that the digital divide, rather than being "binary," encompasses social inclusion as well as digital access. This trend "implies a chain of causality" (Warschauer 2002) preventing development among marginalized groups while rewarding those with proficiency, resulting in a "democratic divide" (Mossberger, Tolbert, and Stansbury 2003). In a similar vein, Jung, Qiu, and Kim's (2001) study measured the quality of Internet use in a "social context" to determine the "Internet Connectedness Index (ICI)." They contend that ICI and accompanying skills are markers in the digital divide that predict gainful employment. Servon (2008) posits that those without requisite IT skills not only lose out to the better trained but also, particularly in times of economic boom, may be placed in positions beyond their competency and that high-growth jobs may not require degrees, while factors such as poverty and race cannot be downplayed.

Regarding access to information and knowledge online, Van Dijk (1999) listed four information-access barriers: "mental access," involving user engagement and anxiety, "material access" to digital tools and the Internet, "skills access," the inability to operate systems because of design or support issues, and "usage access," based on the opportunities for experience. He asserted that the final two barriers require operability of the first two, enabling users to navigate systems competently to find and process information toward task completion, or informational skills with which experience culminates to strategic skills (Van Dijk and Hacker 2003). It is these skills that are critical in a knowledge workforce and are factors that deserver greater attention.

2.2 *(Mobile) Knowledge Workers*

Knowledge workers have long been studied, but those that are mobile less so. In the literature, the terms mobile and nomad are recurrent. Mobility entails transporting work resources (i.e., devices, office artifacts) to generic spaces in order to get work done (Perry 2007). Mobility is related to nomadicity: their differences are highlighted by Ciolfi and de Carvalho (2014) in that the former entails movement across sites for work, while the latter's threshold for productivity is entrenched in greater complexity of resource use.

To date, several scholarship domains have defined knowledge work (e.g., IS, information science, organization science). A collective conception of it entails knowledge production and transmission drawing from intellectual and analytic acuity and theoretical and technical knowledge (e.g., Creplet et al. 2001; Davis 2002). Knowledge work entails not only concrete but also abstract work (Davenport, Jarvenpaa, and Beers 1996), what Rainie and Wellman (2012, p. 173) describe as "atom" work and "bit" work. Davis (2002) emphasizes the dimension of "human mental work." Erickson et al. (2014) stress the "unpredictable" aspects of knowledge work that promotes novel approaches toward tasks and solutions.

Mobile knowledge work can also be defined by its actors' professions and movement. They regularly travel to meet with clients, vendors and colleagues to conduct work knowledge-intensive work (Erickson and Jarrahi

2016) and may moving within one work environment or engage in short or long travel (Dahlbom and Ljungberg 1998). Erickson et al. (2014) conceptualize mobile knowledge work based on three central attributes: "knowledge composition; location independence; and infrastructural engagement." To conduct work in a mobile state and across different locations, mobile knowledge workers navigate and cross multiple forms of boundary, temporal, spatial, infrastructural, or social.

Organization science and IS scholarship tend to focus on the traditional and stationary work contexts. Mobile knowledge workers are, however, different from traditional office workers who are tied to a single locale (Perry et al. 2001). As opposed to conventional office workers who rely on a stable set of resources in their home offices, mobile knowledge workers tend to constantly seek, manage and reconfigure their own work resources to keep their mobile offices operable (Su and Mark 2008). Folding all these pieces together yields a work context that is both dynamically changing and, as yet, not well understood. This information context is fluid, unpredictable, and problem driven, as these workers frequently grapple with a lack of access to centrally organized resource-allocation mechanisms (prevalent in stationary forms of work) and a concomitant requirement to navigate across multiple contextual boundaries.

2.3 *Mobilization and ICTs*

Mobile knowledge workers confront various types of boundaries in their work: temporal, spatial, physical, digital, and social. It is these barriers that can cause personal and professional conflict, disrupting workflows and processes (Orlikowski 2002) stemming from differences in the nature of the portable or mobile office and the traditional work space, which includes human, document, and digital resources (Perry 2007; Su and Mark 2008). Perry emphasizes the lack of familiarity that mobile knowledge workers experience with mobile working spaces, contrasted to their counterparts in traditional workspaces who benefit from a sense of familiarity and certainty that enables the structuring of work practices around known resources. Su and Mark (2008) remark on the added load of making that office space portable, in which resources (e.g., laptops, portable printers, storage drives, and batteries) must be toted from site to site.

Mobile workers adapt to the aforementioned barriers by choosing an array of ICTs; extant research suggests that leveraging ICTs is critical for mobile knowledge work (Ciolfi and de Carvalho 2014; Erickson and Jarrahi 2016). ICTs empower people to better control their mobility, toward performing daily tasks, and they foster locational independence through robust communicative and information retrieval and processing power (Perry et al. 2001). Several lines of scholarship have examined the use of ICTs among mobile knowledge workers (e.g., Chen and Nath 2008; Lilischkis 2003; Perry et al. 2001).

These technological suites enable mobility that, at its best, manifests itself as Weiser's original vision of ubiquitous computing (Weiser 1993), conceptually rooted in the idea of technological environments instilled with an invisible integration. In this scenario, users engaged in a multi-infrastructural environment cannot detect systemic gaps; instead, they engage with seamless technologies. Contrariwise, the reality of mobile workers' use of digital infrastructures is far from the ideal of "seamlessness," while infrastructural arrangements tend to remain persistently visible for this population precisely because of the increased number of barriers and breakdown occurrences that work arrangements afford (Erickson and Jarrahi 2016). Advancing theorization relative to these technological challenges offers significant insight into the work of mobile knowledge workers and can be considered a step toward articulating more inclusive policies. As such, the goal of this chapter is to examine what technical barriers mobile knowledge workers face and how they contend with them by strategizing and deploying ad hoc or recurrent practices to work across technological barriers.

3. Theoretical Framework

For this study, we adopted the analytical framework and vocabulary espoused by Vertesi (2014), which describes both physical and digital infrastructural layers. Vertesi (2014) builds on the studies in technoscience that examine the effects of infrastructure in knowledge production and practice.

In this analytical framework, Vertesi (2014) adopts two vocabularies. The first is heterogeneity that refers to the existence of multiple actors and systems, of diverse classifications and "overlapping" infrastructures or infrastructure layers (p. 268). Her work focuses on the role of cyberinfrastructure in scientific context as a multi-infrastructural space where there are clear gaps across different systems. Second, Vertesi borrows the language of seams from ubiquitous computing, fathered by socio-technical visionary Mark Weiser. Weiser (1993) described this vision as "the nonintrusive availability of computers throughout the physical environment, virtually, if not effectively, invisible to the user" (p. 71). In the last two decades, as technology and its users have evolved, Weiser's vision of seamless integration among devices and digital space has begun to be realized. Indeed, "this language of seamlessness, invisibility, and non-intrusion" in Human-Computer Interaction (HCI) and systems design has instrumented system ecologies (Vertesi 2014: 269). Using the language of seams, seamlessness, and seamfulness, Vertesi (2014) further highlights the integration and tensions among multiple infrastructures. Specifically, she states that they "often collide: their seams are visible in their many edges, endings, and exclusions" (p. 269), in a way that systems are dispersed among infrastructural layers in implicit or explicit compatibility or incompatibility wherein features may overlap. In this sense, each infrastructural layer presents its own point of inclusion and exclusion when it supports or fails to support the work practices of social actors. Vertesi contends that it is less about infrastructural boundaries than it is about patching the seams of heterogeneous systems together to complete tasks, albeit ephemeral, alternate infrastructural alignment and misalignment are in place. Working in this infrastructural complexity, users aggregate tools, forming layers of systems with devices, and system ecologies to engage in seamless or seamful information practices.

In her study, Vertesi witnessed local, dynamic engagement with digital actants and among actors to achieve task completion. Success meant using two distinct systems to complete a task; for example, a scientist aligning systems of voice communications and e-messaging: the first in order to relay critical data and the second in order to expound on mission potentialities. Conflict among device ecologies created "multi-infrastructural torque," (Vertesi 2014) as in a scenario involving a scientist trying to align Mac and PC platforms and media formats in order to present data.

Following this conceptual vocabulary, this chapter examines digital information practices of mobile knowledge workers in order to identify common technological barriers that hamper work practices. These technological barriers in the work of mobile knowledge workers can be understood as seams or disconnect among multiple systems. Seams can also include exclusion points poised by material constraints of an infrastructure. For example, the Wi-Fi signal can be weak in certain locations in a building; meaning that the Wi-Fi infrastructure excludes these places from its coverage due to a lack of rigorous design. Figure 12.1 demonstrates how work practices are hindered by the two forms of seams: 1) infrastructural exclusions (where an infrastructure fails to cover work practices) and 2) infrastructural disconnect (the gap between multiple infrastructures).

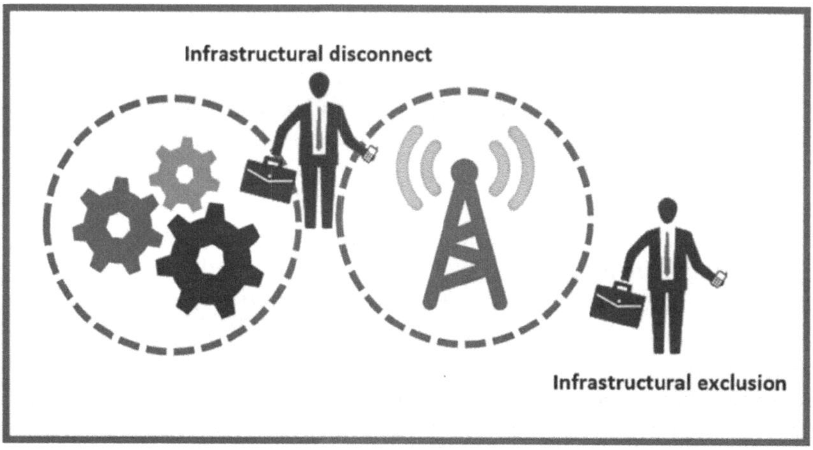

Figure 12.1 Two Forms of Infrastructural Seams.

Such an examination will uncover the scope of seamful and seamless in workers' digital practices that has not yet been addressed in the literature. Applying this framework to mobile knowledge workers uncovers yet unidentified strategies that these workers employ to interact with the technological infrastructures and how they align infrastructures to overcome breakdowns among disparate systems.

4. Methods

This exploratory study aimed to understand mobile knowledge workers' information practices, mobile behaviors, and digital technology use. This chapter shows the relationships between mobile knowledge workers' information practices in lieu of the associated technological infrastructures.

Sixteen mobile knowledge workers comprised the sample for this study. The data was collected through comprehensive, semi-structured interviews. Initially, participants were identified through purposive sampling, while successive participants were found through snowballing techniques. We followed a form of theoretical sampling in which cases are selected to replicate or extend the emergent theory (Eisenhardt 1989), and this strategy generated a pool of study participants that was not random but provided some basis for comparison. Participants were selected for inclusion in the study based on the following criteria: 1) engagement in knowledge work, 2) mobility in their work, and 3) centrality of nomadic work practices. Data gleaned from early sampling identified notable issues and helped us identify better participants to include in our sample. For example, the work practices of the later participants exhibited a higher level of spatial mobility. Participants span a domain of fields, including education, IT, journalism, web and software development, and consulting. Some were self-employed and others were employed by an organization. Relevant details are listed in Table. 12.1.

Before selecting the participants, an interview protocol was created that concentrated on probing the following: the nature of participants' professional domain; their organizational role and responsibilities; their work arrangement, mobility threshold, and environment; the configuration of their professional networks, professional tools, and resources, with emphasis on digital infrastructure, entailing personal or organizational acquirement; mobilization capacity; and their perceptions of resource availability, organizational challenges, and work-life balance affordances. A significant theme that emerged was mobile knowledge workers' interactions with technology.

Interviews ranged from 50 to 90 minutes and were audio recorded. Audio recordings were transcribed verbatim and loaded into the analytical software application NVivo 10. As recommended for qualitative research, data collection, and analysis proceeded concurrently (Miles and Huberman 1994). The analysis involved numerous iterations between data collection

Table 12.1 Participant Demographics (O = Larger Organization; SE = Self-Employed)

Participant	*Gender*	*Role (Arrangement)*	*Work Space(s)*
P1	F	instructor (O)	co-working space
P2	F	web developer (SE)	home office
P3	F	strategy consultant (O)	multiple "offices"
P4	M	knowledge manager (O)	home office
P5	M	web developer (SE)	co-working space
P6	M	columnist (O)	home office
P7	F	business consultant (SE)	multiple "offices," client sites
P8	M	IT consultant (O)	in transit, client sites
P9	M	corporate trainer (SE)	multiple "offices"
P10	M	lawyer (O)	multiple "offices"
P11	M	IT developer/ consultant (O)	multiple "offices," client sites
P12	M	IT consultant (O)	multiple "offices," client sites
P13	M	web developer (O)	home office
P14	M	realtor (O)	in transit
P15	M	IT support (O)	multiple "offices," client sites
P16	F	realtor (O)	in transit

and construction of an emerging understanding about primary technological challenges of mobile knowledge work. Data analysis was inductive, since we were looking for emergent ideas, leads, and issues (Glaser 1978). Data analysis was also guided by the concepts of infrastructural seams. This iterative process of coding, included open coding, initial memoing, focused coding, and integrative memoing.

5. Findings

As indicated, the analysis was aimed at identifying salient seams categorized into common themes: 1) communications and collaboration, which includes knowledge sharing; 2) information and data management, involving data transfer, preservation and security; 3) organizational boundaries, entailing organization restrictions; 4) and spatial constraints. These themes emerged from our inductive analysis and our focus on different barriers (seams) that were rooted in either infrastructural exclusions or infrastructural disconnects (see Figure 12.1). The seams were developed based on key functions of mobile work that the infrastructural seams inhibit (i.e., communication and collaboration, and information and data management) and infrastructural issues that stem from contextual reasons (i.e., organizational and spatial constraints). While there is an overlap between these themes, each experience or statement made by a participant was categorized to the dominant attribute.

5.1 Communication and Collaboration

Participants emphasized the sheer necessity of accessible and adaptive communication platforms. All participants indicated that, as many of their associates were physically distant, online communications were crucial to collaboration. Others who were remote from their organization or collaborated with a network of clients, pinpointed ICTs as a central locus of virtual spaces, where information critical to work completion was transmitted, knowledge disseminated, and a sense of social inclusion imparted. Common technologies included phone, e-mail, IM, text messaging, videoconference, and message boards.

Associated seams are related to communication and knowledge sharing between workers, their co-workers, clients, and partners (contractors). In some ways, they overlap with information and data management as well as spatiality, but issues are more collective and collaborative while seams plaguing data and information management are more personal.

As seen in the interviews, seams involving infrastructural platforms consisted of devices and applications. These were proscribed by user limits, perceived, or actual, such as device limitations in features or capability and application feature deprivation, or a combination of the two. A common complaint among highly mobile participants (i.e., multiple "offices," in transit) was the difficulty of using on-screen keyboards, a mental access barrier (Van Dijk 1999). "I can't write on the phone. I don't use the iPad for that much work either, for the same kind of reason; I don't like typing on glass" (P6). Another participant complained of communication applications that did not work properly on the smartphone, limiting communication capabilities of the primary mobile ICT device. Similarly, a participant needed to carry both her smartphone on top of her laptop, because the latter had such poor audio quality that it made VoIP communications impractical without a headset.

All participants indicated that they use multiple technologies for communication and collaboration, including e-mail, phone, IM services, social media, and text messaging. As many worked remotely from colleagues, they emphasized the importance of open communication channels and the ability to manage multiple documents to complete immediate tasks and extended projects. Most informants emphasized the importance of maintaining open communication channels with devices preferred by their clients and contractors, and even by their colleagues. A common problem is that of heterogeneity of written communication systems and the need for multiplicity of devices to provide access to these mediums.

One participant noted that among close colleagues, using personal e-mail addresses was frequent for work communication. Speaking of his editor, a journalist notes: "I think he knows: [it's] better to send me anything urgent on the [company e-mail]." Another participant underscored the need for access to multiple communication platforms, as his clients regularly communicated on social sites (e.g., LinkedIn, Facebook, and Twitter), even to the relegation of e-mail as a preferred channel,

> Because I think now e-mail is kind of reaching that evolutionary stage where it is almost as inconvenient as paper mail. Because everybody is on Facebook, everybody is on some form of social media. I mean you can look at the statistics. People are checking those types of avenues multiple times a day.
>
> (P8)

We also found that in work situations where multiple e-mail communication systems are used, convenience, habit and absent-mindedness, or social constraints, may trump organizational policies (whether loose or tightly controlled) and practices. These trends underscore the tensions among systems, entailing exclusions in which infrastructural limitations could be visible or invisible to users.

Most participants saw knowledge sharing as a vital but challenging practice. Informants intimated that document sharing involved more than one tool or device. For instance, they stressed the detrimental impact of having to use a diversity of document types and platforms among clients or colleagues. Incompatible formats and system interoperability often caused frustration. Even convenient distribution mediums such as e-mail and cloud storage could cause knowledge-sharing disruptions, particularly in large, often fragmented, groups, or with document sets.

5.2 Information Access and Data Management

Managing data across digital environments was a common theme among participants. Relevant seams involve data loss and dispersion, as seen with work on virtual machines. All participants used a number of ICTs and underscored barriers relative to managing data and information.

Usability of platforms hosted on a virtual private network (VPN) or remote desktop protocol was jeopardized by power and network interruption and slow connection. With shared login practices, a user might be logged off by co-workers (P12). These technological challenges manifested themselves as slowing the pace of work but sometimes losing unsaved work or data when source systems, like virtual machines, went down.

Among participants, we found seams involving integration and distribution of data through systems. One participant mentioned difficulties with accessing online databases via smartphones (P14) and another with difficulties with cloud calendaring service,

> There are issues with having multiple devices and having all the right data in the right place. A lot a people I have to communicate with based on calendaring and meetings. They're on Microsoft Exchange; their

> Exchange system works better for communicating with Android devices than Outlook, or even Outlook in the cloud now. So I would like Outlook in the cloud to be as good as Outlook Exchange.
>
> (P10)

Issues with integration, privacy, and control were common among participants. One participant noted,

> I have problems with calendaring right now, because I'm having trouble with making all the technologies work, using an Android phone and an Android device, Google wants to take over. It wants Google Calendar to be your calendar. Google Calendar doesn't work well for me. It's like I have to change the whole way I think in order to make Google Calendar work for me.
>
> (P3)

As mobile knowledge workers are dependent on multiple devices and generic software, networking data among devices, and applications is vitally important, as compared to their counterparts who are not mobile, they can feel unsupported by their organizations (e.g., no support or inadequate support for devices). Furthermore, whether they are organizational bound or freelance, it may be incumbent on them to resolve these disparities.

5.3 Organizational Boundaries

Participants who were members of a larger organization (as opposed to freelancers) or closely collaborated with one indicated organizational boundaries may impede work practices when they used technologies across intra- and inter-organizational lines. On the other hand, those working with clients or collaborators cross multiple organizational boundaries and coordinate personal and organizational devices. Thus, we concluded that in many cases, organizational rules and policies might not be favorable in helping mobile knowledge workers integrate with others over technological platforms.

Accessing organization servers was problematic when restrictions were imposed on client networks, as this testimonial confirms: "If you're not on a password protected network, you can't get on the [company] network. But we do have a way to access our e-mail outside of the [company] network" (P3). Related to that were policies that restrict network access based on the client's IP address,

> So you have situations where companies have set up constraints that I couldn't connect to their site right here, right now. There are places that have some higher level of security clearances where they need to know what IP your request is coming from.
>
> (P11)

In a similar vein, given the importance of enforcing access to servers, it was not surprising to see software and data restrictions impeding download from and onto organizational devices and cloud platforms. Restrictions may also be by-product of technology limitations imposed by the organization. For example, one participant was tasked with editing video files sent to him by a co-worker, and he had to do that on his personal computer that had the software that his company would not provide. Because of an overly restrictive set of policies, that task proved seamful and intrusive,

> I need to get this movie file between my work and my personal computer, and I was trying to figure out ways to do it. I wish I had Dropbox, because our [company e-mail] inbox, in addition to being our primary mode of communication, is very restricted in terms of size limitation, so we're constantly in an e-mail jail. So I constantly have to move things to my ShareDrive in order to make room on my e-mail.
>
> (P4)

Mobile knowledge workers may also have to change their platform preferences and modus operandi based on organizational policies, often because, as indicated in the previous section, of the lack of IT support.

5.4 Spatial Seams

All participants indicated challenges they faced when physically away from their usual and quasi-regular workspace, performing work tasks in variable environments and conditions. Our findings suggest these spatial seams specifically emanated from barriers involving device use, network connectivity, and resource deprivation. While such seams have some overlap with preceding categories (e.g., spatial constraints determine availability or convenience of ICTs, digital infrastructure access), they reflect barriers that arise from misalignment of technology use and physical space.

Battery life can be an issue, such as on long flights or in cars where AC outlets are not available. The findings indicate preferences for compact devices, issues with battery life and power sources, especially when traveling. One participant noted he had to take two different laptops primarily because of power,

> I've been taking my personal laptop and my work laptop, because typically when I'm traveling for work, I'm constantly having to go to the East Coast, which is a long flight, so just for the battery life sake, I'll bring both.
>
> (P4)

Another participant commented on the frustration of managing work without her forgotten AC adapter, and she and some others complained of the incompatibility among charging devices as well as having to carry so many. Most participants with a high level of mobility also commented on the lack of AC outlets at airports and public places (e.g., coffee shops) at which some might work periodically or meet clients.

Most participants found themselves in situations without Internet or effective network access to organizational resources. As noted earlier, there was a stated need of employing the technologies at hand to get work done but this is not always possible. Network access was proscribed by security limitations by organizational security constraints. For instance, all participants who fly regularly commented on the lack of Wi-Fi connectivity in airports and airplanes, while some public places which provided network access have temporal restrictions: "They have Wi-Fi but they kick you off after 30 minutes sometimes for the day, because they're too busy so they don't want people hanging around there" (P2). As seen through the interviews, mobility suffered because of fragmented access to online resources.

5.5 *Adaptive Strategies*

Participants in this study employed various strategies to overcome the barriers existing within and across technological systems, whether built into the infrastructure or institutionalized by their respective organization. Analysis of these barriers exposes three primary adaptive strategies. Communication, collaboration and organizational barriers are addressed through the adoption of particular technology assemblages: orchestrated infrastructural constellations of technologies that address contextual, temporal and spatial issues (Rossitto, Bogdan, and Severinson-Eklundh 2014). On the other hand, we found that information and data management barriers are primarily remedied by data and service integration: migrating data from one or more sources into a central repository, or vice versa. Finally, spatial barriers are ameliorated by familiarizing oneself with local infrastructure and services and leveraging them as situations arise.

Using Orchestrated Infrastructural Constellations

A theme that emerged in the interviews was the instances of combining platforms for communicatory or collaborative tasks. These could be ad hoc or habitual integrations. These strategies involved using multiple devices or platforms for the same task. Research on media pairings on managers in knowledge work (Leonardi, Neeley, and Gerber 2012; Watson-Manheim and Bélanger 2007) explored how managers choose ICTs respective of the richness of content and that they practice simultaneous media paring, using multiple devices conjunctively in simultaneous and antecedent multi-communicating. Although these practices are ideal for knowledge sharing, resolving conflicts, and persuading colleagues, they can result in attention fragmentation and heightened stress.

We recognize two types of smart portfolio development: concurrent, which entail the use of two or more platforms at a time; and sequential which entails the use of two or more platforms in successive order. This phenomenon of smart portfolio development invariably emerged in technological practices of all self-employed participants as well those with organizational membership and often resulted in pairing multiple tools, because of the richness of the task or the mere availability of the tool. In relation to communication challenges, we found an abundance of collaboration bounced from e-mail to instant message, phone and message boards, wherever situationally practical or promoted social inclusion. Communication media aligned themselves within the situational framework and progressive designs and impromptu needs of communicants.

While it was common for participants to pair devices and applications, what was of continuing interest was how participants objectively stitched together seamless infrastructures. One explained her concurrent-paring strategy for note taking during calls as follows,

> Usually I open up a new mail and I just type there, because it's automatically saved as a draft and everything, and then at the end of the call, if there's anything important I'll send it to myself or if not I'll just delete it.
> (P5)

Following the same strategy of aligning two disparate systems, a support technician created a seamless experience between himself and the client as effective, in his mind, as in-person support,

> So I'm on the phone with the client, and then what I'll do is to give an explanation of the issue. I can dial right into their computer, whether it's from my cell phone or any device. And then I get to see exactly what they're doing, and they can create—recreate that scenario.
> (P8)

We found that in communication practices sequential media pairings aided managers in resolving conflicts and sharing knowledge. This is manifest with several participants' revelations. Participants talked about dialogue bouncing from e-mail, to instant message or phone, given the circumstances and practically responding to the situation, contextually, socially, and spatially.

Integrating Information Sources

These are strategies that participants enacted by nesting information and data or by embracing platforms that provided richer and more convenient access to information sources and foster communication, hence enabling greater resource usability and social inclusiveness.

More specifically, participants exhibited different strategies through centralizing, connecting satellite information sources, and leveraging integrative technologies. For instance, centralizing entails drawing data from diverse systems and sources into a customized manifestation so as to overcome data dispersion barriers. One participant, works with multiple clients and through multiple media, explained her method to aggregate and organize all her e-mail accounts. Her strategy is to centralize all her e-mails by importing them into one Outlook account, creating folders for each e-mail (P3). Another participant manages all incoming documents from clients into one cloud storage system called Box and states that preserving them in one manageable locations aids in information retrieval (P10). This strategy involves transferring data among systems that may not be interoperable. This could be done both manually and automatically. Several participants noted that they migrated data among systems to make it accessible across devices and software platforms. While many used multiple cloud storage services to preserve data for additional security and access points, some employed this strategy to increase availability. For example, data migration across applications and platforms was an important solution adopted by the following organization,

> What we'll do is if we have a deadline in the case, it gets entered into Clio, which then gets pushed to the Google Calendar, then gets pushed to mobile devices or computers, so if people want it locally, it is all nicely integrated. People have multiple calendars; we've got my law firm calendar, my Clio calendar, a family calendar, and those are Google oriented. They all tie together nicely, so I can access those from any device. (P10)

As another strategy, uses of technologies that integrate features, data, and services were seen to aid knowledge mobile workers, enabling information diffusion across systems and minimizing technological conflicts. These are technologies that users employed for efficiency, mobility, and ease of use, and, particularly, for seamless integration of features. Not surprisingly, Google Drive was a popular medium for collaborating in composing documents, sharing documents, and files. Project management web-software package BaseCamp allowed one consultant firm to integrate communication, file storage and task management into one environment (P11, P12). Specifically, it fostered social capital and inclusion by skirting traditional, overused communication models (e.g., e-mail) and via extending usability, that is the creation of seamless channels of discourse that allow users to maneuver among multiple infrastructures to share data and screens as well as knowledge transmission and preservation. However, usability and social inclusion breakdowns occurred, as some clients refused participation or due to organizational budget constraints that prohibited expanding services.

Mobilizing

Adaptive strategies involve planning ahead due to potential loss of infrastructure and given the changing spatial work arrangements. As mobile knowledge workers predict situations in which resources will be unavailable, they circumnavigate these losses in numerous ways.

In mobilizing, participants may, for example, skirt network deprived situations by e-mailing documents before outage occurrences rooted in security restrictions or situational deprivation. Many circumvented device or network privations by e-mailing themselves documents ahead of time (e.g., P4, P16). One participant wrote e-mails and blogs in a word processor while traveling by car or airplane, and then copied them to respective platforms once online (P11).

Mobilization strategies in relation to the use of different devices varied. While one participant toted only a tablet on trips, another took his work and personal laptop, since the work one was restricted by security measures and lacked software he used regularly. Using nomad battery chargers extends battery usage by two fold, indispensable to fieldwork (P13).

Generating local awareness was another key dimension of mobilizing strategies. An IT support professional employed Professional Services Automation software that enables communication and GPS team tracking, promoting social inclusion through verbal and spatial awareness (P8). While on the go, some used voice-based personal assistants like Siri to read them e-mail or to leave voicemail messages as reminders and notifications to themselves (P7, P8). A highly mobile IT support professional working on a campus where Wi-Fi was incohesive between and in buildings used 4G rather than Wi-Fi "to conserve battery life and maintain connectivity" (P15), citing that the phone's Wi-Fi adaptor drained the battery when powered on and persistently searches for connections.

Mobilizing in this sense also entails familiarizing oneself with environmental conditions, adapting to spatial seams that hinder resource access in physical and digital space. As vital as network and Internet connectivity was seen to be for daily work, specific strategies in this regard included accessing information about airport and coffee shop Wi-Fi, VPN hotspots, cellular networks (P9, P11) or clients' infrastructures (P8, P11). To overcome spatial barriers where Wi-Fi was unavailable, participants created hot spots with their phones to VPN on laptops (e.g., when traveling by car) (P9, P11).

6. Conclusion

In this exploratory examination of work practices of a small sample of mobile knowledge workers, digital barriers among participants' infrastructural systems were revealed. By importing the language of seams (Vertesi 2014) into this analytical framework, we shed light on multi-infrastructural torques that lead to recurrent breakdowns. These infrastructural barriers adversely impact how mobile knowledge workers mobilize enterprise resources and collaborate and share knowledge with their colleagues. Breakdowns, therefore, may hold mobile workers back to some degree, hence preventing equal access to technological resources and social inclusiveness.

This investigation further focused on adaptive solutions adopted by mobile knowledge workers. With these adaptive strategies, they improve their mobile situation, enhance their capacity to engage with multiple actors and technologies across time and space, and guard from information overload by discriminating among the number of platforms to use. Participants were seen overlaying or stitching together infrastructural seams, pairing technologies to perform various tasks, ad hoc and habitually; integrating technologies and data into constellations of platforms and services (Rossitto, Bogdan, and Severinson-Eklundh 2014); and taking advantage of resources at their fingertips. Not all participants had free rein to choose their technology, due to organizational, physical, and situational constraints, but those that did constructed what Carroll (2008) describes as portfolios of technology: these portfolios engendered social inclusion, usability of resources across systems and efficient work mobile work practices.

Rather than sole use of organizational tools, as described in the classic "push" models of Hagel and Brown (2008) in which organizations push tools on to workers in a top-down, tightly controlled practice or conventional suites of technologies, our participants actively sought tools that enabled work effectiveness. As independent agents unconstrained by traditional work roles, most participants drew on past experience, taken with situational relevance of the present, and chose approaches that effectively facilitate seamless integration of multiple technological infrastructure. However, works with multiple clients and through multiple mediums, these practices depend on the cultivation of informational and strategic skills (Van Dijk and Hacker 2003) and are fundamental to the skillset of mobile knowledge workers to contend with situational work circumstances and the availability of digital resources. As they do so, mobile knowledge workers continually break through the lines of the digital divide by adapting to digital barriers through incorporating seamful practices with seamless infrastructures.

References

Barley, S. R., and G. Kunda. "Contracting: A New Form of Professional Practice." *The Academy of Management Perspectives* 20, no. 1 (2006): 45–66.

Barzilai-Nahon, K. "Gaps and Bits: Conceptualizing Measurements for Digital Divide/s." *The Information Society* 22, no. 5 (2006): 269–78.

Carroll, J. "Theorizing the IT Artifact for Mobility: A Portfolio, Not a Singularity." In *ICIS 2008 Proceedings*, 2008, 64.

Castells, M. *The Rise of the Network Society*, Vol. 1. Oxford: Blackwell, 2006.

Chen, L., and R. Nath. "A Socio-Technical Perspective of Mobile Work." *Information, Knowledge, Systems Management* 7, no. 1 (2008): 41–60.

Ciolfi, L., and A. F. P. De Carvalho. "Work Practices, Nomadicity and the Mediational Role of Technology." *Computer Supported Cooperative Work (CSCW)* 23, no. 2 (2014): 119–36.

Costas, J. "Problematizing Mobility: A Metaphor of Stickiness, Non-Places and the Kinetic Elite." *Organization Studies* 34, no. 10 (2013): 1467–85.

Cousins, K. C., and D. Robey. "Human Agency in a Wireless World: Patterns of Technology Use in Nomadic Computing Environments." *Information and Organization* 15, no. 2 (2005): 151–80.

Creplet, F., O. Dupouet, F. Kern, B. Mehmanpazir, and F. Munier. "Consultants and Experts in Management Consulting Firms." *Research Policy* 30, no. 9 (2001): 1517–35.

Dahlbom, B., and E. Ljungberg. "Mobile Informatics." *Scandinavian Journal of Information Systems* 10 (1998): 227–34.

Davenport, T. H., S. L. Jarvenpaa, and M. C. Beers. "Improving Knowledge Work Processes." *MIT Sloan Management Review* 37, no. 4 (1996): 53.

Davis, G. "Anytime/Anyplace Computing and the Future of Knowledge Work." *Communications of the ACM* 45, no. 12 (2002): 67–73.

De Carvalho, A. F. P., L. Ciolfi, and B. Gray. "The Making of Nomadic Work: Understanding the Mediational Role of ICTs." In *Handbook of Research on Mobility and Computing: Evolving Technologies and Ubiquitous Impacts*, edited by M. Cruz-Cunha and F. Moreira, 381–96. Hershey, PA: IGI Global, 2011.

Department of Commerce. "U. S. Falling Through the Net: A Survey of the 'Have-Nots' in Rural and Urban America." 1995. Accessed June 15, 2016. www.ntia.doc.gov/ntiahome/fallingthru.html

Eisenhardt, K. M. "Building Theories from Case Study Research." *The Academy of Management Review* 14, no. 4 (1989): 532–50.

Erickson, I., and M. H. Jarrahi. "Infrastructuring and the Challenge of Dynamic Seams in Mobile Knowledge Work." Paper presented at the *Proceedings of the ACM Conference on Computer Supported Cooperative Work & Social Computing*, San Francisco, CA, 2016.

Erickson, I., M. H. Jarrahi, L. Thomson, and S. Sawyer. "More than Nomads: Mobility, Knowledge Work, and Infrastructure." Paper presented at the the *European Group for Organizational Studies Colloquium*, Rotterdam, The Netherlands, 2014.

Florida, R. *The Rise of the Creative Class, and How It Is Transforming Work, Leisure, Community and Everyday Life*. New York: Basic Books, 2002.

Glaser, B. *Theoretical Sensitivity: Advances in the Methodology of Grounded Theory*. Mill Valley, CA: Sociology Press, 1978.

Hagel, J., and J. Brown. "From Push to Pull—Emerging Models for Mobilizing Resources." *Journal of Service Science—Third Quarter* 1, no. 1 (2008): 93–110.

Harmer, B. M., and D. J. Pauleen. "Attitude, Aptitude, Ability and Autonomy: The Emergence of 'Offroaders,' a Special Class of Nomadic Worker." *Behaviour & Information Technology* 31, no. 5 (2012): 439–51.

Hoffman, D. L., T. P. Novak, & A. E. Schlosser. "The Evolution of the Digital Divide: Examining the Relationship of Race to Internet Access and Usage Over Time." In *The Digital Divide: Facing a Crisis or Creating a Myth*, edited by B. M. Compaine, 47–97. Cambridge, MA: MIT Press, 2011.

Jarrahi, M. H., and S. Sawyer. "Theorizing on the Take-Up of Social Technologies, Organizational Policies and Norms, and Consultants' Knowledge-Sharing

Practices." *Journal of the Association for Information Science and Technology* 66, no. 1 (2015): 162–79.
Jung, J.-Y., J. L. Qiu, & Y.-C. Kim. "Internet Connectedness and Inequality Beyond the 'Divide.'" *Communication Research* 28, no. 4 (2001): 507–35.
Leonardi, P. M., T. B. Neeley, and E. M. Gerber. "How Managers Use Multiple Media: Discrepant Events, Power, and Timing in Redundant Communication." *Organization Science* 23, no. 1 (2012): 98–117.
Liegl, M. "Nomadicity and the Care of Place—on the Aesthetic and Affective Organization of Space in Freelance Creative Work." *Computer Supported Cooperative Work (CSCW)* 23, no. 2 (2014): 163–83.
Lilischkis, S. "More Yo-Yos, Pendulums and Nomads: Trends of Mobile and Multi-Location Work in the Information Society." *STAR Issue Report* 36 (2003).
Mazmanian, M., W. J. Orlikowski, and J. Yates. "The Autonomy Paradox: The Implications of Mobile Email Devices for Knowledge Professionals." *Organization Science* 24, no. 5 (2013): 1337–57.
Miles, M. B., and A. M. Huberman. *Qualitative Data Analysis*, 2nd ed. Thousand Oaks, CA: Sage Publications, 1994.
Mossberger, K., C. J. Tolbert, and M. Stansbury. *Virtual Inequality: Beyond the Digital Divide*. Washington, DC: Georgetown University Press, 2003.
Orlikowski, W. J. "Knowing in Practice: Enacting a Collective Capability in Distributed Organizing." *Organization Science* 13, no. 3 (2002): 249–73.
PearnKandola. "A Study: Understanding and Managing the Mobile Workforce." Cisco Systems, 2007. Accessed July 11, 2015. http://newsroom.cisco.com/dlls/2007/eKits/MobileWorkforce_071807.pdf
Perry, M. "Enabling Nomadic Work: Developing the Concept of 'Mobilisation Work.'" Paper presented at the *ECSCW 2007 Workshop: Beyond Mobility: Studying Nomadic Work Limerick*, Ireland, 2007.
Perry, M., K. O'Hara, A. Sellen, B. Brown, and R. Harper. "Dealing with Mobility: Understanding Access Anytime, Anywhere." *ACM Transactions on Computer-Human Interaction (TOCHI)* 8, no. 4 (2001): 323–47.
Rainie, H., and B. Wellman. *Networked: The New Social Operating System*. Cambridge, MA: MIT Press, 2012.
Rodino-Colocino, M. "Laboring Under the Digital Divide." *New Media & Society* 8, no. 3 (2006): 487–511.
Rossitto, C., C. Bogdan, and Severinson-Eklundh, K. "Understanding Constellations of Technologies in Use in a Collaborative Nomadic Setting." *Computer Supported Cooperative Work (CSCW)* 23, no. 2 (2014): 137–61.
Schultze, U., and R. J. Boland. "Knowledge Management Technology and the Reproduction of Knowledge Work Practices." *The Journal of Strategic Information Systems* 9, no. 2 (2000): 193–212.
Servon, L. J. *Bridging the Digital Divide: Technology, Community and Public Policy*. Malden: Blackwell, 2008.
Sørensen, C. *Enterprise Mobility: Tiny Technology with Global Impact on Work (Technology, Work, and Globalization)*. London: Palgrave Macmillan, 2011.
Su, N., and G. Mark. "Designing for Nomadic Work." Paper presented at the *Proceeding of the 7th ACM Conference on Designing Interactive Systems*, Cape Town, South Africa, 2008.
Thomson, L., and M. H. Jarrahi. "Information Practices in the Broader 'Deportment' of Mobile Knowledge Work." Paper presented at the *Proceedings of ASIST'15, Saint Louis*, MI, 2015.

Van Dijk, J. *The Network Society, Social Aspects of New Media*. Thousand Oaks, CA: Sage, 1999.

Van Dijk, J., and K. Hacker. "The Digital Divide as a Complex and Dynamic Phenomenon." *The Information Society* 19, no. 4 (2003): 315–26.

Vertesi, J. "Seamful Spaces: Heterogeneous Infrastructures in Interaction." *Science, Technology & Human Values* 39, no. 2 (2014): 264–84.

Warschauer, M. "Reconceptualizing the Digital Divide." *First Monday* 7, no. 7 (2002).

Watson-Manheim, M. B., and F. Bélanger. "Communication Media Repertoires: Dealing with the Multiplicity of Media Choices." *Management Information Systems Quarterly* 31, no. 2 (2007): 267–93.

Webster, F. *Theories of the Information Society*, 3rd ed. London: Routledge, 2006.

Weiser, M. "Some Computer Science Issues in Ubiquitous Computing." *Communications of the ACM* 36, no. 7 (1993): 75–84.

Wolcott, P., M. Kamal, and S. Qureshi. "Meeting the Challenges of ICT Adoption by Micro-Enterprises." *Journal of Enterprise Information Management* 21, no. 6 (2008): 616–32.

13 Overcoming Obstacles to Activism With ICTs

An Analysis of MoveOn.Org and the Florida Tea Party Movement

Deana A. Rohlinger and Shawn Gaulden

1. Introduction

Many social scientists are concerned with social inclusion, or the ability of individuals and groups to fully participate in society. Academics concerned with social inclusion consider how resource and socio-economic inequality affect the ability of individuals and groups to control their destiny (Stewart 2000; Warschauer 2003; Selwyn 2004). While the widespread availability of ICTs have been heralded by pundits and politicians as a way to improve the social inclusion of citizens, scholars point out that unequal access to ICTs—which can result from a lack of technical knowledge or skill (Mäkinen 2006), geographical location (Warf 2001), and socio-economic factors such as income and education (Norris 2001)—create a "digital divide" that makes it easier for some individuals and groups to participate more fully in society than others. Concerns with the effects of the digital divide on social inclusion are relevant to conventional forms of democratic activism, such as participation in civic groups and party politics. The failure of citizens to use (or lack of knowledge about) established mechanisms of political influence ensures that their ideas and interests are excluded from political consideration (Norris 2001).

Despite the recognition that the digital divide influences political participation, scholars primarily focus on how bridging the information gap is critical to social inclusion (McCaughey and Ayers 2003; van de Donk et al. 2004; Loader 2007). Essentially, access to information online lowers the costs associated with political participation, which, in turn, pulls people into activism (Bimber 1998, 2000; Norris 2000). Indeed, there is reason for optimism. The calls for democratic governance during the "Arab Spring" and the recent demand for recognition that "Black Lives Matter" suggest that individuals can effectively leverage ICTs to make their voices heard in the political arena—a clear indication of social inclusion. That said, the focus on "information costs" alone obscures other basic obstacles that individuals face when it comes to participation in activism—obstacles that may (or may not) be mitigated by ICTs. For example, in his review of the literature on the digital divide, Jan van Dijk (2006) notes that motivational access, or the desire of individuals to access and/or use ICTs, is woefully understudied. While the existing literature examines non-Internet users (van Dijk 2006), motivational obstacles clearly apply to individuals who have access to ICTs but do not engage in democratic activism. Identifying and analyzing potential motivational obstacles is critical to understanding the relationship between activism and social inclusion. The social movement literature is an excellent starting point in this regard. Scholars identify at least two relevant motivational obstacles that individuals face: lack of biographical availability, or the presence of personal, familial, or employment constraints that make participation in activism difficult, and political inexperience with activism (McAdam 1986; Beyerlein and Hipp 2006; Verba, Schlozman, and Brady 1995; Skocpol 2003).

Of course, individuals are only one side of the activism-social inclusion equation. Activist organizations, which target populations to mobilize to action, rarely target broad swaths of the citizenry. Those groups that do try to appeal to mainstream typically have "paper constituents," who write checks but do not attend meetings or otherwise get involved in organizational (or other) politics (McCarthy and Zald 1977). This trend may be shifting in the digital age. Organizations, which vary in form and function, use ICTs to mobilize some segment of the citizenry to action (Postmes and Brunsting 2002; Fisher et al. 2005; Earl and Kimport 2011; Bakardjieva 2012) and get them involved in the political process (Gigler and Bailur 2014; Gibson et al. 2004). That said, organizations that want to maximize citizen participation face at least two obstacles in their efforts: cultivating a collective identity that has broad appeal and creating mechanisms that make supporters feel efficacious, or as though their participation in the organization matters (Lichterman 1996; Polletta and Jasper 2001; Rohlinger, Bunnage, and Klein 2014).

Drawing on 52 interviews with supporters of MoveOn.org (MoveOn) and the Florida Tea Party Movement (FTPM) as well participant observation data of both groups, we discuss these obstacles to social inclusion and explore how SMOs can effectively use ICTs to overcome these barriers to democratic activism. Specifically, we ask the question: *How do activist organizations use ICTs to overcome the motivation and organization obstacles associated with participation?* We find that groups can use ICTs to help individuals overcome motivational obstacles by connecting information to a tangible political activity and creating opportunities for individuals to learn new political skills. Additionally, we find that groups can overcome organizational obstacles by cultivating identities that focus on individuals' roles as citizens (rather than political affiliation) and making supporters feel efficacious by connecting them to a larger political community, as well as giving them "voice" in the organization. While we do note differences in each organization's use of ICTs, this chapter highlights how ICTs can be used to overcome motivational and organizational obstacles in ways that benefit social inclusion as it relates to political participation. We revisit the potential implications of organizational differences in the conclusion, where we also discuss future research on social inclusion and activism.

2. Literature Review

Social scientists acknowledge that ICTs can aid social inclusion via political participation (Katz and Rice 2002). Yet, scholars largely focus on whether (and how) ICTs help individuals overcome the "information costs" associated with activism. Here, we outline two additional obstacles related to democratic activism—motivational and organizational obstacles. We discuss each in turn.

There are at least two, related motivational obstacles that can prevent individuals from getting involved with a movement and are barriers to social inclusion more generally. The first significant barrier to participation is the lack of biographical availability, which refers to the "absence of personal constraints" (McAdam 1986). When deciding whether to get politically engaged, individuals assess the potential costs and benefits of participation given the realities of their daily lives (Klandermans 1984; Klandermans and Oegema 1987). People with limited biographical availability, or those who juggle full-time employment, health problems, or family responsibilities, often regard the costs of participation higher than those who do not have similar demands on their time. These individuals generally are less likely to get involved with a cause or campaign (McAdam 1986; Snow et al. 1986; Wiltfang and McAdam 1991).

Political inexperience is the second obstacle to activism. While there is no shortage of interest groups mobilizing around (upper) middle-class interests in the United States (Meyer and Tarrow 1998), these organizations often ask supporters for money instead of encouraging them to attend local meetings or rallies around town (McCarthy and Zald 1977). Consequently, supporters often learn very little about the best ways to get involved with activism. The organizational focus on finances is compounded by decreased participation in civic and religious organizations in the United States. Individuals typically learn political skills through involvement in civic and religious organizations (Putnam 1995; Schlozman, Verba, and Brady 1999; Skocpol 2003), which means that those with experience with these organizations are more likely to engage in activism (McAdam and Paulsen 1993; Fisher and McInerney 2012). Politically inexperienced individuals, of course, may be moved to get involved in a cause. The point here is that political inexperience generally is a barrier to involvement.

While overcoming the obstacles to initial participation is no small feat, keeping individuals engaged in activism for any period of time also is difficult. There are at least two related obstacles to keeping individuals involved beyond their initial interest in a cause. First, there is the collective identity problem. Collective identity, or a shared sense of "we-ness," connects individuals to a community and a cause larger than themselves, providing a motivation for mobilization and a rationale for continued engagement over time (Polletta and Jasper 2001; Friedman and McAdam 1992; Taylor 1989). Collective identities vary in their inclusivity (Gamson 1997; Bernstein 1997). Some activist groups are very exclusive, while others define themselves in ways that are designed to appeal to a broad swath of the citizenry (Downey and Rohlinger 2008). The benefit of the latter is that groups with broad appeal can mobilize consensus and move people to action more easily (Klandermans 1988) and in ways that make the politicians, who typically are concerned with their re-election prospects, more responsive to the citizenry.

Second, movement groups must overcome the efficacy obstacle. Individuals who feel as though their participation matters are more likely to stay involved in the group, while those who do not are likely to leave (Klandermans, van de Toorn, and van Stekelenburg 2008). A key barrier to efficacy is that there are different criteria by which individuals may gauge the effectiveness of an organization. If the desire to achieve a concrete, political goal motivates participation, individuals will expect a tangible result from their investment in the organization and will consider the success and failure rate of a group before deciding whether or not to stay involved (Klandermans 1997). However, others will decide to stay (or leave) a group based on their evaluations of the organization's ability to create a sense of belongingness or political community (Kanter 1968; Lichterman 1996). While organizations would be hard pressed to keep 100% of their supporters, they must find ways to appease the majority of their members to ensure their representativeness and survival.

This chapter explores how two groups, MoveOn and FTPM, use ICTs to address these obstacles.

3. Data and Methods

How do activist organizations use ICTs to overcome the motivation and organization obstacles associated with participation?

To answer this question we analyze two groups (MoveOn and the FTPM) that share the goal of social inclusion by making the political system more responsive to ordinary citizens, but vary in terms of their targets, organizational form, and ideological orientation. While these differences have consequences for individual engagement in activism over time (Rohlinger and Bunnage 2015), the purpose of this chapter is to explore how ICTs can be used to overcome motivational and organizational obstacles that undermine social inclusion even in groups that are very different. We briefly discuss each group and provide an overview of the data and methods.

MoveOn is one of the "largest and most forceful voices in digital era politics" (Fouhy 2004). The organization was founded in 1998 by Wes Boyd and Joan Blades, two Silicon Valley entrepreneurs, who sent an e-mail petition to about 100 friends calling on Congress to censure President Clinton for his indiscretion with Monica Lewinsky and "move on" to more pressing political issues. Their e-mail petition generated more than 400,000 replies, and the couple formed MoveOn.org: a political action committee designed to affect elections. MoveOn mobilized progressives and moderate independents around a range of issues of concern including global warming, the war in Iraq, health-care reform, and voting rights. The organization primarily employs a top-down approach to do so. Although the day-to-day operations of MoveOn are managed entirely online, the organization is hierarchically structured with a handful of leaders and issue experts disseminating information and opportunities for involvement to its seven million plus supporters. As a result, MoveOn activities and events primarily originate at the national level (other activities require national approval) and, using ICTs, group leaders solicit local activists to "host" events and rallies that they manage from afar.

In contrast, the FTPM is decentralized, which has resulted in the formation of a hybrid organization that is situated online and in the "real" world. The FTPM, in part, was a response to Rick Santelli's rant against President Obama's mortgage rescue plan. In Tallahassee, the first FTPM event took place in March 2009. Anthony, a 32-year-old conservative activist, participated in a Tea Party organized by his friend (and FreedomWorks activist) Brendan Steinhauser outside of the White House. Anthony spearheaded a similar event in Florida's capital and used a Facebook page as an organizing tool. His event was a success with nearly 300 in attendance. He capitalized on the "event buzz" and, using Facebook, grew the number of supporters for the movement and organized another Tea Party the following month on tax day, April 15, 2009. Although this event was also well attended and featured several state legislators, Anthony, who also works full time, found he could not maintain the movement alone. He turned to other local activists for assistance. The result was the creation of three additional groups that support the FTPM banner, but adopt different orientations to politics: Citizens Holding Government Accountable (a fiscally conservative organization), Working for the American Way (a Judeo-Christian organization), and Christians for Responsible Government (a Judeo-Christian organization that works to increase the overall effectiveness of the FTPM). The Facebook page serves as the "communication hub" for the movement, while the local groups appeal to specific constituencies and bridge these constituents to the larger movement.

This chapter draws on three methods. First, we monitored the organizational websites and public forums for the groups daily for three years. Second, we attended dozens of meetings, rallies, and events hosted by MoveOn and local Tea Party Movement groups.[1] Finally, we conducted semi-structured interviews with supporters of MoveOn, four leaders of the FTPM, and supporters of the FTPM in order to assess how activists use ICT to mobilize people and keep them engaged, how individuals understand the role of ICTs in their participation, and why individuals stay involved (or disengage) over time. We used a variety of methods to locate respondents, including e-mail, listservs, online surveys, giving presentations at meetings, handing out flyers at events, and posting flyers in local coffee shops, on TPM Facebook sites, on campus, and in the local progressive and conservative centers. This strategy yielded a total of 19 MoveOn supporters, who were interviewed between October 2006 and April 2007 and again between December 2008 and June 2009, and 33 FTPM leaders/supporters, who were interviewed between August 2010 and March 2011 and again between May 2012 and January 2013. The interviews ranged in length from 25 minutes to three hours. Data was analyzed thematically. The lead author used ATLAS.ti to analyze the data thematically, using theoretically relevant obstacles to democratic activism (e.g., references to biographical availability and efficacy) to assess whether (and how) the organizations used ICTs to combat these barriers. All respondents are identified with pseudonyms.

Table 13.1 provides an overview of the demographics of the respondents. This table only includes the demographics of those individuals whom we formally interviewed, rather than people we spoke to and informally interviewed at events and rallies. Overall, there are not remarkable differences between the supporters of MoveOn and the FTPM. Supporters are diverse in terms of their age, gender, relationship, parental, and employment status but relatively homogenous in terms of their education, race and ethnicity. The racial and ethnic demographics are not completely representative of the Tallahassee area in which 60.42% of the population is White, 34.24% is African-American, 4.19% is Latino, and 2.4% is Asian.

4. Results

4.1 Motivational Obstacles to Activism

Activist groups can use ICTs to help individuals overcome the motivational obstacles associated with activism that can inhibit inclusion in political processes. Organizations can use social media, websites, and e-mail to help individuals overcome their lack of biographical availability and political inexperience by connecting information to a tangible political activity. The most important way groups use ICTs to combat problems associated with biographical availability and political experience is by explicitly connecting information with political action. Signing petitions, donating money, writing letters to politicians, and calling legislators may all be relatively easy to do, but require time—a resource that is not distributed equally across the population. By offering supporters a range of activities on- and offline in which to engage organizations enable individuals to quickly get informed about relevant issues and get involved; a critical first step for social inclusion.

FTPM and MoveOn supporters agreed that organizational leaders leveraged ICTs in ways that made activism convenient enough for them to get involved, despite their overloaded schedules. FTPM relied primarily on its Facebook page to inform supporters about issues and events, and get them involved on- and-offline. For example, Janet, a 49-year-old business

Table 13.1 Demographic Characteristics of the Interviewees

	MoveOn (N = 19)	*FTPM* (N = 33)
Sex		
Male	47%	69%
Female	53%	31%
Race		
White	89%	81%
Asian	11%	0%
Middle-Eastern	0%	3%
Latino	0%	10%
Multi-Racial	0%	6%
Age		
18–35	37%	31%
36–50	21%	31%
51+	42%	38%
Education		
High school	11%	8%
Some college	0%	16%
Bachelor's degree	42%	56%
Graduate degree	47%	20%
Employment Status		
Employed	74%	61%
Unemployed	5%	6%
Retired	0%	27%
Student	21%	6%
Relationship Status		
Single	47%	25%
Partnered	0%	6%
Married	32%	47%
Divorced	21%	19%
Widowed	0%	3%
Parental Status		
No children	47%	38%
One child	16%	22%
Two or more children	37%	40%

owner and mother of four, noted that the FTPM's use of Facebook made it easy for her to get involved because it connected her to political information and action. "I've never physically met with any of the Tea Party Members but I can still be a part of the movement. . . . I can stay informed and connected." While Janet noted that she could not attend any events, she attributed FTPM's Facebook page for helping her to decide which politicians she should write about her concerns and to whom to contribute money. Deborah, a 55-year-old business owner, agreed adding that FTPM's use of ICTs gave her a way to fight for her country,

> [Tea party websites, listservs and Facebook forums] provide a platform for unity and more organized communication [and action]. Being informed alone, being aware of things has increased [my] response. . . . If I don't know, I can't respond. . . . It gives me a way to fight for my country,.

Samantha, a 34-year-old market researcher, made a similar point about MoveOn, explaining that its e-mails, which provided information and opportunities for action, made it easy for her to stay involved after the birth of her daughter,

> If people just knew something they would do something . . . if you make it easy for them, and if you give them an action, they'll do it. And I think MoveOn has been a really good vehicle for me to do that because post child, once you have jobs . . . sometimes it's hard . . . to figure out what to do. . . . Most of what I did, pre-election, going right up to the election and post-election . . . I probably wouldn't have done [without MoveOn].

Many other respondents echoed the same sentiment. Kendra, a 29-year-old outreach coordinator and mother, noted that she appreciated how she could "click through" MoveOn e-mails and quickly engage in an action,

> They make it easy. Just a click of a link and the petition is already made. You can add your extra comments. You can e-mail it on. They spell it all out. It helps to make my individual involvement in the political process a little bit easier. If I was going to express support or dissent about an issue by myself, I would have to research the e-mail address or the contact information for my appropriate legislator or representative. I'd have to compose the letter. I'd have to manually forward it on to friends and family. They have all of the steps already taken care of. It's really clear, clickable links.

Connecting information to a tangible political activity was critical to involvement in activism because the majority of respondents did not have any previous political experience. The organizations' use of ICTs offered them "easy" ways to get involved in the causes they cared about. Kenneth, a 69-year-old ROTC instructor with no experience in activism, attributed his initial involvement in the FTPM to the ease with which he was able to find it online. He explained,

> Well, I began to see these Tea Party e-mails back and forth and I would answer. It would be like a poll and I'd answer the poll and then of course the other issues that I kind of dabbled in. . . . I made some small donations to some of the conservative political candidates that were backed by the Tea Party.

Kenneth noticed meeting announcements on the Facebook page and, before long, began attending meetings and helping local groups plan events. Similarly, Beth, a 26-year-old business owner, went online to learn more about the FTPM after she heard Glenn Beck talk about the movement on the radio,

> I guess what he was saying really spoke to me. It kind of woke me up. I haven't been paying attention to this. I have been really removed from it—from politics—and, you know, in a sense that I thought it didn't impact me. . . . The more I learned about it the more I became interested and the more I became passionate.

Beth explained that she first became very involved online through the Facebook page, but saw an announcement and decided to attend a meeting in the "real world" as well. Meeting other Tea Partiers face-to-face, she noted, made a difference. She began to attend meetings regularly and even helped set up FTPM events.

MoveOn supporters also noted that the organization made it easy for supporters to do something new. Amanda, a 52-year-old social worker, explained that MoveOn made it easy for her to learn the ropes of activism because they emailed her directions and all the necessary materials,

> It's all done for you pretty much . . . the paperwork, the reports, the printouts, flyers . . . They e-mail it to you and you print it out on your own printer and you're set to go . . . They make it easy for the people who don't know what they're doing and have never done this before.

Several supporters also reported receiving e-mails from MoveOn asking them to take on more demanding roles. Amanda shared her story about how MoveOn asked her to be a precinct captain after she had hosted a few events at her home,

> I said "no" when they asked me. I mean, I had absolutely no idea what I was doing. I had never done any kind of door-to-door thing before. But they kept asking. They said they needed another one, so I finally did it.

Amanda ultimately was glad that MoveOn encouraged her to get more involved. "It was interesting. I didn't know what the response was going to be. It was just so new. It was foreign to me and so it was a little scary. But, actually it [the experience] was mostly positive."

In sum, organizations can use ICTs to help individuals overcome the motivational obstacles associated with activism that can inhibit inclusion in political processes. MoveOn and FTPM effectively do so by explicitly connecting information with political action. It is important to note, however, that because they rely on different ICTs (e-mail versus social media) and have different levels of resources available to them, how they move people to action varies. MoveOn, which has a paid staff, relies on clickable links in e-mails to get individuals engaged, monitors their participation, and encourages engaged supporters to get more involved over time. In contrast, FTPM, which is run by volunteers, relies heavily on free social media platforms and does not have the people power to consistently or accurately identify whom to encourage to get involved beyond the Facebook page. FTPM relies on individuals moving themselves from online to offline political engagement. These differences may have long-term effects on inclusion in political processes; a point to which we return in the conclusion.

4.2 Organizational Obstacles to Activism

Organizations can use ICTs to overcome the obstacles associated with collective identity and efficacy, which can undermine inclusion in political processes. First, activist groups can use ICTs to cultivate collective identities that have broad appeal and encourage participation (Bennett and Segerberg 2012). MoveOn and FTPM capitalized on skepticism regarding the ability of the political parties to represent the interests of those "outside the beltway" and called on "concerned citizens" to "hold politicians' feet to the fire" by making them accountable for their legislative decisions. Specifically, both groups called on "average" citizens to use the electoral process and protest to force parties to be more responsive to their demands. These identities increase inclusivity insofar as they unite individuals around their roles as democratic citizens and downplay political differences.

Anthony, the FTPM founder, saw the FTPM as a movement of "the American people" and consequently focused on points of agreement among citizens in order to unite them against politicians who clearly were not serving their interests. He explained,

> The Tea Party has not been focused on social issues or the cultural issues that divide America, but has been focused more on the issues that 70 to 80% of people agree with, [such as] responsible government, accountable elected officials, and balancing the budget.

Similarly, one of MoveOn's founders, Wes Boyd, noted that the group's primary goal is to bring "as much diversity to the power structure as possible. That is, ordinary citizens who can provide the countervailing influence against the notion that some kind of inside-the-beltway elite can make all our decisions" (Bernhard 2004). In short, both groups rejected party lines in order to get citizens involved in democratic activism, regardless of their political affiliation.

ICTs played an important role in creating and maintaining the inclusive identities of these very different groups. The FTPM Facebook page clearly indicated that the movement was not about partisan politics, but about providing a space for citizens to discuss issues, share information, and organize non-partisan, political events. The group description noted,

> This isn't a conservative or liberal thing. This is about the government forking over billions of dollars to businesses that should have failed. This is about taking money from responsible people and handing it over to CEOs who squandered their own.[2]

Citizens, the page added, need to "work together" to take back government from career politicians, who do not operate in the interest of "the people." More importantly, Anthony, the FTPM founder, cultivated and maintained this inclusive identity by monitoring the page, and, when necessary, criticizing or removing flaming posts. Anthony was not alone in this task. Other FTPM supporters helped maintain this inclusive identity by sanctioning "bad posts" and making it clear that the page was not a partisan forum, but a place where concerned citizens could come together and discuss political issues in respectful ways.[3]

MoveOn initially used e-mail to cultivate and maintain an inclusive identity. When a supporter signed up for MoveOn's e-mails, she was asked to choose from a predetermined list of "democratic" issues on which she wanted to receive information and action alerts (e.g., War in Iraq, environment, voters' rights). Individuals concerned about the war in Iraq, for instance, would receive information about petitions and events related to the issues, which they could attend. MoveOn supporters noted that the strength of this approach is that it allows them to connect with one another in a meaningful way while minimizing political differences. Janice, a 29-year-old graduate student, said it best,

> You know, a lot of organizations ask you to take everything and buy in. MoveOn doesn't. It lets you just get in on the issue you want. It gets rid of a lot of conflicts. The Internet makes that much easier that way. You don't have to look someone in the face and say, "I don't agree with you about everything. I only agree with you about one thing." The online structure [pause] minimizes that awkwardness.

This approach works quite well. For example, at a home event we attended regarding substandard health-care treatment for veterans, MoveOn provided a movie for us to watch, questions based on the film for us to discuss, and a letter-writing packet complete with a sample letter, paper, envelops, and the addresses of the relevant politicians—all of which stimulated discussion as well as created a sense of comradery and purpose among the attendees. More recently, MoveOn extended this approach to social media platforms such as Twitter, Facebook, and Tumblr. MoveOn provides a forum for supporters to discuss the political issues of its choosing and encourages them to share their experiences online. For instance, MoveOn, which supported Bernie Sanders's candidacy for the democratic nomination, collected supporters' photos, and stories from the campaign trail so that they could be shared—and commented—on via social media. This virtual photo album visually reminded supporters that even though they worked in separate communities, they were part of a much larger political effort.

Second, organizations can use ICTs to help individuals feel efficacious. This is important because individuals who do not feel as though their participation matters are likely to "exit" an organization (Hirschman 1970); a move that undermines inclusivity. As previously mentioned, the FTPM Facebook page allowed supporters to engage in expressive political speech. Almost all the supporters noted that this was critical to feeling as though they were part of a larger community and that their participation on- and offline mattered. While this may seem inconsequential, the majority of respondents were "surprised" to go online and learn that they were not "alone" in their discontent with government. Katherine, a 47-year-old geographer, explained,

> For me, realizing that so many other people felt the same way was something I didn't know. You certainly didn't get that from the [mainstream] media. You do now. But, we didn't know that before. We thought we were this little minority. We didn't realize there was such a great crowd of people in America who all held that same ideal in their hearts and wanted to see it kept alive. . . . So, I think that is the biggest strength [of the FTPM] by far. People going "Oh my gosh! I'm not alone!" And there are many, many people who feel as strongly as I do about this, and it reinforces your own belief. People have these convictions, and they realize that they're not alone and it strengthens them.

While many supporters engaged in this political community online only, others used FTPM events and the Facebook page to meet with "like-minded folks" in the real world. These respondents, in particular, reported a strong sense of community. For instance, Oliver, who is retired from the Navy, explained that he enjoyed

> being with people of a like-mind and knowing that we [him and his wife] are not alone. So many of our friends are indifferent. Apathetic as we used to be. I found out there are other people that are concerned, and they aren't concerned about politics—which party you're aligned with.

He added that his participation in this community was important because it gave him "hope" that they could work together over time and change the political system. His wife, Vera, echoed this sentiment commenting on the power of "being with people that are united for the cause. . . . That's my draw in factor. I think it really makes me congeal with them [other Tea Partiers]." Bradley, a 51-year-old who works in mergers and acquisitions, concluded: "It feels good to be connected. It feels very good to be a part of it."

MoveOn also used ICTs to make supporters feel efficacious, albeit very differently. MoveOn primarily gave supporters "voice" (Hirschman 1970) in the organization's agenda, which made them feel as though their participation mattered. In fact, the majority of MoveOn supporters noted that they "had a say" in the organization and felt that they could change MoveOn's priorities by participating in the group's surveys, polls, online town halls, and forums or by e-mailing the leadership directly. While these may seem like routine ways to foster voice, respondents noted that their ability to give MoveOn "feedback" through organizational channels kept them engaged even when they did not completely agree with the group or its priorities.

For example, Marcia, a 60-year-old geologist, decided not to leave the group even though she was not happy with MoveOn's agenda. She pointed to the feedback mechanisms online that kept her involved,

> I wish the environment was at the top [of their list] but they went around all the MoveOn people and had them submit this poll about what their priorities are, and the environment wasn't really all that high . . . I've written to them about particular issues I've got and they do respond, so that's good. I think probably right now, the big issue is Iraq, and that's what they're focused on more than anything.

Marcia was still engaged with MoveOn in 2008, even though she was disappointed that the group decided to support Barrack Obama for the democratic presidential nomination. Again, she pointed to the fact that MoveOn had given her a voice on the issue. Marcia explained, "They took a vote. I voted. Most of the group voted. [And the group voted] to support Obama in the primaries. So I was okay with that." In a similar exchange, James, a 56-year-old psychiatrist, expressed some annoyance over MoveOn's avoidance of health-care reform. When asked why he stayed involved with MoveOn, he cited the potential to change the agenda through surveys, e-mails, and town hall meetings. James joked, "I'm trying to persuade the MoveOn group on health-care reform, and they are trying to persuade me on other issues. So, I guess they're working on me, and I'm working on them."

In sum, organizations can use ICTs to overcome the obstacles associated with collective identity and efficacy, which is important for keeping individuals engaged beyond initial mobilization. First, activist groups can use ICTs to cultivate collective identities that have broad appeal. These identities increase inclusivity insofar as they unite individuals around their roles as democratic citizens rather than along partisan lines. Second, activist groups can overcome the efficacy obstacle by using ICTs to create political community and give supporters voice in the organization's agenda. Efficacy matters because it provides supporters personal rewards for participation beyond the promise of electoral change (Lichterman 1996). Of course, it is also clear that FTPM and MoveOn use ICTs differently to achieve these goals. This is in part a function of the ICTs each group uses. FTPM's reliance on social media meant that its leadership had to consistently monitor and manage the conversation on Facebook to ensure that the collective identity remained non-partisan. While this was time intensive for the FTPM's founder Anthony, it effectively created an inclusive identity and political community that kept supporters engaged beyond initial mobilization. In contrast, MoveOn's use of ICTs allowed supporters to opt-in to the political issues of interest and connect with others who also cared about a particular cause. This minimized political differences, making it easier to reinforce an inclusive collective identity that emphasized democratic citizenry. That said, MoveOn supporters did not get a sense of efficacy from the political community. Instead, they pointed to their ability to use ICTs to weigh in on (and potentially shape) the organization's agenda. These differences, particularly in terms of how organizations make supporters feel efficacious, may be consequential over time—a point we return to next.

5. Conclusion

This chapter draws on 52 interviews with supporters of MoveOn and the FTPM as well participant observation data of both groups to explore how organizations can use ICTs to overcome obstacles to democratic activism.

We showed that organizations overcome motivational barriers by using ICTs to connect information to a tangible political activity, such as donating money, signing a petition, sending a letter to a politician, attendance at meetings and rallies, and quickly connecting individuals to one another through online forms. While this may not be extraordinary, it is clear from respondents that ICTs help individuals deal with biographical unavailability and learn new political skills. Additionally, we found that groups can effectively use ICTs to cultivate and maintain inclusive collective identities and give supporters voice in political processes and the groups themselves. In short, these results suggest that ICTs can be used in ways that aid democratic activism and social inclusion by engaging citizens in the political process and making politicians more responsive to their claims.

Clearly, there is much work to do before we truly understand the role of ICTs in democratic activism and social inclusion. First, future research should consider how different organizational structures affect the depth and breadth of individual activism and what this means for social inclusion.

Our analysis suggests that both hierarchical (MoveOn) and decentralized (FTPM) structures can effectively engage citizens across the political spectrum beyond initial mobilization. However, hierarchical organizations may be more effective at engaging individuals in a breadth of activities because they can monitor who is involved, how they are involved, and encourage them to do more over time. Decentralized structures, in contrast, may encourage deeper activism because individuals have to take that first step (e.g., finding a movement's online presence or attending a meeting) on their own without an organizational cheerleader cheering them on. If scholars do find differences, than this has implications for social inclusion insofar it suggests there is a digital divide in how citizens participate. Second, scholars need to assess how important face-to-face interactions are to continued participation. Most of the respondents that stayed engaged in activism over two years had at least one experience in which they were involved in the "real" world. It may be that ICTs are useful for getting and keeping people involved in the short term, but real-world experiences are key to sustained democratic activism. If this is the case, organizations that can help individuals connect in the real world will be the most beneficial to social inclusion.

Of course, the kind of ethnographic research needed to address these issues is rife with methodological obstacles. For example, in the best of circumstances, finding a relatively diverse pool of activists to interview about their experiences with activism can be difficult. This challenge is exacerbated when scholars are reliant on ICTs to find respondents, particularly ones who only engage online. Likewise, the number of platforms available to activist organizations change quickly and, with it, the ways activists engage the citizenry. It is quite possible that some platforms are more effective than others at engaging citizens in activism over time. That said, it is clear that scholars need to pay more attention to individual and organizational obstacles when assessing the ability of ICTs to get people politically engaged. Reducing information costs alone does not drive people to get involved in activism. Without more empirical work on how organizations use ICTs to engage citizens and how individual experience affects organizational effectiveness scholars will not fully understand whose interests are purposefully (or inadvertently) excluded from political consideration and what this means for social inclusion more broadly.

Notes

1 We attended all MoveOn events between 2004 and 2006 and attended all TPM group events and meetings (monthly) 2010 and 2012. In total, we have attended 42 events, rallies, and meetings.

2 Find the full description at www.facebook.com/groups/TallahasseeTeaParty/.

3 This arrangement fell apart over time. Anthony moved out of the area and the Facebook page is now "closed," meaning it is for members only. For a discussion, see Rohlinger and Bunnage (2015).

Bibliography

Bakardjieva, Maria. "Reconfiguring the Mediapolis: New Media and Civic Agency." *New Media & Society* 14, no. 1 (2012): 63–79.

Bennett, Lance, and Alexandra Segerberg. "The Logic of Connective Action." *Information, Communication & Society* 15, no. 5 (2012): 739–68.

Bernhard, Brendan. "Tempest from a Teapot." 2004. Accessed June 2, 2006. www.laweekly.com/general/features/tempest-from-a-teapot/1452/.

Bernstein, Mary. "Celebration and Suppression: The Strategic Uses of Identity by the Lesbian and Gay Movement." *American Journal of Sociology* 103 (1997): 531–65.

Beyerlein, Kraig, and John Hipp. "A Two-Stage Model for a Two-Stage Process: How Biographical Availability Matters for Social Movement Mobilization." *Mobilization: An International Quarterly* 11, no. 3 (2006): 299–320.

Bimber, Bruce. "The Internet and Political Transformation: Populism, Community, and Accelerated Pluralism." *Polity* 31, no. 1 (1998): 391–401.

Bimber, Bruce. "The Study of Information Technology and Civic Engagement." *Political Communication* 17 (2000): 329–33.

Downey, Dennis, and Deana Rohlinger. "Linking Strategic Choice with Macro Organizational Dynamics: Strategy and Social Movement Articulation." *Research on Social Movements, Conflicts and Change* 28 (2008): 3–38.

Earl, Jennifer, and Katrina Kimport. *Digitally Enabled Social Change: Activism in the Internet Age*. New York: MIT Press, 2011.

Fisher, Dana, and Paul-Brian McInerney. "The Limits of Networks in Social Movement Retention: On Canvassers and Their Careers." *Mobilization: An International Quarterly* 17, no. 2 (2012): 109–28.

Fisher, Dana, Kevin Stanley, David Berman, and Gina Neff. "How Do Organizations Matter? Mobilization and Support for Participants at Five Globalization Protests." *Social Problems* 52, no. 1 (2005): 102–21.

Fouhy, Beth. *MoveOn.org Becomes Anti-Bush Online Powerhouse*. The Associated Press, 2004. Accessed May 5, 2007. http://web.lexis-nexis.com.proxy.lib.fsu.edu/universe/document?_m=62436525555f4f656f9566dcd74a683d&_docnum=1&wchp=dGLbVzz-zSkVA&_md5=4db06dcf4903dd59b625017ca18db2ca

Friedman, Debra, and Doug McAdam. "Collective Identity and Activism: Networks, Choices, and the Life of a Social Movement." In *Frontiers in Social Movement Theory*, edited by Aldon Morris and Carol McClurg Mueller, 156–73. New Haven, CT: Yale University Press, 1992.

Gamson, Joshua. "Messages of Exclusion: Gender, Movements, and Symbolic Boundaries." *Gender & Society* 11, no. 2 (1997): 178–99.

Gibson, Rachel, Andrea Römmele, and Steven Ward. *Electronic Democracy: Mobilisation, Organisation and Participation via New ICTs*. Routledge, 2004.

Gigler, Björn-Sören, and Savita Bailur. *Closing the Feedback Loop: Can Technology Bridge the Accountability Gap?* Washington, D.C.: World Bank Publications, 2014.

Hirschman, Albert. *Exit, Voice, and Loyalty: Responses to Decline in Firms, Organizations, and States*. Cambridge, MA: Harvard University Press, 1970.

Kanter, Rosabeth Moss. "Commitment and Social Organization: A Study of Commitment Mechanisms in Utopian Communities." *American Sociological Review* 33, no. 4 (1968): 499–517.

Katz, James, and Ronald Rice. *Social Consequences of Internet Use: Access, Involvement, and Interaction*. Cambridge, MA: The MIT Press, 2002.

Klandermans, Bert. "Mobilization and Participation: Social-Psychological Expansions of Resource Mobilization Theory." *American Sociological Review* 49, no. 5 (1984): 583–600.

Klandermans, Bert. "The Formation and Mobilization of Consensus." In *International Social Movement Research: A Research Annual Volume I*, edited by Bert Klandermans, Hanspeter Kriesi and Sidney Tarrow, 173–96. Greenwich, CT: JAI Press Inc., 1988.

Klandermans, Bert, Jojanneke van de Toorn, and Jacquelien van Stekelenburg. "Embeddedness and Grievances: Collective Action Participation Among Immigrants." *American Sociological Review* 73 (2008): 992–1012.

Klandermans, Bert. *The Social Psychology of Protest*. New York: Wiley-Blackwell, 1997.

Klandermans, Bert, and Dirk Oegema. "Potentials, Networks, Motivations, and Barriers: Steps Towards Participation in Social Movements." *American Sociological Review* 52, no. 4 (1987): 519–31.

Lichterman, Paul. *The Search for Political Community: American Activists Reinventing Commitment Cambridge University Press*. Cambridge: Cambridge University Press, 1996.

Loader, Brian. *Young Citizens in the Digital Age: Political Engagement, Young People and New Media*. New York: Routledge, 2007.

Mäkinen, Maarit. "Digital Empowerment as a Process for Enhancing Citizens' Participation." *E-Learning and Digital Media* 3, no. 3 (2006): 381–95.

McAdam, Doug. "Recruitment to High Risk Activism: The Case of Freedom Summer." *American Journal of Sociology* 92 (1986): 64–90.

McAdam, Doug, and Ronnelle Paulsen. "Specifying the Relationship Between Social Ties and Activism." *The American Journal of Sociology* 99, no. 3 (1993): 640–67.

McCarthy, John, and Mayer Zald. "Resource Mobilization and Social Movements: A Partial Theory." *American Journal of Sociology* 82, no. 6 (1977): 1212–41.

McCaughey, Martha, and Michael Ayers. *Cyberactivism: Online Activism in Theory and Practice*. New York: Routledge, 2003.

Meyer, David, and Sidney Tarrow. *The Social Movement Society: Contentious Politics for a New Century*. New York: Rowman and Littlefield Publishers, Inc., 1998.

Norris, Pippa. *A Virtuous Circle: Political Communications in Postindustrial Societies*. New York: Cambridge University Press, 2000.

Norris, Pippa. *Digital Divide: Civic Engagement, Information Poverty, and the Internet Worldwide*. Cambridge: Cambridge University Press, 2001.

Polletta, Francesca, and James Jasper. "Collective Identity and Social Movements." *Annual Review of Sociology* 27 (2001): 283–305.

Postmes, Tom, and Suzanne Brunsting. "Collective Action in the Age of the Internet." *Social Science Computer Review* 20, no. 3 (2002): 290–301.

Putnam, Robert D. "Bowling Alone: America's Declining Social Capital." *Journal of Democracy* 6, no. 1 (1995): 65–78.

Rohlinger, Deana, and Leslie Bunnage. "Connecting People to Politics over Time? Internet Communication Technology and Retention in MoveOn.org and the Florida Tea Party Movement." *Information, Communication & Society* 18, no. 5 (2015): 539–52.

Rohlinger, Deana, Leslie Bunnage, and Jesse Klein. "Virtual Power Plays: Social Movements, ICT, and Party Politics." In *The Internet and Democracy: Voters, Candidates, Parties and Social Movements*, edited by Bernard Groffman, Alex Trechsel and Mark Franklin. New York: Springer, 2014.

Schlozman, Kay Lehman, Sidney Verba, and Henry E. Brady. "Civic Participation and the Equality Problem." In *Civic Engagement in American Democracy*, edited by Theda Skocpol and Morris Fiorina. Washington and New York: Brookings Institution Press and Russell Sage Foundation, 1999..

Selwyn, Neil. "Reconsidering Political and Popular Understandings of the Digital Divide." *New Media & Society* 6, no. 3 (2004): 341–62.

Skocpol, Theda. *Diminished Democracy: From Membership to Management in American Civic Life*. Oklahoma: University of Oklahoma Press, 2003.

Snow, David, E. Burke Jr. Rochford, Steven Worden, and Robert Benford. "Frame Alignment Processes, Micromobilization, and Movement Participation." *American Sociological Review* 51 (1986): 464–81.

Stewart, Angus. "Social Inclusion: An Introduction." In *Social Inclusion: Possibilities and Tensions*, edited by Peter Askonas and Angus Stewart, 1–16. Houndmills: Macmillan, 2000.

Taylor, Verta. "Social Movement Continuity: The Women's Movement in Abeyance." *American Sociological Review* 54 (1989): 761–75.

Van de Donk, Wim, Brian Loader, Paul Nixon, and Dieter Rucht. *Cyberprotest: New Media, Citizens and Social Movements*. New York: Routledge, 2004.

Van Dijk, Jan A. G. M. "Digital Divide Research, Achievements and Shortcomings." *Poetics* 34, nos. 4–5 (2006): 221–35.

Verba, Sidney, Kay Lehman Schlozman, and Henry E. Brady. *Voice and Equality: Civic Voluntarism in American Politics*. Cambridge, MA: Harvard University Press, 1995.

Warf, Barney. "Segueways into Cyberspace: Multiple Geographies of the Digital Divide." *Environment and Planning B: Planning and Design* 28, no. 1 (2001): 3–19.

Warschauer, Mark. *Technology and Social Inclusion: Rethinking the Digital Divide*. Cambridge, MA: MIT Press, 2003.

Wiltfang, Gregory, and Doug McAdam. "The Costs and Risks of Social Activism: A Study of Sanctuary Movement Activism." *Social Forces* 69, no. 3 (1991): 987–1010.

14 Social Inclusion, Farmer Resignation, and the Challenges of Information Technology Implementation

Ranjan Vaidya

1. Introduction

Social exclusion refers to the phenomenon of exclusion of individuals or groups from the usual activities of society. An important question within social exclusion discussions is "exclusion from what?". Zheng and Walsham (2008) provide a capability deprivation model of social exclusion, and describe social exclusion as capability deprivation. Earlier studies have identified different social activities from which people can be excluded. For example, Burchardt (1999) identifies five categories of activities from which people can be excluded. These are *consumption, savings, production, politics,* and *social activities*. Exclusion can be based on individual attributes as well as on collective attributes, such as caste, gender or class (Brown 1995; Bryson 1996). Also, the type of social exclusion faced by a person or group may depend on the domain of the study. For example Murray (2007) while studying the social exclusion of the children of prisoners identifies seven types of social exclusions including linguistic, administrative, and stigmatic.

Irrespective of these discussions, social exclusion has physical, affective, behavioral, psychological, and cognitive impacts. It thwarts the meeting of the basic human needs of belongingness and has negative impact on self-esteem of individuals (Kim et al. 2012). Experimental studies from psychology indicate that social exclusion has a direct impact on the self-regulation abilities of individuals. Socially excluded people are less likely to regulate their behavior, which further makes them vulnerable to social exclusion (Baumeister et al. 2005). In other words, once excluded the path back to social inclusion is extremely difficult (Gamian-Wilk 2013).

Social exclusion is more than a matter of sole concern of an excluded individual. The outcomes of social exclusion are equally threatening for groups and communities. Social exclusion is known to reduce the altruism propensity of individuals and lowers their ability to empathize (DeWall and Baumeister 2006). According to DeWall et al. (2009: 57), "Excluded people see the world through blood-coloured glasses" and have a higher sense of hostility. In work place situations, social ostracism—such as bullying—results in withdrawal of people and makes them less compliant. It also impacts their cognitive abilities (Gamian-Wilk 2013). This situation gets further acerbated in light of the fact that there is an overlap between real-world social exclusion and online social exclusion, and there is evidence available to indicate that these two fields often overlap (Helsper 2012). All these characteristics are extremely detrimental for social progress and harmony, and hence social exclusion is a matter of grave concern. Given this background of social exclusion, the focus of this study is on understanding the context within which the farmers use the withdrawal form of strategy to cope with social exclusion. To do so, the study looks at the computerization of agricultural marketing board in India. The project involved four stakeholder groups, namely farmers, traders, government officers, and private partners implementing the project.

The remainder of the paper is organized as follows. The next section presents the findings of the literature review on social exclusion. This is followed by the section on Bourdieu's theory of practice. Next, the section on research methods is presented. The section on case study presents the case of the agricultural marketing board where the information technology project was implemented. This is followed by a section on the findings of the study. Lastly, the discussion and conclusions are presented.

2. Literature Review

2.1 Conceptualization of Social Exclusion and Types

Based on the literature review, some outlines can be drawn that can help in the conceptualization of social exclusion. Social exclusion refers to the exclusion of individuals or groups from the normal activities that a society performs. Authors have grouped these activities in different categories, such as consumption, political, production, and social activities (Scharf et al. 2005). Another approach to social exclusion is based on Amartya Sen's capability framework whereby social exclusion is described as depriving individuals of their potential capabilities (Zheng and Walsham 2008). There is also a transactional form of social exclusion that views social exclusion as an outcome of transactions between individual and groups. For example, in individual transactions, some people may be perceived as selfish and unfair, and may be excluded from networks. Similarly, people may be denied access to certain commodities because of their group affiliations. In such transactional form of social exclusion, the role of boundary agents becomes highly critical and they are described as "gatekeepers" (Vranken 2001). Two important issues may be kept in mind while conceptualizing social exclusion. First, its relationship with poverty. Though studies indicate that social exclusion is more than poverty, there are also studies that suggest that social exclusion is in fact an effort to deny the presence of poverty in society (Vranken 2001). In other words, poverty plays a significant role in determining social exclusion (Levitas et al. 2007). Second, social exclusion has definite structural roots. For example, the study by Shucksmith (2004) suggests that while in a modern world young individuals are increasingly held responsible for their conditions, the fact remains that their conditions are determined by factors, such as race, gender, access to transportation, and the rural or urban nature of their neighborhood. Though the structural roots of exclusion are well agreed upon, there are discussions about the individual attributes that relate to social exclusion (Brown 1995). For example, an individual's adaptability to change (Scharf et al. 2005), trust and reciprocity have been identified as important criteria that relate to social exclusion (Morrow 2001). In any case, there are various definitions of social exclusion thathave been criticized due to not being operationally implementable (see Levitas et al. 2007 for a list of definitions).

Social exclusion can be of various types. On the basis of the dimensions impacted, social impact can be wide, deep, or concentrated. There are possibilities that, in a given social exclusion study, some people are impacted on few exclusion indicators or activities, and others are excluded on many indicators (Levitas et al. 2007). Social exclusion can take aggressive and passive forms. When exclusion is direct and explicit, it results in the outright rejection of individuals. On other occasions, exclusion may be passive and take symbolic forms (Molden et al. 2009). Similarly, Lee and Shrum (2012) distinguish between the "ignoring" and "rejecting" forms of social exclusion. Each of these impacts different types of human needs. Ignoring impacts the efficacy needs of people and results in what authors called "conspicuous consumption." On the other hand, the rejection form of exclusion impacts people's needs of socialization and results in more prosocial behavior.

2.2 Behavioral Characteristics of Socially Excluded People

An important theme in social exclusion studies relates to the behavioral characteristics of socially excluded people. This has been a consistent research theme since the '90s. For example, earlier studies have indicated that excluded people spend more time with people whom they do not know well, and give less time to their kith and kin, they are more susceptible to feelings of dejection and envy, and have lower self-esteem (Leary 1990). Studies have indicated that social exclusion results in the lowering of self-control, makes people vulnerable to temptations, and negatively impacts their cognitive abilities (Gamian-Wilk 2013). Experimental studies from psychology corroborate these findings, and indicate that social exclusion results in self-destructive behaviors such as over eating and smoking (Baumeister et al. 2005). Socially excluded people are found to be good at mimicking people, and also in the judging of the cues related to social exclusion and acceptance. For example, they are more able to differentiate the genuine smiles from the deceptive or portrayed smiles (Bernstein et al. 2010). Another set of studies indicates that once socially excluded, the path back to inclusion is extremely difficult (Gamian-Wilk 2013). The two arguments taken together suggest that, although socially excluded people are better in their cognitive abilities, through the use of which they can reclaim inclusion, yet they find it difficult to become socially accepted. This again indicates that social exclusion has structural roots rather than agency roots. This takes us to another important theme in social exclusion studies, namely that of the strategies that excluded people employ to counter social exclusion.

2.3 *Social Exclusion Coping Strategies*

Molden et al. (2009) present three types of strategies through which social exclusion is countered. The first strategy is based on self-reflection and introspection and results in increased awareness about one's own personality traits that perhaps led to exclusion. Under this approach, people try to reclaim the lost territory with renewed efforts for social acceptance, and people start becoming compliant and adopt social norms of behavior. The second strategy relates to the outward justification of exclusion whereby the cause of social exclusion is located in the surrounding environment or people's behavior. Under this approach, aggression and opposition mechanisms are used. These mechanisms may be directed toward particular people or toward the society in general. At times, the extent of the anger toward society may be so high that socially excluded people completely abandon self-regulation (Baumeister et al. 2005: 590). The third type of response strategy is withdrawal whereby agents may lose all hope for inclusion and may refrain from participation in any efforts through which they can gain acceptance. The withdrawal form of social exclusion has also been cited in other studies in organizational situation (Gamian-Wilk 2013). It is well known that social exclusion has bodily manifestations. For example, Allen (2004: 487–8) quotes the works of many social geographers to suggest that mismatching social environments (such as rural people subjected to urban environments) yield a feeling of dislike for one's own habitus, and often result in a withdrawal type strategy. Studies have also suggested that withdrawal strategies are a natural outcome of neoliberalism because they ignore the existence of group dependency by emphasizing individualism too much. In fact, studies have directly attributed what is termed "amoral familyism" to the withdrawal strategies that are an outcome of neoliberalism (Neilson 2015). Furthermore, the withdrawal form of strategy is also highlighted in the work of Pierre Bourdieu(Swartz 1997). The focus of this study is on the implications of this form of strategy. Before discussing the implications in the section of findings, the sections on theory of practice, research method and case study are presented.

3. Theory of Practice: Overview

Why people behave as they do has been an interesting question for sociologists. Empirical studies present consistent evidence that, although people may think that they control their lives, in truth their status in society is determined by group attributes, such as class or gender (Shucksmith 2004). Subjectively speaking, an individual may feel in full control of his own fate. Yet, objectively speaking, there seems to be no doubt that the elite class—economically or culturally—have better placed off springs, as also shown by Bourdieu in his study of the French education systems. In other words, while we all know empirically that if we had the highest assortment of capital, we would perhaps be better placed in the society, yet paradoxically, we claim that neoliberalism has given us more choices, and that we are better off in deciding our future. This contradiction between objective facts and subjective perceptions is explained by Bourdieu's theory of practices. Certain specific terminologies that are used by Bourdieu and will help in the understanding of theory of practices are discussed next.

3.1 Habitus

Habitus is a specialized term that refers to the dispositions and mannerisms of people. It can be understood through the terms of mimicking and mimeses. As kids, we all mimic our parents or teachers—beings who have had a mark or influence on us. We would perhaps wear their shoes, eyeglasses, or perhaps we would speak like them. This mimicking continues for a long time, even when adults, as it also helps us to strengthen our position in the society. Initially, this is at a conscious level, untill mimicking gets transformed into mimesis—the unconscious mimicking of our class, rank or group affiliations in the society (McKay 2001). Habitus is similar to mimesis—a sum of our tastes, dispositions that are peculiar to one's group affiliations. It is also the internalization of the external world.

3.2 Field

Field refers to social spaces or arenas of struggle where agents participate to acquire capital and hence to strengthen their position. According to Bourdieu, fields represent the interests of agents, the social space where these interests are contested present the idea of field (Bourdieu 1990: 88). The social field can be described using many variables, including the volume and type of capital possessed. Each of these variables can be represented along a different axis, and the position of agents can be described along various coordinates resulting in different theoretical classes. However, it is important for a researcher to be aware of the fact that such a categorization of classes is a theoretical construction, and not a real-world situation (Bourdieu 1985). It is a first level objectification of the reality, and a true picture of reality can be obtained when there is a second level of objectification—that of the researcher himself. Authors studying agriculture have conceptualized field as a network of relationships. For example, Raedeke et al. (2003) conceptualize the farming field as a relationship between the economic, family, and rental relationship of farmers with their landlords.

3.3 Capital

Bourdieu discusses four different types of capital, namely economic, social, cultural, and symbolic. Economic refers to capital that has a material or economic value. Social refers to capital that can be accrued on the basis of social networks. For Bourdieu, social capital is a heuristic tool that can provide guidance to empirical work. Bourdieu's version of social capital provides better explanations to social capital usage but has only recently started receiving attention (O'Brien and Ó Fathaigh 2005). Cultural capital refers to the tastes and manners that have an exchange value, and are highly revered by a community (Carolan 2005). Symbolic refers to a capital that has a symbolic value. For example, Olsen and Neff (2007) discuss the case of Indian farmers to show that education has a symbolic value among the poor farmers. These forms of capital are inter-convertible and their value is field-specific. A capital that is valued in one field may not be of value in another. Also, the struggle over capital is of three types. Firstis the struggle over the volume of capital. Second is the struggle over the conceptualization of capital—i.e., what constitutes the capital. Lastly, since forms of capital are inter-convertible, there is also a struggle over controlling the relative advantages that may be obtained through the conversion of capital (Chopra 2003).

3.4 Practices

Practices refer to the improvizations that happen over a period of time untill certain behavioral mechanisms are mastered. Practices have a historical component as well as an operational component—i.e., "opus operatum" and "modus operandi" (Carolan 2005: 389). The practices of people are an outcome of their habitus, the field characteristics, and the volume of capital possessed by the agents. This is summarized through the following relationship taken from Bourdieu (1984: 101):

(Habitus)(Capital) + Field = practices

According to Bourdieu, for the agents to be successful in the field, there needs to be harmony between the habitus of the agents and the requirements of the field. For example, the stout manners of farmers, while highly sought in the farming field, are not worthwhile in the occupations of fine arts. In interactions between the habitus and field requirements, three situations emerge. First, when there is a synergy between the two, the social structure is maintained. Second, when there is some synergy between them, there are modifications in the social structure. And third, when there is a complete mismatch between them, either there is revolt or the agents, with a feeling of hopelessness, resign or perhaps succumb to their exploitation (Swartz 1997). Our study presents the evidence of the *resignation* outcome as it goes on to demonstrate that in absence of a level playing field, the powerless farmers had chosen to resign to their exploitation.

4. Research Method

The data for the present study was collected using semi-structured interviews. The field visits were carried out in 2009, 2010, and 2012. A total of 23 interviews were conducted across four stakeholder groups: farmers (eight), traders (three), government officers (eight), and employees of the private vendor company (five). These groups were identified on the basis of importance-influence criteria (Bailur 2007). Past studies have suggested that farmers, traders and government officers are important in agricultural marketing (Meenakshi and Banerji 2005). Studies have also suggested that traders are one of the most influential groups in agricultural marketing (Molony 2008). Also, studies on information technology implementation have identified that the role of private vendors is extremely important for the success of information technology projects (Best and Kumar 2008). The interviews with the farmers were recorded and transcribed. The other stakeholder groups were not comfortable with the recording of the interviews and hence notes were taken. These notes and the transcribed interviews formed the data corpus for analysis. The term "data corpus" is used on the basis of the guidelines for conducting thematic analysis (Braun and Clarke 2006). The data was analyzed using the constant comparison guidelines suggested by Boeije (2002). Under this, the codes and interview summaries are constantly compared until an overarching thematic relationship is obtained between the constructs. The key stakeholder themes obtained are presented in Table 14.1. Clearly, the results indicate that stakeholder strategies was the most discussed theme among the stakeholders.

There are many theories that explain the stakeholder practices including classical theories, such as rational choice theory and stakeholder theory. We have chosen Bourdieu's theory of practice, as this theory has the scope for

Table 14.1 Major Themes From Stakeholder Interviews

Themes Summary	*Sources*	*References*
Stakeholder strategies	19	188
Trust issues among stakeholders	11	44
Negative perceptions about other stakeholders	10	27
Government employee issues	8	72
Context of farmers	14	98
Context of trade	9	22
Major problems in yards	8	34
Major problems in IS implementation	11	59
General perceptions about Indian society[7]	2	3

analyzing the symbolic systems and artifacts against the mere analysis of material resources (Mosse 1997). Furthermore, the theory has been found to be extremely useful in explaining the sociology of agriculture and has been used in studies based on India as well as in developed countries (Carolan 2005; Moritz 2010; Olsen and Morgan 2010; Olsen and Neff 2007; Vasan 2007). In fact, a major focus of Bourdieu's work has been on rural peasants, and the theory has developed through explaining the peasantization phenomenon (for example, see Bourdieu and Nice 2008).

5. Case Study

The present case is a case of agricultural marketing IS implementation in one of the Indian States. Like other parts of India, the state is home to considerable cultural and regional diversity represented through the different dialects, religions, ethnicities, and agro-climatic regions. Also, the state is home to eight different tribal populations, and has the largest tribal population in India. The state is a leading producer in India of soya bean, gram, oilseeds, pulses, linseed, and maize. The total population of the state is over 70 million, and the primary occupation of the people is agriculture. The farmers of the state can sell their agricultural commodities through various channels, such as selling to local village or town based merchants, government food storage centers, private companies, or traders, and through government controlled agricultural market yards. Of these various methods, the government-controlled market yards are most frequently used by farmers as these use the auction method against the fixed price methods used by other channels. These yards are controlled by a state-level government organization referred to as the agricultural marketing board. The current case study relates to one such agricultural marketing board of one of the states of India. The board was formed in 1972 with the objective of providing a marketing platform to the farmers. The board controls over 200 yards of the state where farmers and traders accumulate for agricultural trade. The government acts as an intermediary and certifies the trade transaction for a service fee.

5.1 Processes

The farmers bring their agricultural commodities to the yard. At the yard, a series of documents are issued which eventually lead to the completion of a trade transaction. At each successive step of the trade transaction, documents are generated, such as entry slips, auction slips, weight slips, payment slips, and service charge paid certificates. In the manual system, these documents are reconciled periodically by manual matching of each document. Manual reconciliations resulted in data quality issues, and the farmers did not enjoy fair prices for their commodities, as this data was not shared with other yards. It is a known that the farmers also faced some other problems in the yard,s including cheating in commodity weighing, delays in the payments, etc. (Anupindi and Sivakumar 2007).

5.2 Computerization of Yards

It is against this background that the board decided to computerize the yard processes in 2001. Under the new project, it was envisaged that all the yards of the state will be interconnected through the use of modern day information technology. This would enable the sharing of auction data across the yards. Also, IT was to be used to automate the internal yard processes of entrance, auctioning, weighing, service charge payments, etc. This was done through the use of computer terminals, handheld devices for capturing auction prices and installation of electronic weighing machines and bridges. As an example, Figure 14.1 presents a digital device, through which the auction data was captured. A pilot project was launched in 2001, and after its success, as part of the first project phase, the project was rolled over to more than 60 yards of the state. A private vendor company was selected for the implementation of the project. This private company brought the technology, and shared a percentage income from the total yard earnings.

The project continued successfully for the first few years. This is indicated in the department of information technology's report that concluded that the project was successful (DIT 2008: 92). However, by 2008–2009,

Figure 14.1 Digital Device to Capture Auction Data.

conflicts between the private vendors started to emerge. The government blamed the private vendor company for not fulfilling their contractual obligations, and the private vendors blamed the government for corruption. In 2011, the conflict surfaced as the government income tax department conducted several raids on the offices of private partners for income tax evasions. By 2012, the project was completely abandoned. By this time, however, the government officers were dependent on the computers for the generation of yard reports. Following the abandonment of the project, the government officers within each yard hired contractual personnel who continued to provide their services. Our correspondence with the additional director of the board in 2016 indicated that the government has launched a new project that looks exclusively at only one of the yard processes, namely that of the generation of no objection certificate. This project has been rolled over in 50 yards of the state and is called e-anugya project (in the local Hindi language, anugya refers to a "no objection certificate"). Also, the central government in India has rolled over the National Agriculture Market project that will connect electronically many yards of the state with those of other states. In other words, despite the failure of this project, both the state government and the national government are convinced that ICTs can play an important role in streamlining the otherwise complex agricultural trade (Vaidya 2016).

6. Study Findings

Past studies indicate that communities, especially those that migrate, construct an "ethnoscope" which provides a reference point for comparing their practices. The reference point can be the practices of the migrating community or the indigenous communities. That migration also changes the landscapes of both these communities (McKay 2001). This seems to be particularly true of the farmer community. Farmers travel long distances, to cities where the yards are usually located, for selling their commodities. This travel provides them with an opportunity to compare their lives with urban residents'.

6.1 Uncertainties of Agriculture

First of all, farmers' travel to cities is full of uncertainties. The village roads are rugged, and often the vehicles end up broken. For example, in a meeting with a farmer, the farmer was late by about an hour as his vehicle got broken in the way (in the words of the farmer: "There was a little problem with my tractor and so I was late"). Upon reaching the yards, their experience of trade transactions are also not pleasant. First, they are already exhausted through long travel, and the yards lack basic amenities, such as clean drinking water or a place where they can take some rest. Second, they have to interact with other groups that are more powerful, such as the traders and government officers. Furthermore, there is no certainty of the final outcome of the yard processes. The queue for auction may be long and, by the time their commodity is auctioned and weighed, it may be already late evening, the banks may be closed, and the traders may be unable to make payments. The drudgeries faced are clearly described in the following words of a farmer,

> Coming to the yard is not very easy. The roads are inadequate and also there is a lot of dust so it is full of stress. The farmers come to this yard from quite a distance. Sometimes it may be as high as 200 kilometers. Also, there might be problems with their vehicle on the way. Often he may arrive late in the evening, when the yard has closed. So he is already late, exhausted, and worried. The farmer has to travel early morning or even previous night to be at the yard. Furthermore, there are no facilities in the yard. Those who come in the night sleep for the whole night in the yard amidst the mosquitoes.

6.2 *Physical Hardships of Agriculture*

It is not that if farmers were not traveling, there would be any lessening of their drudgeries. Agriculture in India is not a mechanized activity. Right from plowing of land until the harvesting, most of the work is done manually. Also, there are issues such as supply of electricity to the farms and villages. This is evident from the statement of a farmer who mentioned,

> Agriculture is one of the most laborious professions; if we do not work hard then we are not able to do agriculture. You go to any farm in our state, and you will observe that the electricity is available only in the night. So the farmer works all night to water his field. Also, a farmer needs to do a lot of hard work. Each of these bags that you see here is of one quintal, and the farmer has to lift it many times to bring it finally to the yard. We first cut the soybean in the farm, and then we put it in bags so that it can be transported from the farm to a drying area. After drying the commodity is again put in the bags. Then the bags are transported in the lorry to the yard. So there is a lot of effort.

The overall perception of the farmers about agriculture is that it is a labor intensive activity. So when the farmers reach the yards, they have already passed through a physically challenging process.

6.3 *Exploitation by Stakeholders*

Not only agriculture is physically challenging but, according to the farmers, it is also ridden with exploitative practices of groups such as local traders and government officers. For example, a government officer, who was also a farmer, perceived exploitation to be a part and parcel of the farmer's life. According to him,

> A farmer is in problems everywhere, in India. Whether it is the yard, the society or anywhere, the farmer is in problems. He is exploited everywhere and hence makes no progress. Let's take the case of Seed Corporation. The basic function of the corporation is to provide seeds to the farmers, but they do not provide the seeds to the farmers easily. They harass the farmers. They ask them to come again and again. So, farmers feel dejected and depressed. Even in the yard, the traders do not make payments timely. I will give another example of farmer exploitation. The farmer produces milk and sells it to the local supplier. The local supplier sells the milk forward so that it reaches the city. The supplier is the middleman; he adulterates the quality of milk by adding water to it. This increases his profit. If someone complains about the milk quality, the middleman deducts the amount payable to the farmer.

This excerpt clearly indicates that the farmers approach the yard with a deep sense of exploitation.

6.4 Habitus of Exploitation

This sense of exploitation becomes such an imbedded part of a farmer's personality that further exploitation in yards is taken for granted. Discussions with farmers indicated that during various yard processes and despite knowing that they were being exploited, they had chosen to remain silent. For example, one of the farmers mentioned that the payment process was particularly prone to exploitation and mentioned,

> the farmer keeps his patience in dealing with the traders. He tolerates all the treacheries of the traders. He cannot help it because if he chooses to fight or argue, the trader will delay his payments. So the farmer tolerates everything in the yard.

Another context in which farmers use this strategy is the weighing process. Here, they tolerate the exploitation both through the hands of the laborers who do the weighing as well as the traders. One farmer described this exploitation and mentioned,

> As per the rules, it is the duty of laborers to weigh the commodity. They are required to lift the commodity from the vehicle and put it on the weighing machines. They are also required to remove it from the machine, but they do not do the entire work. We have to help them. Yet they charge full payments from us . . . So we have to exercise patience in all such situations.

6.5 Patience as a Form of Withdrawal Strategy

Studies on social exclusion suggest that people use different strategies to cope with social exclusion. These include introspection and improvizations, opposition, and withdrawal. Bourdieu's theory of practice also indicates similar conclusions in the "socialisation model of habitus." As mentioned by Swartz (1997: 212), "habitus tends to reproduce those actions consistent with the conditions under which it was produced." Bourdieu uses the term *doxa* to suggest that within each grouping of people (community, class, rank, etc.) there are orthodoxies that are regarded objective truths, and any deviations are not acceptable by the group (Chopra 2003). In other words, as *habitus* becomes formed, its formation happens within the limits of *doxa*. Also, Bourdieu mentions that in an interaction between the field and habitus, there are three outcomes. First, when there is conformity between the habitus and field—i.e., the requirements for success in a field conform to the habitus of the participants—the existing social structures are consolidated. Second, when there is some mismatch between the two, the habitus tends to adapt to the field requirements. Incidentally, one can see among the farmers in the yards some effort to match their *hexis* (another Bourdieuvian term that represents the physical dress up, mannerism, etc.) to those of the other yards participants. They do so by wearing dresses that suit the urban environment (such as trousers, shirts, and shoes) against wearing their traditional village attire. The third situation is a situation in which there is a complete mismatch between the field requirements and the offers of the habitus. According to Bourdieu, in such situations, there is a possibility of either revolt or resignation—i.e., behavioral patterns that are non-adaptive (Swartz 1997: 212–14). The case of the yards clearly suggests that farmers use patience as a withdrawal strategy. They are convinced that they are powerless people, exploited everywhere in the society, and that no one would come to their help, as mentioned by one of the farmers: "they do not help the farmers, so farmers feel dejected and depressed."

7. Discussion and Conclusion

The study findings clearly reveal that agricultural uncertainties, physical hardships, and the exploitative context of the farmers' work become ingrained in the habitus of the farmers so much that they believe that exploitation is a fait accompli. This is also corroborated from past studies that indicate that stigma acerbates social exclusion as stigmatized people start believing their stigma, and hence start behaving in a stigmatized manner. In other words, stigma contributes to the habitus of people, as it makes people behave in a stigmatized way (Murray 2007). The fact that social exclusion can stigmatize people is also supported from past studies. Through six experimental studies, Baumeister et al. (2005) demonstrate that social exclusion reduces self-regulation. Similarly, another experimental study suggested that social exclusion results in a lowering of the ability to empathize (DeWall and Baumeister 2006). Studies on social exclusion reveal that the socially excluded individuals are less altruistic than those who are socially included. Incidentally, more recent studies on trust indicate that trust formation bears a strong relationship with altruism propensities, and with other normative constructs such as humility (Foddy, Platow, and Yamagishi 2009; Sandeep and Ravishankar 2014; Song 2009). In other words, social exclusion impacts the trusting propensities of farmers, increases their sense of hostility for the other groups, and makes them more socially excluded. Trust and reciprocity are constructs that both relate to social exclusion (Morrow 2001).

In the yards, the other stakeholder groups particularly the traders and government officers blame the farmers for manipulating their commodities. According to these groups, the farmers show low quality sample during the auctions, but when the commodity is seen during the weighing process, the quality is much lower. Even the farmers agreed that this did happen occasionally in the yards. The farmers, with a deep sense of helplessness, often end up manipulating their commodities. Upon such manipulations, they are further stigmatized, and this in turn makes them further behave in a stigmatized manner—i.e., they manipulate the commodity qualities without remorse or compunction. Such manipulations perhaps provide a temporary relief to them, as they develop a sense that rather than always being at the receiving end of exploitation, they are sometimes at the opposite end. This perhaps is their improvisation of their practices—an effort to adapt to the field situations. However, when such commodity discrepancies get unearthed, the entire trade transaction is canceled, and the whole trade process starts again. Thus, farmers' improvizations to yard situations—those that can relive them of the feeling of exploitation—yield no results, and farmers feel more helpless. According to Bourdieu's model of action, this results in what is called "avoidance strategies", whereby the agents refuse to even be exposed to information that may further consolidate a feeling of crisis (Swartz 1997: 212). As an outcome, the only option left to the stakeholders is to withdraw from the situation and succumb to the exploitation of the powerful actors.

In developing countries that are marred by such inequalities, information technology projects need to address the challenges of exclusion. One way to study these challenges is through the framework provided by Bourdieu's theory of practices. Many agriculture related studies have used it to study the sociology of agriculture. For example, Olsen and Morgan (2010) and Olsen and Neff (2007) apply it to study what they call "aspiration paradox." Using Bourdieu's theory, of practice, the authors show how poor farmers over commit themselves to unsustainable debts to acquire cows that perhaps has a symbolic value in rural areas of India. Another example of the application of Bourdieu's theory is found in Moritz (2010) who studied the differences between the strategies of mixed and pastoral farmers of Africa and detailed how pastoral farmers have a broad range of strategies and can activate much more varied forms of capital than mixed farmers. The same theory has also been used to compare the differences between two fields before and after an intervention. In fact, Carolan (2005) uses the term "field work" to suggest that the field analysis should include "field work," an analytical exercise that lays the change expectations in the new field. This exercise can be very fruitful for information technology projects. Apart from comparing the fields, the theory has been used to compare the habitus of different stakeholders (for example, in Olsen and Neff 2007). As a starting point, the theory can be applied by drawing the details of the field that will be impacted due to new information technology implementation. For example, Raedeke et al. (2003) conceptualized the agriculture field as a field of relationships between the economic field, the social and family life of the farmers, and their relationship with their landlords. Past studies indicate that Bourdieu's theory of practice provides a robust tool for studying agricultural sociology (Moritz 2010). The current study supports this and suggests that Bourdieu's theory of practice can be used for studying information technology projects, particularly projects that are conceptualized toward inclusive development—such as "emancipatory information systems" (Kanungo 2004) or "public sector information and communications technology (PICT)"(Sandeep and Ravishankar 2014). The key challenge of such information technology implementation is to address the mismatch between the field requirements and the habitus of stakeholders.

References

Allen, Chris. "Bourdieu's Habitus, Social Class and the Spatial Worlds of Visually Impaired Children." *Urban Studies* 41, no. 3 (2004): 487–506.

Anupindi, Ravi, and S. Sivakumar. "Supply Chain Reengineering in Agri-Business." In *Building Supply Chain Excellence in Emerging Economies*, edited by H. L. Lee and C.-Y. Lee, Vol. 98, 265–307. New York: Springer, 2007.

Bailur, Savita. "Using Stakeholder Theory to Analyze Telecenter Projects." *Information Technologies & International Development* 3, no. 3 (2007): 61–80.

Baumeister, Roy F., C. Nathan DeWall, Natalie J. Ciarocco, and Jean M. Twenge. "Social Exclusion Impairs Self-Regulation." *Journal of Personality and Social Psychology* 88, no. 4 (2005): 589.

Bernstein, Michael J., Donald F. Sacco, Christina M. Brown, Steven G. Young, and Heather M. Claypool. "A Preference for Genuine Smiles Following Social Exclusion." *Journal of Experimental Social Psychology* 46, no. 1 (2010): 196–9.

Bernstein, Michael J., Steven G. Young, Christina M. Brown, Donald F. Sacco, and Heather M. Claypool. "Adaptive Responses to Social Exclusion Social Rejection Improves Detection of Real and Fake Smiles." *Psychological Science* 19, no. 10 (2008): 981–3.

Best, M., and R. Kumar. "Sustainability Failures of Rural Telecenters: Challenges from the Sustainable Access in Rural India (SARI) Project." *Information Technologies and International Development* 4, no. 4 (2008): 31–45.

Boeije, Hennie. "A Purposeful Approach to the Constant Comparative Method in the Analysis of Qualitative Interviews." *Quality & Quantity* 36, no. 4 (2002): 391–409. doi: 10.1023/a:1020909529486

Bourdieu, Pierre. *Distinction. Translated by Richard Nice.* Cambridge, MA: Harvard Univer, 1984.
Bourdieu, Pierre. "The Social Space and the Genesis of Groups." *Theory and Society* 14, no. 6 (1985): 723–44.
Bourdieu, Pierre. *The Logic of Practice.* Stanford, CA: Stanford University Press, 1990.
Bourdieu, Pierre, and R. Nice. *The Bachelors' Ball.* Chicago: University of Chicago Press, 2008.
Braun, V., and V. Clarke. "Using Thematic Analysis in Psychology." *Qualitative Research in Psychology* 3, no. 2 (2006): 77–101.
Brown, Phillip. "Cultural Capital and Social Exclusion: Some Observations on Recent Trends in Education, Employment and the Labour Market." *Work, Employment & Society* 9, no. 1 (1995): 29–51.
Bryson, Bethany. "'Anything But Heavy Metal': Symbolic Exclusion and Musical Dislikes." *American Sociological Review* 61, no. 5 (1996): 884–99.
Burchardt, Tania, Julian Le Grand, and David Piachaud. "Social Exclusion in Britain 1991–1995." *Social Policy & Administration* 33, no. 3 (1999): 227–44.
Carolan, Michael S. "Barriers to the Adoption of Sustainable Agriculture on Rented Land: An Examination of Contesting Social Fields." *Rural Sociology* 70, no. 3 (2005): 387–413.
Chopra, R." Neoliberalism as Doxa: Bourdieu's Theory of the State and the Contemporary Indian Discourse on Globalization and Liberalization." *Cultural Studies* 17, nos. 3–4 (2003): 419–44. doi: 10.1080/0950238032000083881
DeWall, C. Nathan, and Roy F. Baumeister. "Alone But Feeling No Pain: Effects of Social Exclusion on Physical Pain Tolerance and Pain Threshold, Affective Forecasting, and Interpersonal Empathy." *Journal of Personality and Social Psychology* 91, no. 1 (2006): 1.
DeWall, C. Nathan, Jean M. Twenge, Seth A. Gitter, and Roy F. Baumeister. "It's the Thought that Counts: The Role of Hostile Cognition in Shaping Aggressive Responses to Social Exclusion." *Journal of Personality and Social Psychology* 96, no. 1 (2009): 45.
DIT. *Draft Report on Impact Assessment of e-Governance Projects.* New Delhi: Department of Information Technology, Government of India, 2008.
Foddy, Margaret, Michael J. Platow, and Toshio Yamagishi. "Group-Based Trust in Strangers: The Role of Stereotypes and Expectations." *Psychological Science (Wiley-Blackwell)* 20, no. 4 (2009): 419–22. doi: 10.1111/j.1467-9280.2009.02312.x
Gamian-Wilk, M. "Does Bullying Increase Compliance?" *Social Influence* 8(2–3) (2013): 131–48. doi: 10.1080/15534510.2012.756429
Helsper, Ellen Johanna. "A Corresponding Fields Model for the Links Between Social and Digital Exclusion." *Communication Theory* 22, no. 4 (2012): 403–26.
Kanungo, Shivraj. "On the Emancipatory Role of Rural Information Systems." *Information Technology & People* 17, no. 4 (2004): 407–22.
Kim, Nam Jung, Junghoon Moon, Jaeseok Jeong, and Minghao Huang. "Social Exclusion Online: A Literature Review and Suggestions for Future Research." Paper presented at the *AMCIS*, 2012. http://aisel.aisnet.org/amcis2012/proceedings/SocialIssues/8
Leary, Mark R. "Responses to Social Exclusion: Social Anxiety, Jealousy, Loneliness, Depression, and Low Self-Esteem." *Journal of Social and Clinical Psychology* 9, no. 2 (1990): 221–9.

Lee, J., and L. J. Shrum. "Conspicuous Consumption Versus Charitable Behavior in Response to Social Exclusion: A Differential Needs Explanation." *Journal of Consumer Research* 39, no. 3 (2012): 530–44. doi: 10.1086/664039

Levitas, Ruth, Christina Pantazis, Eldin Fahmy, David Gordon, Eva Lloyd, and Demy Patsios. "The Multi-Dimensional Analysis of Social Exclusion." 2007.

McKay, Deirdre. "Migration and Masquerade: Gender and Habitus in the Philippines." Paper presented at the *Geography Research Forum*, 2001.

Meenakshi, J. V., and A. Banerji. "The Unsupportable Support Price: An Analysis of Collusion and Government Intervention in Paddy Auction Markets in North India." *Journal of Development Economics* 76, no. 2 (2005): 377–403. doi: 10.1016/j.jdeveco.2004.02.001

Molden, Daniel C., Gale M. Lucas, Wendi L. Gardner, Kristy Dean, and Megan L. Knowles. "Motivations for Prevention or Promotion Following Social Exclusion: Being Rejected Versus Being Ignored." *Journal of Personality and Social Psychology* 96, no. 2 (2009): 415.

Molony, Thomas. "Running Out of Credit: The Limitations of Mobile Telephony in a Tanzanian Agricultural Marketing System." *The Journal of Modern African Studies* 46, no. 4 (2008): 637–58. doi:10.1017/S0022278X08003510

Moritz, Mark. "Crop—Livestock Interactions in Agricultural and Pastoral Systems in West Africa." *Agriculture and Human Values* 27, no. 2 (2010): 119–28.

Morrow, Virginia. "Young People's Explanations and Experiences of Social Exclusion: Retrieving Bourdieu's Concept of Social Capital." *International Journal of Sociology and Social Policy* 21(4/5/6) (2001): 37–63.

Mosse, David. "The Symbolic Making of a Common Property Resource: History, Ecology and Locality in a Tank-Irrigated Landscape in South India." *Development and Change* 28, no. 3 (1997): 467–504.

Murray, J. "The Cycle of Punishment: Social Exclusion of Prisoners and Their Children." *Criminology and Criminal Justice* 7, no. 1 (2007): 55–81. doi: 10.1177/1748895807072476

Neilson, David. "Class, Precarity, and Anxiety Under Neoliberal Global Capitalism: From Denial to Resistance." *Theory & Psychology* 25, no. 2 (2015): 184–201. doi: /10.1177/0959354315580607.

O'Brien, Stephen, and Mairtin Ó. Fathaigh. "Bringing in Bourdieu's Theory of Social Capital: Renewing Learning Partnership Approaches to Social Inclusion." *Irish Educational Studies* 24, no. 1 (2005): 65–76.

Olsen, W. K., and J. Morgan. "Aspiration Problems in Indian Microfinance: A Case Study Exploration." *Journal of Developing Societies* 26, no. 4 (2010): 415–54. doi: 10.1177/0169796X1002600402

Olsen, W. K., and D. Neff. "Informal Agricultural Work, Habitus and Practices in an Indian Context." *Global Poverty Research Group Working Paper GPRG-WPS-079*, 2007.

Raedeke, Andrew H., John J. Green, Sandra S. Hodge, and Corinne Valdivia. "Farmers, the Practice of Farming and the Future of Agroforestry: An Application of Bourdieu's Concepts of Field and Habitus." *Rural Sociology* 68, no. 1 (2003): 64–86.

Sandeep, M. S., and M. N. Ravishankar. "The Continuity of Underperforming ICT Projects in the Public Sector." *Information & Management* 51, no. 6 (2014): 700–11.

Scharf, Thomas, Chris Phillipson, and Allison E. Smith. "Social Exclusion of Older People in Deprived Urban Communities of England." *European Journal of Ageing* 2, no. 2 (2005): 76–87.

Shucksmith, Mark. "Young People and Social Exclusion in Rural Areas." *Sociologia Ruralis* 44, no. 1 (2004): 43–59.

Song, Fei. "Intergroup Trust and Reciprocity in Strategic Interactions: Effects of Group Decision-Making Mechanisms." *Organizational Behavior and Human Decision Processes* 108, no. 1 (2009): 164–73.

Swartz, David. *Culture and Power: The Sociology of Pierre Bourdieu*. London: The University of Chicago Press, Ltd., 1997.

Vaidya, Ranjan. "Trust Formation in Information Systems Implementation in Developing Countries: The Role of Emancipatory Expectations." *Journal of Information, Communication and Ethics in Society* 14, no. 2 (2016): null. doi:10.1108/JICES-10-2015-0032

Vasan, Sudha. "Timber Access in the Indian Himalaya: Rethinking Social Capital in Public Policy." *Journal of Development Studies* 43, no. 7 (2007): 1215–33.

Zheng, Yingqin, and Geoff Walsham. "Inequality of What? Social Exclusion in the e-Society as Capability Deprivation." *Information Technology & People* 21, no. 3 (2008): 222–43. doi:10.1108/09593840810896000

15 Smartphones Adoption and Usage by 50+ Adults in the United Kingdom

Jyoti Choudrie, Sutee Pheeraphuttharangkoon, and Uchenna Ojiako

1. Introduction

Over the years, the capabilities of ICTs have significantly advanced. Advances in ICT devices, such as laptops, tablets, and smartphones, have made them important to the business, educational, and personal lives of individuals by providing information that is expedited, easily accessible and manageable (Line, Jain, and Lyons 2011). Smartphones began to take off after 1996 with developments offered by the novel forms of mobile phones leading them to become one of the fastest evolving technologies in the mobile phone market.

Benefits of smartphones include the provision of information and knowledge on entertainment, travel, finance, health care, and social networks (Xu et al. 2011). Since its introduction, there has been an exponential increase in the consumer market. For instance, in the year 2015, statistics showed that there were approximately two billion smartphone users in the consumer market and this number was further estimated to increase to approximately three billion by 2020 (Statista 2017). In the United Kingdom, the numbers of smartphone owners have increased continuously with present statistics revealing that smartphone ownership was in four out of five adults (81%) or 37 million individuals having a smartphone in the period up to June 2016. In penetration terms, there were rises by just 7% in 2016, compared to 9% in 2015, 13% in 2014, and 19% in 2013, and it is anticipated that user numbers will peak and have very modest rises in the coming years (Deloitte 2016).

In demographics terms, a global trend that is evident is an aging population. Due to medical advances and better quality of life, individuals live longer (UN DESA 2009). In the United Kingdom, individuals aged 65 or over increased by 3.9% between 1974 and 2014, from 13.8% to 17.7%. This proportion is projected to increase by a further 6.6% of the UK population by 2039 (OFS 2016). Therefore, individuals belonging to the older population group are increasing. In this research, the term silver surfer is defined as an individual who is 50 years old and above (Netlingo 2010). Due to improved quality of life and better economic conditions within families, some older adults are still working or becoming entrepreneurs; thereby owning and managing enterprises (Meyer 2013). It is within this context that smartphones may also provide assistance.

Older adults are usually at risk of being lonely and socially excluded (Stewart et al. 2013, 15). Previous studies suggest that smartphones can significantly improve personal relationships, reduces loneliness and hence offer pathways to social inclusion (Park, Han, and Kaid 2013). Moreover, Cho (2015: 350) found that using smartphone apps plays a vital role in aiding social inclusion and thus, improving the quality of people's lives. Further, smartphones may provide benefits in terms of health care for retired individuals (Joe and Demiris 2013), and a reduction in loneliness by connecting older adults with their friends and family (Blažun, Saranto, and Rissanen 2012).

Additionally, the older adult population is a likely group of adopters of smartphones, but there are still many who have not adopted such devices; thereby offering a motivation to pursue this research. For this purpose, the aim of this chapter is *to identify, examine, and explain the adoption and usage patterns of smartphones in the United Kingdom within the 50 years old and above population.*

The contributions of this research include enhancing the understanding of adoption and use of mobile telephony within the United Kingdom's older adult population. For practitioners, this research identifies factors that will encourage or inhibit the acceptance of smartphones among this group. For policy makers, this research is beneficial as it forms an understanding of smartphones devices, which can inhibit or encourage more interaction with government and other organizations such as smartphone providers and developers. To familiarize readers, the following section provides the literature review and a conceptual framework related to older adults, smartphones and smartphone adoption research. This is followed by a presentation of the research model, the research methodology, and the results. Finally, the chapter closes with a discussion of the results, their implications for other research and some conclusions.

2. Literature Review

When considering the theoretical foundation of older adults and smartphones, gaps that exist within the areas of older generation, digital divide, mobile phones, and smartphones research were initially identified.

2.1 Older Adults and Digital Divide

The divisions between individuals, society groups, and nations in terms of their associations with ICTs are varied, but are widely known as the "digital divide" (Tsatsou 2011). The following definitions are those widely agreed to capture the nature of the phenomenon of the digital divide. The digital divide is commonly defined as the divide between "those who have access to a particular technology and those who do not" (Curwen and Whalley 2010: 210). It is also posited that "the digital divide is the 'uneven diffusion' or 'gap' or 'disparities' between different socio-economic levels or across countries or between developed and developing nations in terms of 'access to' and 'use (usage)' of ICTs" (Hwang 2006: 19). When considering "the digital divide", a "typical" description refers to Internet access, but the term has been broadened to include other ICTs and a wide range of adoption parameters that go beyond access (Yusuf 2010).

The digital divide has provoked immense debates that have resulted in it being considered in a variety of contexts including, socio-economic status, gender, age, race, region or geography (Tsatsou 2011). One significant component of the digital divide is age (Selwyn et al. 2003). Having lived many years in the world without the Internet older adults tends to perceive the Internet as "non-essential." Additionally, age-related problems, such as declining vision, cognitive, and chronic diseases, are posed as major challenges to overcome. This has resulted in a significant age-based divide between young and old, with Internet use largely decreasing in the older group (Greengard 2009).

Over the years, many researchers have examined older adults' applications of and benefits from novel technologies (Vroman, Arthanat, and Lysack 2015). Previous studies of ICT adoption and usage patterns among older adults brought to light various aspects. One aspect included the digital divide where the gap between individuals who have used ICTs and those who have not used ICTs was examined (Barnard et al. 2013). Concurring with these studies is a study that found the existence of a digital divide and the gap is not likely to close anytime soon (Kim 2011). Digital divides can occur due to older adults facing difficulties when adopting novel technologies (Lee, Chen, and Hewitt 2011). In a study of a 55 years old and above population in Finland, it was found that around one-third of the respondents did not use the Internet (Vuori and Holmlund-Rytkönen 2005). In Australia, among silver surfing individuals, the Internet is used five times less than in the under 30 years age group (Willis 2006).

Several studies have attempted to investigate this issue and identify the factors leading to the age-related digital divide. Factors included the perceived lack of benefits (Heart and Kalderon 2013), lack of interest or motivation (Fu 2013), lack of knowledge (Peacock and Künemund 2007), lack of access (Peacock and Künemund 2007), cost (Carpenter and Buday 2007), and physical limitations (Saunders 2004).

2.2 *Mobile Phones and Smartphones*

As smartphone technology continues to advance, research findings on the role of demographic factors such as age and gender are continuously emerging (Aldhaban 2012). For instance, a 2000 study of digital divide in mobile phones and the Internet revealed the existence of a digital divide in terms of age, gender, income, work status and education (Rice and Katz 2003). Furthermore, similarities in the adoption and use of mobile phones and the Internet were apparent (Rice and Katz 2003). Research was also conducted on the health-related information differentials in gender terms and revealed that within silver surfer females, age is a significant factor given that older adults are less aware of novel technologies (Xue et al. 2012). In 2011, a study of health and caregiving in the silver surfer population identified that 79% of the silver surfers owned mobile phones, but only 7% adopted smartphones. Also, within this age group, approximately half of the silver surfers used or intended to use mobile technology for health-related matters. In terms of using this technology for health purposes, 11% of the sample population used it for basic health matters, such as weight, blood sugar, and blood pressure measurements (Barrett 2011). Such studies assisted this research to recognize the benefits of smartphones for the older population and identified the gaps in adoption studies.

2.3 Theory Building

The core concepts of this research are adoption and use. In terms of the IS discipline and adoption research, it was identified that research in this area has matured, but studies related to adoption are still developing. The main theories applied in adoption studies are the Diffusion of Innovation (DoI) (Rogers 2003); UTAUT (Venkatesh, Thong, and Xu 2012), Technology Acceptance Model (TAM) (Davis 1989), and Theory of Reasoned Action (TRA) (Ajzen and Fishbein 1980). Additionally, the factor of Perceived Enjoyment was adopted from previous studies and applied in this research (Song and Han 2009).

To determine the theories for this research, a review of various IS adoption theories was conducted. It was found that TAM is the most popular, followed by UTAUT and TRA (Aldhaban 2012). However, there was also a pattern of combining two or more IS theories. For instance, DoI and TAM were combined to explain the adoption of smartphones in a logistic industry (Chen, Yen, and Chen 2009). This pattern also applied in medical studies examining the adoption of smartphones among medical practitioners (Park and Chen 2007). UTAUT and enjoyment were combined to examine the importance of enjoyment in mobile services (Song and Han 2009). Based on this, the study applied this pattern of combining theories to provide a better understanding of the research problem.

2.4 Research Model

The proposed conceptual framework assumed that the dependent variable of behavioral intention to use smartphones is influenced initially by Observability and Compatibility that have been drawn from DoI (Rogers 2003). The second group of constructs include, social influence, facilitating conditions, performance expectancy and effort expectancy that are drawn from UTAUT (Venkatesh, Thong, and Xu 2012). Third, Perceived Enjoyment (Chtourou and Souiden 2010) is also integrated in the model. Finally, the dependent variable actual use (ACU) is influenced by the intention to use smartphones. Usage was measured using the features of a smartphone, such as e-mailing, browsing, using social media, taking a photo, and playing games.

DoI: Observability

An innovative product is defined as a new product where the features are novel or improved significantly from the predecessors. The contemporary features may develop using innovative technologies, knowledge or materials currently available (Rogers 1998). Therefore, smartphones can be considered to be an innovative product because first, they were introduced in 2007 with advanced designs and sophisticated technologies such as, an iPhone (Honan 2007). Second, they had applications and immense advanced features compare to a feature phone. Therefore, Rogers's DoI is applied to this framework.

Observability (OBS) is defined as the degree that smartphones are visible to silver surfers. Previous studies related to smartphones also identify that Observability is important for technology adoption (Koenig-Lewis, Palmer, and Moll 2010). Therefore, from DoI, this research posits that there is more likelihood of silver surfers adopting smartphones when they see a smartphone being used. Thus the following hypothesis is proposed.

H1: *Observability has a positive influence on the behavioral intention of smartphone adoption among silver surfers.*

DoI: Compatibility

Compatibility (COM) is also drawn from DoI, which is defined as the degree that a smartphone is compatible with silver surfers' lifestyles (Rogers 1998). Smartphones can benefit users in many ways, such as for business and personal communication and information on health issues (Chang et al. 2016). Therefore, smartphones may be compatible with the silver surfers' lifestyles, which led to the following hypothesis being proposed.

H2: *Compatibility has a positive influence on the behavioral intention of smartphone adoption within silver surfers.*

UTAUT: Social Influence

Social influence (SOC), one of the factors drawn from UTAUT, can be defined as the degree to which an individual perceives that other individuals important to the person, such as, family, friends or other close peers believe that the person should use the new system (Venkatesh 2012). It was found that silver surfers adoption of new technologies is normally influenced by other individuals, particularly, those who are close to them; for instance, their family and friends (Berner et al. 2015). Previous studies associated with smartphones also show that SOC is important for technology adoption (Bouwman and Reuver 2011). Therefore, the following hypothesis is proposed.

H3: *Social Influence has a positive influence on the behavioral intention of smartphone adoption among silver surfers.*

UTAUT: Facilitating Conditions

Facilitating conditions (FC), drawn from UTAUT, can be defined as the degree to which an individual believes that an organizational and technical infrastructure exists to support the use of a smartphone (Venkatesh 2012). This factor is described as older adults having necessary resources, such as knowledge, time, and money to adopt smartphones (Zhou, Lu, and Wang 2010). However, as with any novel technology, users who want to adopt a smartphone will need to have some understanding when using a new device. Additionally, in terms of cost, if a fee for using the smartphone is affordable and viewed as most beneficial to the silver surfers, then a positive attitude may occur and the older adults can use this technology. From previous research on mobile acceptance, the construct FC is viewed as one of the main factors leading to acceptance or adoption (Zhou, Lu, and Wang 2010). Therefore, the following hypothesis is proposed.

H4: Facilitating Conditions have a positive influence on the behavioral intention of smartphone adoption among silver surfers.

UTAUT: Performance Expectancy

Also drawn from UTAUT, Performance Expectancy (PE) is defined as the degree to which an individual believes that using the system will help him or her to achieve completion of their jobs or tasks (Venkatesh 2012). Theory also revealed that performance is one of the factors that affect user behavioral intention (Venkatesh 2012). UTAUT identifies a user's perception of the infrastructure functions of a smartphone, such as mobility, Internet connection, and application. If older users recognize the potential benefits that a smartphone provides, then they are likely to adopt and use a smartphone. Therefore, the following hypothesis is proposed.

H5: Performance Expectancy has a positive influence on the behavioral intention of smartphone adoption among silver surfers.

UTAUT: Effort Expectancy

Effort Expectancy (EE) is defined as the degree of ease associated with the use of a system (Venkatesh 2012). EE reflects the perceived effort construct when users adopt a new system. This factor is comparable to the perceived ease-of-use construct of TAM and the complexity construct from DoI (Venkatesh et al. 2003). It explains a user's perception of the difficulty associated with using a smartphone. If using a smartphone is considered a difficult task, then fewer older adults will adopt and use it. This research postulates that older adults may have different perspectives that need to be studied. Therefore, the following hypothesis is proposed.

H6: Effort Expectancy has a positive influence on the behavioral intention of smartphone adoption among silver surfers.

TAM3: Perceived Enjoyment

Perceived Enjoyment (ENJ), drawn from TAM3, is defined as the extent to which the activity of using a specific system is perceived to be enjoyable in its own right, aside from any performance consequences resulting from system use (Venkatesh 2012). A smartphone, which has additional capacities, such as connecting older adults with friends and family, playing music, watching videos, installing and playing games, and surfing some entertaining content, can be a device that provides enjoyment to older adults. ENJ was found to significantly affect the intended use of new technology (Rouibah, Lowry, and Hwang 2016). This factor was studied in both the contexts of using software in smartphones (Verkasalo et al. 2010) and using mobile Internet especially for shopping (Agrebi and Jallais 2015). Thus, this research believes that older people may find smartphones enjoyable in many ways. Therefore, the following hypothesis is proposed.

H7: Perceived Enjoyment has a positive influence on the behavioral intention of smartphone adoption within silver surfers.

Behavioral Intention / Use Behavior

From UTAUT (Venkatesh 2012), Behavioral Intention (INT) is the level to which a person has formulated a conscious plan to use a device in the future. It is the middle factor between the dependent variables and use behavior. INT is considered to influence the adoption or usage of smartphones in this research. Previous research applying UTAUT presented a relationship between the dependent variables and INT (Venkatesh et al. 2003).

Also, studies suggest that old people fear using unfamiliar technology, such as mobile phones (Kurniawan 2008). However, with stronger intentions, the

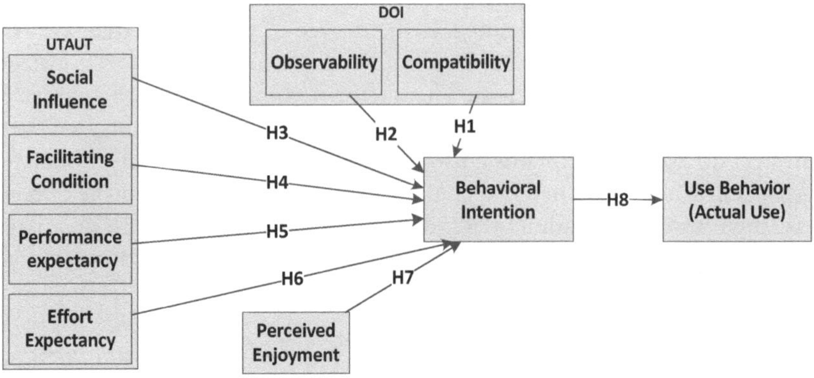

Figure 15.1 Research Model and Hypotheses.

larger benefits of smartphones for old people, especially health care (Joe and Demiris 2013), appropriate learning time and environments, can lead to the acquisition of technology knowledge similar to the younger generation (Chaffin and Harlow 2005). Therefore, the following hypothesis is proposed.

H8: Behavioral Intention has a positive influence on the smartphone usage of silver surfers.

To illustrate and understand the combination of factors, their relationships, and the formed hypothesis, a structural model was formed, as shown in Figure 15.1.

This research also recognizes the moderating variables of gender, age, experience and voluntariness in usage of smartphones (Venkatesh 2012). However, adding variables increases the number of hypotheses. Therefore, this research will discuss three of the main moderator variables that are gender, age and experience in the results section (Nysveen, Pedersen, and Thorbjørnsen 2005). Moreover, as aging occurs, adults face health problems, which led to the inclusion of the gender, age, and experience variables. Excluded from this research study is the voluntariness of technology use, "the degree to which use of the innovation is perceived as being voluntary, or of free will" (Moore and Benbasat 1991: 195). This is because the focus was on general users who have the freedom to use smartphones or not.

3. Research Methodology

For data collection, an online questionnaire survey was developed and posted on surveymonkey.com. The questionnaire consisted of two sections.

The first section examined the demographics and background details of participants. The second section sought to ascertain whether respondents used or did not use smartphones. If the respondents were current smartphone users, the questionnaire continued to seek reasons for using smartphones including the question related to the main constructs of the research model.

The research site was northern London, which was selected due to the well-developed mobile coverage infrastructure offered in the vicinity compared to other areas in the United Kingdom (Ofcom 2013). In 2011, North London has a population of 1,880,852, with 474,873 older adults of 50 years old and above; in other words, 25.25% of the overall population are 50 years and above (OFS 2011).

A leaflet to the survey location was sent to 19,760 households. This activity was completed between December 2013 and January 2014. The link was open for 3.5 months. This resulted in 1,030 completed responses, of which 984 usable responses were from silver surfers, making a response rate of 5.21%. This low response rate was attributed to the researchers having no control over the selection of the households that consisted of silver surfers.

Data Analysis

For data analysis, this research applied the Partial Least Square (PLS) technique with the help of SmartPLS version 2.0M3 (Ringle, Wende, and Becker 2005) and Structural Equation Modeling (SEM). SEM is appropriate to explain complicated variable relationships among hierarchical structural models (Gefen, Straub, and Boudreau 2000), which is the case in this study. For this study, SEM was used after data cleansing for verification and validation.

4. Research Findings

The survey responses revealed that 702 respondents had adopted smartphones, 134 respondents planned to have a smartphone and 148 respondents did not plan to have a smartphone. While 52.24% of the responses were from males, 47.76% were from females. In age terms, the majority of the respondents (56%) were from the 50–59 age group, 34.45% was from the 60–69 age group, 7.52% from the 70–79 age group, 1.63% from the 80–89 age group, and 0.2% was over 90 years old. Among smartphone adopters, the majority (64.10%) was from the 50–59 age group and 30% was from the 60–69 age group. However, 49.32% of the respondents who did not plan to have a smartphone were from the 60–69 age group. Moreover, respondents in the 60–69 age group who planned to have a smartphone were mo than the respondents in the same group who were current users of a smartphone device. In terms of employment status, 32.83% of the respondents were in full-time employment, 19.61% were pensioners (65+), and 12.60% were self-employed. Further, 6.5% were unemployed respondents, 3.15% entrepreneurs, 1.12% disabled, and 0.81% housewives.

4.1 Uses of Smartphones

As the usage of smartphones was of interest to this research, a question on smartphone usage was included in the survey. For this variable, as shown in Table 15.1, 15 questions on smartphone usages were included using a Likert scale measurement ranging from 1–7, where 1 is never and 7 is many times per day.

From Table 15.1, it can be concluded that people tended to send SMS more frequently than making a phone call. For the advanced features, 89.60% of the respondents indicated that the phone was used for browsing, and 85.47 % used their phone's e-mail facility. Mapping or navigation was used by 78.77% of the respondents with a 3.21 mean frequency. Managing appointments and calendars was used by 72.36% of the respondents with 3.52 mean frequency, while 68.66% read online news or magazines with a

Table 15.1 Smartphone Usages

Usages of a smartphone (Scales range from 1 to 7)	*Average*	*Total (n = 702)*	
		number	*%*
1. Making a phone call	4.76	687	97.86
2. SMS, Text messaging	5.19	689	98.15
3. E-mailing	4.19	600	85.47
4. Taking a photo	3.58	647	92.17
5. Filming a video	2.37	454	64.67
6. Browsing-surfing website(s)	4.35	629	89.60
7. Playing games	2.89	420	59.83
8. Watching videos for example YouTube	2.45	426	60.68
9. Mapping, Navigator such as Google Map, Tom-Tom, Copilot	3.21	553	78.77
10. Taking notes such as shopping lists or task that I need to do	2.95	472	67.24
11. Managing my appointment on my calendar	3.52	508	72.36
12. Using social network such as Facebook, Twitter	3.26	440	62.68
13. Reading online News and online Magazines	3.15	482	68.66
14. Using Facetime, Skype, ooVoo, Google Talk, Viber, Fring	2.22	322	45.87
15. Using to contact government authorities—NHS, Jobcentreplus, UKBA	1.80	243	34.62

3.15 mean frequency. Taking notes, filming a video, using social networks such as Facebook, watching videos and playing games involved more than half of the users. The frequency of using social media was the highest in this group at 3.26. There was also a question seeking information on Voice over Internet Protocol (VoIP) usage, or video calling applications, such as, Facetime, Skype, or Viber, followed by using smartphones to contact government authorities such as NHS or job center, with a low frequency of 2.22 and 1.80, respectively.

4.2 Adoption of Smartphones

Analysis of the previous findings led to obtaining 27 observed items over eight (8) latent constructs, as shown in Table 15.2. After the first analysis, some observed items (SOC4 and FC4) with loading factors below 0.8 were removed.

Table 15.2 illustrates the results of reliability and consistency. Composite reliability (CR), which measures the internal consistency, exceeds the 0.7 thresholds for all constructs, thus ensuring their reliability. Next, the items loaded well on their respective factors, exceeding 0.7. Furthermore, all the constructs' AVE was above or almost above 0.5. Finally, according to Fornell and Larcker (1981), when identifying discriminant validity, the square root of AVE for all the constructs needed to exceed all the other cross-correlations. This criterion was also satisfied for the overall constructs. As such, the model also exhibited satisfactory discriminant validity.

Following assessment of the reliability of the results, the next step was to identify the strengths of the various constructs. The results from applying SmartPLS are shown in Figure 15.2 and Table 15.3, respectively. According to the R-square, the model shows 76.0% of the variance belonged to INT, and 20.8% of the variance to the ACU of smartphones. The path coefficients (β) and t-values from the bootstrap and PLS algorithm were applied to explain the results. Thus, ENJ (H7) had the strongest factor influence of INT to use smartphones within the silver surfers with $\beta = 0.380$, t-value = 10.447 and a significant level of $(p) < 0.01$. COM (H2) and PE (H5) were strong factors, with $p < 0.01$, $\beta = 0.252$, t-value = 5.116, and $\beta = 0.235$, t-value = 5.765, respectively. FC (H4) and EE (H6) were considered significant ($p < 0.05$) with $\beta = 0.089$, t-value = 2.014 and $\beta = 0.083$, t-value = 2.511. Importantly, the INT for the overall sample population appears to have an important effect on ACU ($\beta = 0.456$, t-value = 12.490 and $p < 0.01$) However, OBS (H1) and SOC (H3) were considered not significant, with t-value = 0.480 and 1.032, respectively. Therefore, of the eight hypotheses, six were supported as shown in Table 15.3.

Table 15.2 Cross-Correlations, Item Loadings, Average Variance Extracted (AVE), Composite Reliability (CR), R-square and Cronbach's Alpha (CA) of the Research Model.

	Cross-correlations								*Item loadings*	*AVE* > 0.50	*CR* > 0.70	R^2	*CA* > 0.70
	COM	*EE*	*ENJ*	*FC*	*INT*	*OBS*	*PE*	*SOC*					
COM	***0.94***								*0.92–0.95*	*0.87*	*0.95*		*0.93*
EE	*0.61*	***0.97***							*0.96–0.97*	*0.93*	*0.97*		*0.93*
ENJ	*0.66*	*0.65*	***0.98***						*0.98–0.98*	*0.96*	*0.98*		*0.96*
FC	*0.73*	*0.66*	*0.54*	***0.86***					*0.84–0.88*	*0.74*	*0.90*		*0.83*
INT	*0.77*	*0.66*	*0.78*	*0.66*	***0.88***				*0.84–0.91*	*0.78*	*0.91*	*0.76*	*0.86*
OBS	*0.55*	*0.36*	*0.33*	*0.55*	*0.42*	***0.95***			*0.95–0.95*	*0.90*	*0.95*		*0.90*
PE	*0.75*	*0.57*	*0.61*	*0.61*	*0.74*	*0.43*	***0.88***		*0.85–0.90*	*0.77*	*0.91*		*0.85*
SOC	*0.53*	*0.28*	*0.40*	*0.40*	*0.43*	*0.46*	*0.51*	***0.86***	*0.82–0.88*	*0.74*	*0.89*		*0.82*
									-			*0.21*	

Note: The Diagonal Elements in Bold in the Cross-Correlations Matrix Are the Square Root of AVE.

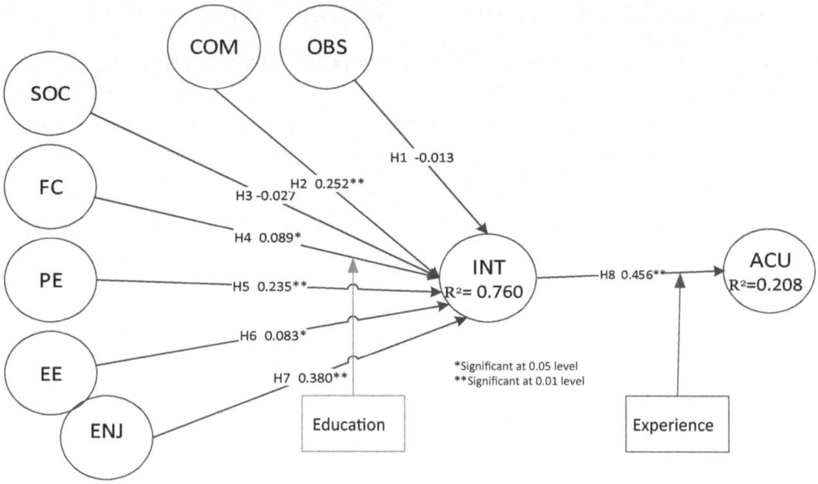

Figure 15.2 Hypothesis Testing Results.

Table 15.3 Hypothesis, Path Coefficients, T-value, Significant, and Hypothesis Support

Hypothesis	*Path coefficients (β)*	*t-value*	*Significant (p)*	*Supported*
1. Observability -> Behavioral intention	−0.013	0.480	-	NO
2. Compatibility -> Behavioral intention	0.252	5.116	< 0.01	YES
3. Social Influence -> Behavioral intention	−0.027	1.032	-	NO
4. Facilitating -> Behavioral intention	0.089	2.014	< 0.05	YES
5. PE -> Behavioral intention	0.235	5.765	< 0.01	YES
6. EE -> Behavioral intention	0.083	2.511	< 0.01	YES
7. Perceived Enjoyment -> Behavioral intention	0.380	10.447	< 0.01	YES
8. Behavioral intention -> smartphone usage	0.456	12.490	< 0.01	YES

4.3 Results on Adopting Smartphones With Moderator Variables

Also determined were the moderator variables drawn from the original UTAUT in order to gain more understanding of the study's contributions. According to UTAUT, the moderating variables affect relationships between independent and dependent variables (Venkatesh et al. 2003). The original moderating variables from UTAUT are gender, age, experience, and voluntariness of use. Experience in this study is defined as the experience of using smartphones. Moreover, since this research is related to older adults, health is selected as a moderator variable. Education is also often used as a moderator variable in technology adoption research (Park, Yang, and Lehto 2007). Five moderators were examined that are gender, age, experience, health, and education (Park, Yang, and Lehto 2007). The data was analyzed using the process of Lowry and Gaskin (2014) and a formula provided by Chin (2000). For gender, the dataset was divided into male and female. For age, the dataset was separated between 50–9 and 60–79. User experience was divided to under two years and more than two years of using smartphones. Health is a self-assessment variable where three choices were available: poor, good, and excellent. For moderator analysis, the good and excellent expressions were grouped against poor. Education levels were higher degree, and First Degree against High Diploma, Diploma, A level, and O level. The sub-groups were analyzed using SmartPLS and Chin's (2000) formula to find t-values or significant values. Only the significant results are shown in Table 15.4 and Figure 15.2.

The results illustrate that education moderated the relationship between FC and INT while experience moderated the link between INT and ACU at a significant level (p) < 0.01. This was understood as FC having a stronger positive effect on INT for the higher education than the lower education. Similarly, for those who have used smartphones for a long period of time, INT has a stronger positive effect on ACU.

Table 15.4 Significant Moderator Variables

Moderating Model-Experience										
	Less than 2 years (n = 238)				*More than 2 years (n = 464)*				*Compare*	
Hypothesis	β	t-value	Mean	STERR	β	t-value	Mean	STERR	t-value	p-value
INT->ACU	0.525	9.342	0.5232	0.0562	0.352	7.079	0.3502	0.0497	2.159	0.031
Moderating Model-Education										
	Low(n = 405)				*High(n = 282)*				*Compare*	
Hypothesis	β	t-value	Mean	STERR	β	t-value	Mean	STERR	t-value	p-value
FC->INT	0.199	3.687	0.1997	0.054	−0.088	1.320	−0.087	0.0666	3.366	0.001
INT->ACU	0.404	7.923	0.4027	0.051	0.523	9.847	0.5233	0.0531	1.600	0.110

4.4 Smartphone Diffusion and Communication Channels

For diffusion, the questions began by seeking information of the features considered when purchasing smartphones. These questions were asked to both the users and those who plan to use a smartphone in order to investigate the attitudes of these two groups.

From Table 15.5, it can be concluded that those who plan to use a smartphone have less experience with smartphones because they did not own the devices. Within this group, 58.21% were concerned with the brand of the smartphone that they planned to purchase, 43.28% were concerned with the price of the smartphone and 38.81% demonstrated concerns with the battery life. Furthermore, 35.82%, 35.07%, 31.34%, and 30.60% were interested in the operating system of the smartphones, screen size, appearance of the phone and camera, respectively. This group was less bothered about speed, screen resolution, weight, sizes of memory, and voice clarity (less than 20%).

Meanwhile, Table 15.5 shows that the adopters who had some experience or knowledge displayed diverse interests. They were highly concerned with the price (66.10%), while the screen size and battery life were the second concern (64.39%). Then 61.54% of the adopters were concerned with the brand. Next, 56.55% were interested in the operating system, and 48.01%,

Table 15.5 The Smartphone Features Considered When Buying

Consideration when buying a smartphone	*Adapted (n = 702)*		*Plan to use (n = 134)*		*Total*	
	Number	*(%)*	*Number*	*(%)*	*Number*	*(%)*
1. Appearance (such as color or material)	284	40.46	42	31.34	326	39.00
2. Camera	337	48.01	41	30.60	378	45.22
3. Operating System (such as iOS, Android or Windows Mobile)	397	56.55	48	35.82	445	53.23
4. Brand (such as Apple, Samsung, Nokia, or Blackberry)	432	61.54	78	58.21	510	61.00
5. Price	464	66.10	58	43.28	522	62.44
6. Operating Speed	290	41.31	22	16.42	312	37.32
7. Screen Size	452	64.39	47	35.07	499	59.69
8. Screen Resolution	215	30.63	20	14.93	235	28.11
9. Weight	245	34.90	22	16.42	267	31.94
10. Battery Life	452	64.39	52	38.81	504	60.29
11. Size of Memory	268	38.18	20	14.93	288	34.45
12. Voice Clarity	171	24.36	21	15.67	192	22.97

41.31%, and 40.46% of the group considered the camera, operating speed and appearance of the phone, respectively. The adopters were less concerned with voice clarity, screen resolution, and weight of the smartphones.

To determine the communication channel that silver surfers used when adopting or using smartphones, a question relating to communication channels was added. The communication channels included word of mouth by friends and family; high street stores; media, such as TV, radio, and newspapers; magazines; online social networks; professional technology review websites; peer technology reviews; and a sales person (shown in Table 15.6). This question was asked of both the adopters and those planning to use. Overall, both groups received smartphone information using the word of mouth from friends and family (65.07%). However, the group of those planning to use a smartphone (76.87%) relied more on the word of mouth compared to the adopters (62.82%). The professional technology review websites and high street stores followed this at 30.38% and 29.90%, respectively. The channels that both groups utilized less were peer technology reviews, online social networks, and magazines. The percentages in both groups were similar except for the group of those planning to use a smartphone, as it was being more reliant on high street stores when compared to the adopters group.

In the adopters and non-adopters groups there were diverse views, with those planning to use a smartphone being highly concerned with the brand

Table 15.6 Communication Channels to Get Information About Smartphones

Where do you get information on a smartphone	*Adopters (n = 702)*		*Those planning to use (n = 134)*		*Total (n = 836)*	
	Number	*(%)*	*Number*	*(%)*	*Number*	*(%)*
1. Word of mouth by friends and family	441	62.82	103	76.87	544	65.07
2. High street stores	192	27.35	58	43.28	250	29.90
3. Media (e.g., TV, radio, and newspapers)	157	22.36	36	26.87	193	23.09
4. Magazines	85	12.11	20	14.93	105	12.56
5. Online social networks	70	9.97	12	8.96	82	9.81
6. Professional technology review website (e.g., CNET.co.uk, Trustedreviews.com)	215	30.63	39	29.10	254	30.38
7. Peer technology review (e.g., unboxing video on YouTube)	66	9.40	8	5.97	74	8.85
8. Sales person	153	21.79	31	23.13	184	22.01

and the price, while the adopters group was greatly interested in the price, screen sizes, and brand. Moreover, after experiencing a smartphone, adopters were more likely to consider more of the smartphone's features.

5. Discussion

As shown in Figure 15.2, Performance Expectency was the strongest factor leading to the adoption of smartphones. This result is similar to a mobile shopping study (Agrebi and Jallais 2015), mobile application in Finland (Verkasalo et al. 2010) or a study in Canada and France on mobile devices used for surfing the Internet (Chtourou and Souiden 2010). However, the word "enjoy" or "fun" is interpreted differently among older adults; that is, the silver surfers did not consider action games as enjoyable, but viewed using their smartphone to capture their moments to share with friends or family as "enjoyment" or "fun." Therefore, smartphone or network providers may need to consider such factors when increasing smartphone adoption. By sharing photographs, the smartphone can assist older adults in memory sustainability and keeping in touch with peers and family. As Addis et al. (2010) found, "Older adults remember the good times well, because the brain regions that control the processing of emotions act in concert with those that control the processing of memory, when older adults experience positive events." In terms of academia, these results confirmed that for smartphones adoption, enjoyment is one of the important factors. Therefore, if smartphone providers are seeking increases in smartphones numbers, the enjoyment an older user can gain when using of asmartphone should be noted.

Facilitating condition, which is one of the main constructs applied in the research was measured using four items (FC1, FC2, FC3, and FC4) in the survey. However, the item FC4, which sought to ascertain whether silver surfers required help from someone when using a smartphone, was removed from the structural model. This was because 26.78% responses suggested very strongly that they did not need help from anyone, implying the majority of the respondents wanted to know more about how to use smartphones, but rarely had someone to assist them. From the questions on Effort Expectency (H6), smartphones are considered easy to use since the mean in both questions was 5.67 and 5.54, with 1.41 and 1.46 SD. Thus, policy makers, smartphone providers, and software developers may need to place an emphasis on providing not only easy-to-use devices but also easy-to-understand demonstrations and assistance.

In terms of usage of the various smartphone functions, the silver surfers' behavior appeared to suggest that they used only the basic features of a smartphone, such as making a phone call and SMS. However, new features, such as e-mailing and browsing, were also used with a high frequency. Some advanced functions, such as managing appointments on calendars or taking notes, were adopted less than the basic functions. Watching videos such as YouTube was not popular among the silver surfers, which this study attributed to the smartphone screen sizes. This issue appeared to be a concern as illustrated by more than half of the responses. In some cases, older adults may have physical limitations, such as problems related with the vision or cognition. Therefore, smartphones with large screens may be more compatible with older adults' lifestyles.

For smartphone diffusion, similar to the other age groups, price, and branding were the two most concerning factors. Moreover, for those who had smartphone use experience, higher expectations of newer versions of smartphones were present. Another issue highlighted in our study is the issue of the battery life of smartphones. The most important communication channel appeared to be the "word of mouth" by friends and family, which was closely followed by professional technology reviews that could be used to encourage adoption within computer professionals, such as hardware and software developers, programmers, and testers. Simultaneously, sales personnel in high street shops may need to be trained in a diverse manner in order to provide guidance to silver surfers purchasing or using smartphones. Regarding the research model, compatibility (H2) between smartphones and users should be noted. Matters of importance are, for example, how smartphones can help or improve an older adult's lifestyle. The functions (PE, H5) of a smartphone that older adults could use needs to be paid attention to. This is not restricted to business matters, but also to connecting with family and friends, or taking care of own health. However, the phone features and performance should be explained in a very simple way, so that a smartphone will be considered an easy-to-use device (EE, H6)

6. Conclusion

This study examined the parameters of adoption of smartphones within the United Kingdom's older adult population. To investigate adoption, a research model was developed based on UTAUT, DoI, and PE from TAM 3. Through an online survey, this study received 984 complete and usable responses. Of these, 702 respondents had adopted a smartphone, 134 respondents planned to have smartphones and 148 did not adopt and did not plan to have smartphones. The analysis results from the sampmle population revealed that 76% of the theoretical factor of the behavioural intention construct to use smartphones could be explained by the data.. Additionally, perceived enjoyment was the strongest factor influencing behavioural intention with significant factors in the conceptual model being performance expectancy, compatibility, facilitating conditions, and effort expectancy.. Education was a moderator variable in the link between facilitating conditions and behavioural intention, while experience of using smartphones moderated the affect between behavioural intention and the actual use of smartphones. However, social influence and observability were not significant for this research. What this suggests is that older adults will accept smartphones that are easy to use, lead to improvements in productivity and speed, are aligned to their daily activities and lifestyles as well as offering some technical assistance when needed or required. What is also found is that enjoyment, which is a diverse concept when referring to odler adults could be a leading factor for the adoption of smartphones in older adults.

In terms of contributions, this study provides a research model specifically for 50+ years old adults and their adoption of smartphones. Due to the differences in behavior and attitudes between younger and older adults, this research provided some examples of the digital divide occurring as a result of ICTs' existence. For policy makers, this study offers a view on factors that can encourage smartphone adoption. For the industry, this research offers strengths and weaknesses relating to the adoption of smartphones, which could help promote an increase in sales of smartphones to older adults.

However, this research has some limitations. In applying a quantitative method, this research may not capture additional views apart from the identified factors. Future studies would benefit by verifying the research framework and by seeking to increase the number of respondents. The references in this chapter are also limited, as smartphones were introduced to all memebers of the public quite late in time and research has offered to date rather limited accounts on smartphones and older adults. This led the team to apply reports from other related fields, such as marketing which are much more up-to-date. For theory, this research focused on adoption theories. Moreover, this research did not distinguish between different user cultures; therefore, the results may be used as a guideline for different cultures or geographic areas.

References

Addis, D. R., C. M. Leclerc, K. A. Muscatell, and E. A. Kensinger. "There Are Age-Related Changes in Neural Connectivity During the Encoding of Positive, But Not Negative, Information." *Cortex* 46, no. 4 (April 2010): 451–67.

Agrebi, S., and J. Jallais. "Explain the Intention to Use Smartphone for Mobile Shopping." *Journal of Retailing and Consumer Services* 22 (2015): 16–23.

Ajzen, I., and M. Fishbein. *Understanding Attitudes and Predicting Social Behavior*. Englewood Cliffs, NJ: Prentice-Hall, 1980.

Aldhaban, F. "Exploring the Adoption of Smartphone Technology : Literature Review." In *PICMET '12*, Portland, OR, 2012, 2758–70.

Barnard, Y., M. D. Bradley, F. Hodgson, and A. D. Lloyd. "Learning to Use New Technologies by Older Adults: Perceived Difficulties, Experimentation Behaviour and Usability." *Computers in Human Behavior* 29 (2013): 1715–24.

Barrett, L. *Health and Caregiving Among the 50 + : Ownership, Use and Interest in Mobile Technology*. Washington, DC,: AARP Research, 2011.

Berner, J., M. Rennemark, C. Jogreus, P. Anderberg, A. Skoldunger, M. Wahlberg, S. Elmstahl, and J. Berglund. "Factors Influencing Internet Usage in Older Adults (65 Years and Above) Living in Rural and Urban Sweden." *Health Informatics Journal* 21, no. 3 (2015): 237–49.

Blažun, H., K. Saranto, and S. Rissanen. "Impact of Computer Training Courses on Reduction of Loneliness of Older People in Finland and Slovenia." *Computers in Human Behavior* 28, no. 4 (2012): 1202–12. doi:10.1016/j.chb.2012.02.004

Bouwman, H., and M. D. Reuver. "Mobile TV : The Search for a Holy Grail that Isn't." In *EuroITV'11*, 2011, 185–93.

Carpenter, B. D., and S. Buday. "Computer Use Among Older Adults in a Naturally Occurring Retirement Community." *Computers in Human Behavior* 23, no. 6 (2007): 3012–24. doi:10.1016/j.chb.2006.08.015

Chaffin, A. J., and S. D. Harlow. "Cognitive Learning Applied to Older Adult Learners and Technology." *Educational Gerontology* 31, no. 4 (2005): 301–29. doi:10.1080/03601270590916803

Chang, S. E., W-C Shen, and A. Y. Liu. Why mobile users trust smartphone social networking services? A PLS-SEM approach. *Journal of Business Research* 69, no. 11 (2016): 4890–95.

Chen, J. V., D. C. Yen, and K. Chen. "The Acceptance and Diffusion of the Innovative Smart Phone Use: A Case Study of a Delivery Service Company in Logistics." *Information & Management* 46, no. 4 (2009): 241–8. doi:10.1016/j.im.2009.03.001

Chin, W. W. "Frequently Asked Questions—Partial Least Squares & PLS-Graph." 2000. Accessed September 5, 2014. http://disc-nt.cba.uh.edu/chin/plsfaq.htm

Cho, J. "Roles of Smartphone App Use in Improving Social Capital and Reducing Social Isolation." *Cyberpschology, Behavior, and Social Networking* 18, no. 6 (2015): 350–5. doi:10.1089/cyber.2014.0657.

Chtourou, M. S., and N. Souiden. "Rethinking the TAM Model: Time to Consider Fun." *Journal of Consumer Marketing* 27, no. 4 (2010): 336–44. doi:10.1108/07363761011052378

Curwen, P., and J. Whalley. *Mobile Telecommunications in a High-Speed World : Industry Structure, Strategic Behaviour and Socio-Economic Impact*. Surrey: Gower Publishing Limited, 2010.

Davis, F. D. "Perceived Usefulness, Perceived Ease of Use, and User Acceptance of Information Technology." *MIS Quarterly* (September 1989): 319–40.

Deloitte. "There's No Place Like Phone." 2016. Accessed February 4, 2017. www.deloitte.co.uk/mobileuk/

Fornell, C., and D. Larcker. "Evaluating Structural Equation Models with Unobservable Variables and Measurement Error." *Journal of Marketing Research (JMR)* 18, no. 1 (1981): 39–50. http://search.ebscohost.com/login.aspx?direct=true&profile=ehost&scope=site&authtype=crawler&jrnl=00222437&AN=5015357&h=OgKSbKqDCpGkI8AwgRGFG2iwXDSLbfHaoONj4lNyF/ie1kWdZWEsraVFlCqScEazb5DlbbVC5fbHP0Y+H8uz8Q==&crl=c

Fu, J. "ICT in Education: A Critical Literature Review and Its Implications." *International Journal of Education and Development using Information and Communication Technology* 9, no. 1 (2013): 112–25.

Gefen, D., D. W. Straub, and M. C. Boudreau. "Structural Equation Modeling and Regression : Guidelines for Research Practice." *Communications of AIS* 4 (August 2000): 1–79.

Greengard, S. "Facing an Age-Old Problem." *Communications of the ACM* 52, no. 9 (2009): 20. doi:10.1145/1562164.1562173

Heart, T., and E. Kalderon. "Older Adults: Are They Ready to Adopt Health-Related ICT?" *Journal of Medical Informatics* 82, no. 11 (2013): e209–31.

Honan, M. *Apple Unveils iPhone*. Macworld, 2007. Accessed August 26, 2013. www.macworld.com/article/1054769/iphone.html

Hwang, J. *Deconstructing the Discourse of the Global Digital Divide in the Age of Neo-Liberal Global Economy*. Pennsylvania: The Pennsylvania State University. Doctoral Dissertation, 2006. https://etda.libraries.psu.edu/files/final_submissions/3156.

Joe, J., and G. Demiris. "Older Adults and Mobile Phones for Health: A Review." *Journal of Biomedical Informatics* 46, no. 5 (2013): 947–54. doi:10.1016/j.jbi.2013.06.008

Kim, S. H. "Moderating Effects of Job Relevance and Experience on Mobile Wireless Technology Acceptance: Adoption of a Smartphone by Individuals." *Information & Management* 45, no. 6 (2008): 387–93. doi:10.1016/j.im.2008.05.002

Kim, S. H. "The Diffusion of the Internet: Trend and Causes." *Social Science Research* 40, no. 2 (2011): 602–13. doi:10.1016/j.ssresearch.2010.07.005

Koenig-Lewis, N., A. Palmer, and A. Moll. "Predicting Young Consumers' Take Up of Mobile Banking Services." *International Journal of Bank Marketing* 28, no. 5 (2010): 410–32. doi:10.1108/02652321011064917

Kurniawan, S. "Older People and Mobile Phones: A Multi-Method Investigation." *International Journal of Human-Computer Studies* 66, no. 12 (2008): 889–901. doi:10.1016/j.ijhcs.2008.03.002

Lee, B., Y. Chen, and L. Hewitt. "Age Differences in Constraints Encountered by Seniors in Their Use of Computers and the Internet." *Computers in Human Behavior* 27, no. 3 (2011): 1231–37. doi:10.1016/j.chb.2011.01.003

Line, T., J. Jain, and G. Lyons. "The Role of ICTs in Everyday Mobile Lives." *Journal of Transport Geography* 19, no. 6 (2011): 1490–99. doi:10.1016/j.jtrangeo.2010.07.002

Lowry, P. B., and J. Gaskin. "Partial Least Squares (PLS) Structural Equation Modeling (SEM) for Building and Testing Behavioral Causal Theory: When to Choose It and How to Use It." *IEEE Transactions on Professional Communication* 57, no. 2 (2014): 123–46. doi:10.1109/TPC.2014.2312452

Meyer, H. "Older, Healthier and Working: Britons Say No to Retirement." *The Guardian*, 2013. Accessed August 25, 2013. www.theguardian.com/society/2013/aug/24/working-britons-retirement

Moore, G., and I. Benbasat. "Development of an Instrument to Measure the Perceptions of Adopting an Information Technology Innovation." *Information Systems Research* 2, no. 3 (1991): 199–222. http://pubsonline.informs.org/doi/abs/10.1287/isre.2.3.192

Netlingo. "Silver Surfer." 2010. Accessed November 27, 2013. www.netlingo.com/word/silver-surfer.php

Nysveen, H., P. E. Pedersen, and H. Thorbjørnsen. "Explaining Intention to Use Mobile Chat Services: Moderating Effects of Gender." *Journal of Consumer Marketing* 22, no. 5 (2005): 247–56. doi:10.1108/07363760510611671

Ofcom. "Infrastructure Report 2013." 2013. http://d2a9983j4okwzn.cloudfront.net/downloads/infrastructure-report-2013.pdf

Ofcom. "Three in Five Adults Had a Smartphone in Q1 2014." 2014. Accessed October 27, 2014. http://stakeholders.ofcom.org.uk/market-data-research/market-data/communications-market-reports/cmr14/telecoms-networks/uk-5.81

Office for National Statistics (OFS). "The 2011 Census for England and Wales." 2011. Accessed April 2, 2014. www.ons.gov.uk/ons/guide-method/census/2011/index.html

Office for National Statistics (OFS). "Overview of the UK Population." February 2016. Accessed January 5 2016. www.ons.gov.uk/peoplepopulationandcommunity/populationandmigration/populationestimates/articles/overviewoftheukpopulation/february2016

Park, J., S. Yang, and X. Lehto. "Adoption of Mobile Technologies for Chinese Consumers." *Journal of Electronic Commerce Research* 8, no. 3 (2007): 196–206.

Park, K., S. Han, and L. Kaid. "Does Social Networking Service Usage Mediate the Association Between Smartphone Usage and Social Capital?" *New Media & Society* 15, no. 7 (2013): 1077–93. doi: 10.1177/1461444812465927

Park, Y., and J. V. Chen. "Acceptance and Adoption of the Innovative Use of Smartphone." *Industrial Management & Data Systems* 107, no. 9 (2007): 1349–65. doi:10.1108/02635570710834009

Peacock, S. E., and H. Künemund. "Senior Citizens and Internet Technology." *European Journal of Ageing* 4, no. 4 (2007): 191–200. doi:10.1007/s10433-007-0067-z

Rice, R. E., and Katz, J. E. "Comparing Internet and Mobile Phone Usage: Digital Divides of Usage, Adoption, and Dropouts." *Telecommunications Policy* 27, nos. 8–9 (2003): 597–623. doi:10.1016/S0308-5961(03)00068-5

Ringle, C. M., S. Wende, and J. M. Becker. "Next Generation Path Modeling." 2005. Accessed February 12, 2013. www.smartpls.de

Rogers, E. M. *Diffusion of Innovations*. New York: Free Press, 2003.

Rogers, M. *The Definition and Measurement of Innovation*. Melbourne: Melbourne Institute Working Paper No. 19/98, 1998.

Rouibah, K., P. Lowry, and Y. Hwang. "The Effects of Perceived Enjoyment and Perceived Risks on Trust Formation and Intentions to Use Online Payment Systems: New Perspectives from an Arab Country." *Electronic Commerce Research and Applications* 19 (2016): 33–43.

Saunders, E. J. "Maximizing Computer Use Among the Elderly in Rural Senior Centers." *Educational Gerontology* 30, no. 7 (2004): 573–85.

Selwyn, N., S. Gorard, J. Furlong, and L. Madden. "Older Adults' Use of Information and Communications Technology in Everyday Life." *Ageing and Society* 23, no. 5 (2003): 561–82. doi:10.1017/S0144686X03001302

Song, Y., and J. Han. "Is Enjoyment Important? An Empirical Research on the Impact of Perceive Enjoyment on Adoption of New Technology." In *2009 International Conference on Information Management, Innovation Management and Industrial Engineering*, IEEE, 2009, 511–14. doi:10.1109/ICIII.2009.582

Statista. "Number of Smartphone Users Worldwide from 2014 to 2020 (in Billions)." 2017. Accessed February 2, 2017. www.statista.com/statistics/330695/number-of-smartphone-users-worldwide/

Stewart, J., L. Bleumers, J. V. Looy, I. Mariln, A. All, D. Schurmans, K. Willaert, F. D. Grove, A. Jacobs, and G. Misuraca. "Digital Games and Gaming for Empowerment and Social Inclusion (DGEI)." In *The Potential of Digital Games for Empowerment and Social Inclusion of Groups at Risk of Social and Economic Exclusion: Evidence and Opportunity for Policy*, edited by Clara Centeno, 15–25. Luxernbourg: Publications Office of the European Union, 2013.

Tsatsou, P. "Digital Divides Revisited: What Is New About Divides and Their Research?" *Media, Culture & Society* 33, no. 2 (2011): 317–31. doi:10.1177/0163443710393865

UN DESA. *World Population Ageing 2009*. New York: United Nations, 2009.

Venkatesh, V. "Technology Acceptance." 2012. Accessed November 24, 2012. www.vvenkatesh.com/it/organizations/Theoretical_Models.asp

Venkatesh, V., M. G. Morris, M. Hall, G. B. Davis, and F. D. Davis. "User Acceptance of Information Technology : Towards a Unified View." *MIS Quarterly* 27, no. 3 (2003): 425–78.

Venkatesh, V., J. Y. L. Thong, and X. Xu. "Consumer Acceptance and Use of Information Technology: Extending the Unified Theory of Acceptance and Use of Technology." *MIS Quarterly* 36, no. 1 (2012): 157–78.

Verkasalo, H., C. López-Nicolás, F. J. Molina-Castillo, and H. Bouwman. "Analysis of Users and Non-Users of Smartphone Applications." *Telematics and Informatics* 27, no. 3 (2010): 242–55. doi:10.1016/j.tele.2009.11.001

Vroman, K. G., S. Arthanat, and C. Lysack. " 'Who over 65 Is Online?' Older Adults Dispositions Toward Information Communication Technology." *Computer in Human Behavior* 43 (2015): 156–66. doi:10.1016/j.chb.2014.10.018

Vuori, S., and M. Holmlund-Rytkönen. "55+ People as Internet Users." *Marketing Intelligence & Planning* 23, no. 1 (2005): 58–76. doi:10.1108/02634500510577474

Willis, S. "Beyond the "Digital Divide": Internet Diffusion and Inequality in Australia." *Journal of Sociology* 42, no. 1 (2006): 43–59. doi:10.1177/1440783306061352

Xu, Q., Z. M. Mao, A. Arbor, J. Erman, F. Park, A. Gerber, . . . S. Venkataraman. "Identifying Diverse Usage Behaviors of Smartphone Apps." In *IMC'11*, Berlin, Germany, 2011, 329–44.

Xue, L., C. C. Yen, L. Chang, H. C. Chan, B. C. Tai, S. B. Tan, . . . M. Choolani. "An Exploratory Study of Ageing Women's Perception on Access to Health Informatics via a Mobile Phone-Based Intervention." *International Journal of Medical Informatics* 81, no. 9 (2012): 637–48. doi:10.1016/j.ijmedinf.2012.04.008

Yusuf, D. "Digital Divide." In *International Encyclopedia of Civil Society*, edited by H. K. Anheire and S. Toepler, 604–9. N.Y: Springer, 2010.

Zhou, T., Y. Lu, and B. Wang. "Integrating TTF and UTAUT to Explain Mobile Banking User Adoption." *Computers in Human Behavior* 26, no. 4 (2010): 760–7. doi:10.1016/j.chb.2010.01.013

Appendix 15.1
Factor Loading

Construct Measure	*Mean*	*SD*	*Construct Measure Definition*
Social Influence (SOC1)	4.43	1.95	1. People important to me think I should use a smartphone (For example, friends and family)
Social Influence (SOC2)	3.81	1.95	2. People who influence my behavior think that I should use a smartphone
Social Influence (SOC3)	4.60	1.88	3. It is expected that people like me will use smartphones (For example, similar age or position people).
Social Influence (SOC4)	3.00	1.90	4. I want to use a smartphone because my friends do so.
Observability (OB1)	5.51	1.64	5. I have had many opportunities to see smartphones being used.
Observability (OB2)	5.39	1.68	6. It is easy for me to observe others using smartphones. (For example, I saw my friends use smartphones)
Compatibility (COM1)	5.91	1.37	7. I believe that using the smartphone is suitable for me.
Compatibility (COM2)	5.61	1.60	8. I believe that using the smartphone will fit my lifestyle.
Compatibility (COM3)	5.59	1.66	9. I think that using the smartphone fits well with my lifestyle or my work.
Facilitating Condition (FC1)	5.79	1.43	10. I have the resources necessary to use the smartphone. (For example, time and money)
Facilitating Condition (FC2)	5.86	1.35	11. I have the knowledge necessary to use the smartphone.
Facilitating Condition (FC3)	5.66	1.51	12. The operation costs of a smartphone do not prevent the use of it (such as, price of a smartphone or monthly fee).

Construct Measure	*Mean*	*SD*	*Construct Measure Definition*
Facilitating Condition (FC4)	3.63	2.20	13. I have a person available to assist me when using my smartphone.
Performance expectancy (PE1)	5.77	1.45	14. I feel a smartphone is useful. (e.g., with my lifestyle, my daily routine, and my work)
Performance expectancy (PE2)	4.69	1.92	15. Using a smartphone enables me to finish my personal tasks or work more quickly.
Performance expectancy (PE3)	4.99	1.92	16. Using a smartphone increases my productivity (e.g., to receive or reply e-mails faster).
Effort Expectancy (EE1)	5.67	1.41	17. I find that using the smartphone is easy.
Effort Expectancy (EE2)	5.54	1.46	18. Learning how to use a smartphone is easy for me.
Enjoyment (ENJ1)	5.37	1.62	19. I think it is fun to use a smartphone.
Enjoyment (ENJ2)	5.20	1.73	20. I find a smartphone fun (I had fun using a smartphone).
Behavioral intention (IN1)	5.28	1.69	21. I intend to use a smartphone as much as possible.
Behavioral intention (IN2)	6.18	1.23	22. I intend to continue using a smartphone in the future.
Behavioral intention (IN3)	5.53	1.61	23. Whenever possible, I intend to use a smartphone in my daily lifestyle or job.
Actual use (ACU)	5.87	1.49	Usage frequency of your smartphone
N = 702	The question used Likert scale 1–7(1 = strongly disagree, 7 = strongly agree)		

16 Literacy and Identity Links Forging Digital Inclusion?

Critical Reflections and Signposts From a Qualitative Study

Panayiota Tsatsou, Gillian Youngs, and Carolyn Watt

1. Introduction

Digital technologies, the Internet and especially the World Wide Web (WWW or the Web) have covered a remarkably long distance of innovation in only a few decades. While the Web made its scene changing appearance in the early 1990s (Berners-Lee and Fischetti 1999; Youngs 2007), it has moved from a cognition and text-based system (i.e., Web 1.0 in the 1990s) to a complex landscape of social interactive media (i.e., Web 2.0 in the 2000s). Also, currently Web 3.0 signals the shift to deeper embedding of collaboration and global exchange based on interactive knowledge-based networks and movements (e.g., Wikipedia, Open Source movement) (Fuchs 2008: 127).

Core to these and other digital developments are questions such as what is happening in the interplay between digital technologies (hardware and software) and human interfaces? What are the meanings of contrasting individual paths into and through the web or digital world? What are the implications for social inclusion of different levels of engagement with the digital world through varied forms of use, manipulation, and creation of technology and for the achievement of personal, social, technical, or other ends? Such questions probe socio-technical and ontological dimensions of digital developments, suggest a complex analytical approach to digital transformations beyond the simple good or bad, and call on us to think about the socio-technical implications of the "digital" in nuanced ways.

These remarks are particularly valid for contemplations over digital inclusion. Digital inclusion put simply is "the use of technology either directly or indirectly to improve the lives and life chances of disadvantaged people and the places in which they live" (Digital Inclusion Team 2007: 5). The literature has highlighted the links between digital and social inclusion/exclusion and has suggested that digital inclusion is "a practical embodiment of the wider theme of social inclusion" (Selwyn 2004: 343). Scholarly work has increasingly attempted to theoretically and empirically showcase the complex relationship between digital and social inclusion/exclusion. Helsper (2012) has proposed the "corresponding fields model,' according to which, the influence of offline exclusion on digital exclusion is mediated by social impact mediators (specifically, individuals" access, skills and attitudes), while the influence of digital on offline exclusion is mediated by digital impact mediators (specifically, the relevance, quality, ownership, and sustainability of different types of digital engagement). Although we acknowledge the significance of digital inclusion and its complex links with social inclusion, this chapter presents reflections on people's experiences and practices with/via digital technologies in order to identify some of the links between literacy and age identity in particular and thus to shed light on the polysemy, nuanced content, and value of digital inclusion.

More specifically, in this chapter, we argue that discourses about digital inclusion should go beyond the arguably direct social inclusion benefits of digital technologies and recognize the multidimensional role and implications of technology as well as the ramifications of technology adoption at the micro, individual level. Specifically, we argue that one way of advancing the study of digital inclusion is to revisit the concepts of "literacy" and "identity" and to place more focus on their links and relationship. We contend that digital literacy that cannot be fully captured through a "skills" perspective, as it involves individualized dimensions of people's interaction (or non-interaction) with the "digital" alongside structural or other parameters (such as education, as well as access to and understanding of technology). Thus, our guiding principle in thinking about digital literacy is "interactivity," which inherently involves people's identity and enjoys a dynamic relationship with self-driven and reflexive dimensions of identity in particular.

Our arguments are informed by the results of qualitative interviewing of young adult and older users of digital technologies who reflected on their digital attitudes, experiences, and various aspects of their relationships to the "digital." These qualitative interviews allowed us to address the following research question (RQ): *To what extent are digital users' literacy and identity linked to one another and how might such links account for aspects and traits of digital inclusion?* For the purpose of this study, we focused on the age dimension of digital users' identity. Also, we defined digital technologies as devices and environments that enable the viewing, creation, manipulation, transmission, storage, and sharing of (moving) images, text, and sound. This includes, but it is not limited to desktop/laptop computers, digital still and video cameras, mp3 recorders and players, mobile phones/smartphones/PDAs, tablets, gaming devices/consoles, and the Internet.

Next, we present the theoretical background and our conceptual propositions. Then we present the study methodology and the findings that address the RQ.

2. Theoretical Background

2.1 From Digital Divides to Digital Inclusion

Debates surrounding macro-micro dimensions of digital economy (Youngs 2007, 2013) have increasingly moved from discussions of the "digital divide" to discussions of "digital inclusion," and their focus has, respectively, shifted from technology availability and demographics to skills and literacy (Tsatsou 2011a).

In the 1990s and early 2000s, there was a major focus on questions of access to technology and connectivity. Most researchers articulated arguments in favor of technology innovation and diffusion and considered limited access to technology, socio-demographics, and financial cost common drivers and parameters of the digital divide (Angwin and Castaneda 1998; Katz and Aspden 1998; Katz and Rice 2002; Morrisett 2001: ix; Rice 2002; 106; Rogers 1995, 2001; Walton 1999; Wilhelm 2001). In the 2000s, scholars began to redefine the phenomenon incorporating qualitative elements of technology access and usage (Norris 2001). A new wave of literature gradually emerged that departed from a digital divide/exclusion agenda and drew attention to digital inclusion and "gradations' of inclusion (Livingstone and Helsper 2007).[1]

Specifically, the literature shifted in the 2000s to more complex, layered, and all-embracing approaches to digital inclusion. For instance, Van Dijk (2005: 22) projected a "cumulative and recursive model of successive kinds of access to digital technologies" that consists of motivational access (i.e., motivation), material access (i.e., equipment), skills access (i.e., skills and knowledge) and usage access (i.e., number and diversity of applications used, usage time). Others referred to "usage gaps"—namely, differentiations in how people use the Internet (e.g., Hargittai and Hinnant 2008; van Dijk 2005; van Dijk and Hacker 2003; Zillien and Hargittai 2009), as well as to variation in the purposes, the degree of autonomy, and the effectiveness of use, among disparities in the availability of social support (e.g., DiMaggio et al. 2004). In turn, Bradbrook and Fisher (2004) proposed the "5 Cs" of digital inclusion—"connectivity" (access and its various forms), "capability" (skills and employability), "content" (quality or community focused content), "confidence" (self-efficacy or motivation), and "continuity" (continuous and ongoing usage)—so as to highlight the intermingling of material, cultural, psychological, educational, and social capital in determining whether technologies are used as well as the purposes and effectiveness of their use.

In addition, the literature started in the 2000s to increasingly stress the role of skills and capabilities in determining people's self-efficacy or actual efficacy in the use of digital technologies (van Dijk and Hacker 2003; Livingstone 2007; Punie et al. 2009). Hargittai (2002) considered online skills and related skills gaps critical for the emergence of a "second-level digital divide." Notley defined digital inclusion as "the ICT capabilities people require to participate in society in ways they have most reason to value" (2009: 1212).Van Dijk (2005) categorized skills into instrumental, which allows people to operate technologies; informational, which enables people to search, select, process, and apply information through digital technologies; and strategic, which enables people to strategically use such information to serve purposes. Later on, van Deursen and van Dijk (2010; see also van Deursen, van Dijk and Peters 2011) pointed out that effective and goal-directed use of the Internet is highly dependent on the user's ability to handle technology (i.e., medium-related skills) and to find and use the needed content (i.e., content-related skills).

2.2 The Role of Literacy in Digital Inclusion

The emphasis on skills as a prerequisite for digital inclusion has gone hand in hand with the conceptualization of skills as the foundation of digital literacy. Hargittai and Hinnant suggested in this respect: "We focus on differences in general digital literacy, arguing that the way in which people utilize the Internet is at least in part driven by their online skills" (2008: 605).

A volume of research has addressed digital inclusion from a skills perspective, classifying literacy as skills of various types (e.g., Brandtweiner, Donat, and Kerschbaum 2011; Gui and Argentin 2011; Min 2010; van Deursen and van Dijk 2010; van Deursen, van Dijk and Peters 2011). Quite early on, Carvin (2000) argued that users should be "information literate"—able to identify quality and/or appropriate content, "adaptively literate"—able to develop new skills while using ICTs, and "occupationally literate"—able to apply skills to business, education, or domestic activities. Gui and Argentin (2011) defined digital literacy alongside theoretical/knowledge, operational skills, and evaluation skills. On the one hand, the literature has shifted emphasis from narrow technical skills to informational, strategic, creative and critical skills (van Dijk 2005; Helsper 2008), as well as to users' ability to use technology efficiently (Appel 2012). On the other hand, the prevalent argument is that literacy is to be understood through a skills lens and that skills themselves lie with the user who bears responsibility for their attainment (Thoman and Jolls 2005).

All in all, literacy has gained growing prominence in accounts of drivers and statuses of digital inclusion: "Everywhere, it seems, we hear of cyber-literacy, digital literacy, computer literacy, media literacy, Internet literacy, network literacy, and so on" (Livingstone 2008: 53). However, there is yet the challenge of refining existing conceptualizations of literacy, as literacy has mostly been interpreted alongside competencies, capabilities and skills (Livingstone 2008; 54). We argue that while the skills debate has expanded, it still constrains the approach to digital literacy, which would benefit from a more socio-technical focus that rests on the notion of "interactivity." The notion of "interactivity," in turn, involves user identity, and we discuss it later in the chapter.

2.3 Identity and Digital Inclusion

Identity is an extensively studied concept, viewed through socio-demographic and socio-cultural perspectives: "[Identity is] the process of construction of meaning on the basis of a cultural attribute, or a related set of cultural attributes, that is given priority over other sources of meaning." (Castells 2010: 6). What am I (socio-demographic approach), who am I (psychological approach) and where do I belong to (socio-cultural approach) are some of the usual questions one can pose in relation to identity. This is to say that identity embraces macro-micro linkages and is made up of individual and collective components, with socio-demographic and socio-cultural aspects being perceived and integrated at the psychological level in a number of ways. Identity is also dynamic and varied, containing stable, evolving, and changing elements: "There may be a plurality of identities. Yet, such a plurality is a source of stress and contradiction in both self-representation and social action" (Castells 2010; 6).

The existing literature has examined how socio-culturally driven aspects of identity and associated thinking and acting can explain adoption of digital technologies, patterns of adoption, as well as self-exclusion (Baron and Segerstad 2010; Erumban and Jong 2006; Kvasny 2006; Lenhart and Horrigan 2003; Reisdorf 2011; Robinson and Martin 2009; Selwyn 2006; Stanley 2003; Thomas et al. 2005; Tsatsou 2010, 2011b, 2011c; van Dijk 2005). Also researchers have examined socio-demographic aspects of identity such as age identity to make sense and explain digital inclusion and exclusion. For instance, theey have referred to the young generation as the "net generation" or "digital natives" (Prensky 2001; Tapscott 1998)—namely, "native speakers of the digital language of computers, video games and the Internet" (Prensky 2001: 1).

However, more recently, researchers have viewed age identity as just one of the aspects of identity that matter for digital inclusion. For example, Hargittai and Hinnant (2008) found that, although young adults are broadly perceived as the most highly connected age group, those with higher levels of education and of a more resource-rich background (e.g., longer experience, autonomous Internet use) make more beneficial use of the Internet, or, as the authors suggest, engage with more "capital-enhancing" activities on the Internet. Similarly, Helsper and Eynon (2010) found that "generation" is only one of the predictors of people's likelihood to be "digital natives," since factors such as experience, gender, and educational level also matter.

Such shifting evaluations of the role of age in digital inclusion go hand in hand with shifting conceptualizations of "age identity." Broadly speaking, the concept of age identity consists of an ontological dimension, captured predominantly through biological age, and a socially constructed dimension, which is shaped through processes of social construction and has largely created the "elderly stereotype" (Cuddy et al. 2005). This stereotype and related discourses have paved the way for the development of aging studies or gerontology that explores the question of aged people's social inclusion, among others (e.g., Clarke and Warren 2007; Macnicol 2006; Nelson 2002; Reed et al. 2006). Aging studies takes an "ageism" perspective on the role of age in digital divides, revisiting the "generational digital divides" thesis of digital inclusion research (e.g., Millward 2003; Olphert and Damodaran

2013). In the last few years, aging studies and digital divides research have both been placing more emphasis on psychological aspects of age identity and on emerging intra-generational diversity. Such research has assigned antithetical attributes to older people, such as that of "eternally youthful" and "frail needy seniors" (Loos 2013), while problematizing Prensky's distinction between "digital natives" and "digital immigrants" (2001) and particularly the existence of a unified group of "digital natives" (Helsper and Eynon 2010; Robinson 2014; Selwyn 2009).

We argue that identity involves the ways and extent to which socio-demographic and socio-cultural factors are perceived and integrated at the psychological level. Thus, we view age identity—which is the focus of our study—not only as biological and socially constructed but also as self-driven at the micro-level. On the ground of this presumption, our study aims to shed light on and unpack generational differences as well as intra-generational diversity concerning digital inclusion.

2.4 Literacy and Identity Links Forging Digital Inclusion?

While scholars have developed a multidimensional approach to digital inclusion that increasingly opens up literacy and identity-centered considerations, literacy, and identity have mostly been looked at separately, with literacy being understood as skills and with identity being conceived mostly at the socio-cultural and socio-demographic levels. There are few examples of literature linking literacy and identity, such as Gee (2003) who links identity development and new media literacies in the study of young video games players.

The links between literacy and identity can be revealed when one looks at the interaction of the user with technology, as interaction is influenced by broader systemic conditions while building on user identity. In this respect, Livingstone's proposition (2004, 2008) to define and study literacy beyond "skills" and through people's "relations with different media rather than . . . independently of them" (Livingstone 2004: 8) is a useful one. What Livingstone proposed is a conceptual shift "from an exclusive focus on the viewer to a focus on the interaction between text and reader or between inscribed and actual viewer/user" (2004: 8). For Livingstone, the "mutuality between text and reader" is central to understanding how users interpret, diverge from, confirm to or recreate meanings in the process of engaging with digital media (2008: 55). On the other hand, the mainstream skills-based interpretation of literacy focuses on user ability/capability rather than on the complex interactivity the user develops with technology.

Interactivity is a useful concept in its own right and has been theorized in disciplines that explore human-technology relationships. While for some it is understood as a structural element of the medium/technology (Manovich 2001) or a perception in the mind of the user (Lee 2000; Wise, Hamman, and Thorson 2006), interactivity is a multidimensional concept that refers to more than one actor/agent and practice/situation. Specifically, Szuprowicz (1995) identified three dimensions of interactivity: user-to-user, user-to-documents (or user-to-content/services), and user-to-computer (or user-to-technology). Similarly, McMillan (2002) talked about social interactivity (interactivity among users), textual interactivity (interactivity between users and documents/texts/services), and technical interactivity (interactivity between the user and system/medium/technology), while also suggesting attention to "the locus of control." In turn, Rafaeli and Sudweeks (1997) argued that interactivity is not a characteristic of the medium; it is a process-related construct concerning communication, a continuum, a variable, not just a condition, that can lead to multi-directional communication and various forms of engagement and cooperation. On the other hand, Jensen (1998: 201) drew upon "technology ability" and "user influence" and understood interactivity as "a measure of a media's potential ability to let a user exert an influence on the content and/or form of the mediated communication." For Jensen, interactivity is as an integral feature of media technologies that enables user—content and user—user interaction but via the interaction between user and technology (Carpentier 2015: 17).

The conceptual shift from "skills" to "interactivity" that we suggest here underscores the significance of user-to-technology, user-to-user, and user-to-content/service interaction for literacy development, with interaction involving not only skills but also learning, thinking, and various modes of practicing. This assigns to literacy multiple layers of development, application and effect at the perceptual, knowledge, practical, and behavioral levels. This allows the adoption of the plural—namely, "multi-literacies"—while also pointing to links between literacy and identity in the framework of people's use and appropriation of digital technologies.

Identity and associated meanings of self and self's position at the socio-cultural, socio-demographic, and psychological levels matter for the course and features of "interactivity," for how the user goes about discovering and using digital tools and services as well as for how resistant or receptive the user is to them. Identity matters for digital literacies as it affects the user-to-technology, user-to-user, and user-to-content/service interaction, and related learning, thinking, and practicing. At the same time, user identity is manifested or even altered through human-technology interaction, as interaction effects can take various forms (e.g., reflections, experiences, lessons, knowledge, ideas, and so on), all, many or some of which can interfere with the wide range of socio-culturally driven, socio-demographic, and individuated trends of meaning that constitute user identity at large.

Hence, it can be argued that "multi-literacies" developed through and throughout the user-to-technology, user-to-user, and user-to-content/service interaction challenge the prevalent perception of literacy as a static realm of skills, capability, and expertise that determines people's use of and experiences with technology before technology use even begins. At the same time, "multi-literacies" highlight the importance of focusing on aspects of user identity, such as age, for explaining literacy trends and changes. This proposition invites research to examine digital inclusion as a continuously moving landscape through the user-to-technology, user-to-user, and user-to-content/service interaction and under the influence of the complex mosaic of user identity and age identity in particular.

3. Methodology

This study aimed to operationalize the conceptual propositions presented earlier and address the study's RQ through in-depth semi-structured interviews.

Specifically, we interviewed 12 digital users: 6 older users (55+ years) and 6 younger users (18–24 years). The selection of these two age categories related to our aim to gain insight into the nuances of digital inclusion of young adults—who are often considered "digital natives"—and middle-aged and elderly people—who are broadly assumed to be "digital immigrants." Also, we purposefully looked for a socio-demographic spread of users in each age category (e.g., males and females, working and non-working people). At the same time, both age groups consisted of users who made substantial use of a range of digital tools and services, whom we called "advanced users," as well as of users who used a limited number of digital tools and services, whom we called "minimal users."[2]

The interviews comprised semi-structured discussions with digital users about how they see themselves in relation to the digital so as to unpack the possible connections of their identity in general and their age identity in particular with their digital experiences, including their digital futures. The interview questions were piloted with each age group and adjusted slightly. The analysis of the interview data combined a bottom-up thematic and a top-down deductive approach. On the one hand, we allowed for themes to emerge from the data, while we looked at contradictions and conflict as well as similarities and invisible agreed assumptions within and between interviewees of different age, gender, and use backgrounds. On the other hand, we used NVivo to structure and code the data alongside the four underpinnings of digital inclusion that also comprised the four main topics of inquiry in the interviews (see Figure 16.1)—digital experiences (i.e., range and types of use), evaluation of the digital (i.e., perceptions, views, etc.), digital effects (i.e., effectiveness and consequences of use), and future directions (i.e., future-looking attitudes, digital aspirations, etc).

The concept of digital inclusion and its four underpinnings were then analyzed in the context of the broader research framework of the study (see Figure 16.2). The research framework includes the concept of digital users and its two binary categorizations (i.e., older vs. younger uses; minimal vs. advanced users), as well as the concepts of (age) identity and literacy, with their role in digital inclusion being examined from a digital user perspective.

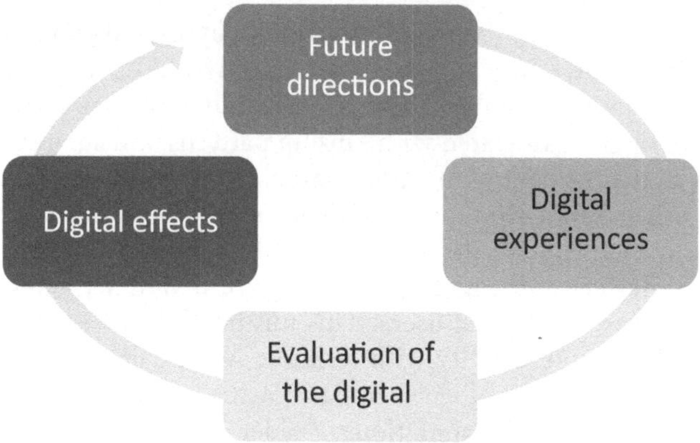

Figure 16.1 Digital Inclusion and Its Underpinnings.

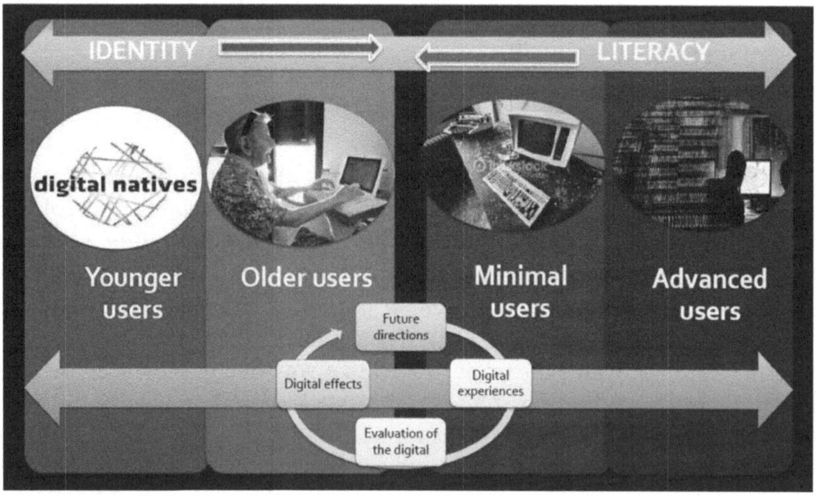

Figure 16.2 Research Framework.

4. Findings

4.1 Digital Experiences

In general, all participants were users of multiple digital devices, smartphones, tablets, etc. Many were working across several devices simultaneously, engaging in a number of activities throughout use.

Older Users

Older users claimed that their decision to take up digital technologies was in order to facilitate family communication and keep up with the interests of the younger family members. They clearly indicated that paths to digital inclusion are tightly associated with kinship patterns and age-related family roles, such as that of parent and grandparent.

On the other hand, they perceived themselves as late adopters facing technological challenges. Although some of them were interacting with many digital devices and engaging in a range of digital activities, they perceived themselves as minimal users. This mismatch between self-perceived and demonstrated interactivity with technology (i.e., literacy) we interpret as the result of the psychological impact of the "elderly stereotype" and other socially constructed assumptions of older people as lagging behind the young generation of "digital natives." In turn, self-perceived literacy seemed to have certain implications on their interaction with technology—namely, on their factual literacy, as it provoked fear, anxiety, and mistrust when technology malfunctioned: "DMUM2: I grew up in an age where you relied on electricity. So, I sort of think, well, what happens if the plug comes out of the computer. That's my fear."

Young Users

Friends, family, and social networks were the main drivers of use for young users, demonstrating the importance of age-related activities and roles at the personal, family, and social levels. Specifically, younger users raised an observer sense of watching what goes on in friends' and other people's lives, as well as a preoccupation with self in providing others with updates on themselves and their lives. Also, they considered digital technologies very important for study and career purposes, in a way being influenced by both biological and socially constructed age affordances. For instance, YUF1, who had just finished college, placed much emphasis on the importance of digital technologies for university study and her career development as a primary school teacher: "I'll be doing primary school teaching, so I'll be using things like Smart Boards . . . it saves a lot of time pre-writing and planning your lessons onto a Smart Board."

Most young users were familiar with the latest gadgets and digital services. In contrast to older users, they demonstrated a high degree of confidence with technology and welcomed prospects of learning through experimenting with and adopting new forms of technology, thus happily undertaking challenging avenues of interaction with technology,

> I just changed to Mac and I don't know what I'm doing; everything is different and it will take me a lot of time to get used . . . I quite enjoy that; it's like learning; it's a challenge.
>
> (YUF2)

Similar to older users though, their overall attitude toward literacy development largely derived from socially constructed perceptions of the importance of communication/networking, study, and career affordances of their biological age.

At the same time, YMM1 and YMM2, two unemployed male users, were minimal users of digital technologies and mostly had an interest in online games, appearing to lack the incentive to enhance their digital use. Such exceptions in the young user groups remind us of the need to refrain from uncritically assigning the "digital natives" label to young people and seriously consider aspects of intra-generational diversity in relation to digital inclusion.

4.2 Evaluation of the Digital

Older Users

Older users and especially the advanced users tended to be utilitarian in their approach to the digital. In assessing technology and making decisions on their present and future digital inclusion, they emphasized how much and in what ways their interaction with technology serves the purposes of "convenience" and "practicality."

On the other hand, minimal users indicated that age and associated thinking, mentality and life traits set limits on their interaction with digital technologies, demonstrating the roles that (self-enforced at the micro-level) socially constructed and biological age constraints play in their interaction with and consequent evaluation of digital technologies,

> I don't really want to know too much about it . . . I went for about two weeks without using my laptop and I forgot all my passwords . . . I'm just at that age . . . It's a bit of a turnover; it's a different mentality.
>
> (DMUM2)

Also, socially driven perceptions of generation gaps as well as family and other experiences at the micro-level seemed to affect older users' attitudes to the role of digital technologies in younger people's lives. Especially minimal users were concerned that the younger ones were obsessed with the digital and counter-proposed the empowering alternative of digital choices that fit life patterns, thus dismissing the social pressure for keeping up with technological development in order not to be left behind,

> Technology is progressing . . . and before long, people are not going to have . . . freedom. Cause they are so hooked on technology that they actually rely on it. So what happens when someone pulls the plug?
>
> (DMUM1)

At the same time, some older users viewed digital technologies as important in order not to be left out, resonating their anxiety over retirement and socially driven mechanisms of social exclusion due to aging. In particular, those who were enthusiastic adopters of digital technology had placed technology at the center of their lives, such as DUF2, a 67-year-old retired female user, who stated that since she got retired and started to spend most of her time at home, she could not live without her digital interactions.

Younger Users

Similar to older users, younger users appreciated digital technologies on the grounds of "convenience" and in accordance with their age-related life trajectories. Unlike older users though, most young users adopted digital buoyancy and largely developed positive approaches to digital applications and possibilities. They stressed in particular the role of digital media in achieving age-related study and career aspirations, demonstrating the role that socially constructed and biological aspects of their age identity play in the interaction patterns they develop with technology and their consequent evaluations of the digital.

At the same time, younger users viewed digital technologies as a two-edged sword: on the one hand, they acknowledged that technology has a practical, positive role in study, work and other activities, which are socially defined as important for them; on the other hand, they were conscious of the tendency to become technologically dependent and anxious about the dominant role of technology in their lives,

> I was on a holiday in Italy and when we were out and about there was no internet connection and incredibly dodgy phone reception. At first, I got panicked thinking: "Oh my God, no one can contact me!" But after a while, I actually felt quite free thinking that "I can't be contacted, so I can completely switch off" . . . it was just like a weight off.
> (YUF3)

In this respect, younger users did not idealize digital technology. Although they acknowledged its increasing importance, they did not think that digital experiences can replace offline experiences. Similar to older users, they experienced social pressure over using as many technologies and as much as possible and they reflected on how rapid technological growth will influence their future living: "I'm kind of worried as to where technology is to go next and how quickly it'll get there . . . I worry whether we're gonna be able to keep up with it" (YUM3). In this sense, they seemed to experience a growing realization of digital risks and they were marked by a similar anxiety to that of older users about the effects on those coming up behind them,

> My neighbor who is three years old sits quite happily using the iPad, knowing how to use it exactly, knowing how to use iPhones . . . it's

> insane. Such young kids are growing up in a world which is driven by technology and this is scary.
>
> (YUF1)

This illustrates that age identity might create a sense of "generation gap" even for young users who are anxious about the "naturalized" role of digital technologies in the lives of the even younger ones, thus challenging socially defined age notions such as those of "digital natives" and "e-generation."

4.3 Digital Effects

Older Users

Digital use appeared to enhance older users' family communication and, in this respect, older users linked digital effects to family drivers of digital use. Likewise, negative effects of digital technologies were placed in older people's age-dependent family-life context. For instance, DMUF1 was concerned about children's safety on the Internet (e.g., cyber-bullying, pedophilia) and highlighted how her role as a mother and family-oriented woman affected her feelings about digital risks and associated harm.

On the other hand, older users found themselves under immense (social) pressure to take up an increasing range of digital possibilities (e.g., e-shopping, e-banking) in order to ease life challenges that come along with aging (e.g., immobility, poor health, quieter lifestyle). In a way, the interplay between social constructions around aging (e.g., marginalization and exclusion) and life conditions that come alongside biological aging (e.g., health-related difficulties) appeared to make older users experience psychological stress over the effects of non-use of digital technologies and services. At the same time, older users expressed a concern about the effects of (social) pressure for more and extended adoption of technology on younger people and they reflected on resulting generational differences in values and fashion, and on their own perceptions of new generation trends and practices.. For instance, DMF2 remarked that young people are losing interest in other subject areas, not truly understanding the science or technology behind the digital equipment they use. In particular old users making extensive use of digital technologies raised concerns about young people's dependency on technology and their inclination to buy the latest gadgets and devices, with DUM1 likening such behaviors to the desire of young people to follow fashion.

Younger Users

Similar to older users, younger users evaluated digital effects on the basis of their drivers of use. Specifically, they assessed digital technologies as practically easing life and enhancing career performance. For instance, YUF3, a

24-year-old law teacher, used smartphones and clips in her teaching in order to enhance students' learning experience. Also, digital technologies appeared to facilitate young people's complicated patterns of social and personal communication: "I was in a long distance relationship and having the tools to be able to communicate while being apart made things easier" (YUF3). Overall, younger users characterized digital technologies as a window on the outside world, a "communal clip board" that enables them to connect, exchange information and become familiar with different places and cultures.

On the other hand, younger users expressed critical viewpoints about how technology reduces immediacy in human communication and impacts adversely on their relationships,

> I was on holiday and my girlfriend was texting me; I was texting her back but my Wi-Fi cut out cause there was a thunder storm above. She thought I was ignoring her cause I didn't text back for about 7 minutes.

Similar to older users, they felt pressured to be digitally connected so as to respond to the emerging culture of instantaneity. The instantaneous broadcasting of news through microblogging and other platforms made them (rather inevitably) a type of celebrity in their social circle and their personal details and emotions were thrust into the public eye. As a result, they thought that having digital breaks could alleviate the social pressure felt over digital technologies. For YUF1, for instance, not checking Facebook and text messages when on a weeklong holiday was refreshing, although she admitted it is too difficult to do this on a day-to-day basis. Another example is YUF3 who had no Internet connection for three months when she moved into her new flat and gradually developed a sense of freedom having escaped from the constant barrage of information and contact.

In addition, younger users talked about digital risks by mostly referring to the younger generations and children. YUM3 argued that 12–13-year-old children post pictures onto Facebook looking for popularity and self-esteem. He also referred to kids and teenagers who are under pressure to acquire the latest game consoles, phones, laptops and the like in order not to experience peer dismissal. YUF1 was concerned that children were unaware of the privacy risks in online social networks and so vulnerable to privacy violation, stalking, and abuse online. Such views not only challenge conventional discourses around age definers such as "young" and "old" but also showcase how individual points of view and experiences shape subjective understandings of age and its association with digital effects.

4.4 *Future Directions*

Older Users

Older users were content with the range of digital technologies they were using and appeared selective with the new technologies they might decide to use in the future. For instance, DUM1 was unlikely to expand his digital use to online shopping or e-banking due to familiarity with face-to-face transactions and anxiety over online fraud. Minimal users in particular had limited digital aspirations and associated age not only with the lack of desire or need to expand their digital use but also with limitations on their capability to learn and improve use in future: "There's such a lot to learn and, being older, it doesn't go in as easy as when you're younger . . . It goes in and comes straight out!" (DMUF3). This suggests that social constructions on how aging constrains learning and personal development influence negatively older users' patterns of interaction with digital technologies as well as their future prospects of digital inclusion.

Older users were aware of the availability of training, but only a couple were planning to take advantage of it, buy new devices and advance their digital use in the future. Those willing to undertake training appeared to cope quite well with age-related anxieties. As DMUF3, an 80-year-old female minimal user, stated,

> I'm learning . . . when you're older and you first start you're frightened that you're going to press the wrong button . . . But it's amazing, I mean, I have got used to a hell of a lot now.

On the other hand, some referred to a mismatch between training provision and their own individual needs, and adopted a utilitarian approach to training For instance, DUM3, a 69-year-old male user with an engineering background and a desire to develop his digital literacy, considered external training pressured and not quite tailored to his needs. Older users' critical evaluation of existing training on the grounds of psychological and self-driven aspects of their identity (e.g., needs, preferences) showcases that bracketing older people as "digital immigrants" can be misleading or tell a partial story, also pointing to intra-generational diversity.

Younger Users

The majority of younger users were content with the technologies they were using and would consider future changes depending on their life circumstances and individual needs. On the one hand, they did not consider it crucial to buy the latest device and were quite reflective on how technology is marketed. On the other hand, they had a self-driven approach to age-appropriateness of technology that influenced their digital aspirations. For instance, YUF3, aged 24, was a Facebook user but said she would use online networking platforms that emerged more recently, such as Twitter or Instagram, only if she was younger. Such views point out the role of rapid technological development and its links to self-perceptions of age. Another female user suggested she would reduce Internet use in the future when traditional and sustainable sociability platforms, such as work and family/kinship, would acquire a more central place in her life—namely, when she would transit from the formative world of a young person to the fully consolidated life of a more mature person.

Overall, younger users did not appear to have a great desire to use more or different technologies in the future. This was explained not only by their satisfaction with current use and the associated lack of a need for new digital experiences but also by their attachment to technologies they perceived as current and suitable for "their" generation, with technology having its own "age identity."

5. Concluding Discussion

In addressing the RQ of the study, the interview findings illustrate that age identity and literacy influence digital inclusion separately as well as jointly, with age identity influencing both actual and perceptual literacy gaps. More specifically, age-related life-roles, interests, and preoccupations affect people's decision to use digital technologies as well as the range of technologies and services they use. Importantly, age identity appears to have some bearing on self-perceptions of literacy—especially among older users—impacting psychologically and practically on users' interaction with technology. In evaluating digital technologies, users rely heavily on age-related convenience and practicality while age identity appears to influence their attachment to technologies they grew up with, their inclination for more or less, basic or advanced use of digital technologies in the future, and subsequently their will to undertake training for interactivity (i.e., literacy) advancement. At the same time, the presence of both minimal and advanced users in the two age groups, as well as the existence of digital literacy discrepancies and differential digital inclusion prospects within the same age group challenge conventional discourses around generation gaps. Specifically, the study demonstrates that considerations of age identity in relation to the digital are subjective and relational, disrupting assumptions about youth as "digital natives" and older people as "digital immigrants" andinviting more extensive study of concrete aspects of intra-generational diversity in relation to digital inclusion.

At the conceptual level, these findings support our suggestion to move from a "skills" to an "interactivity" perspective on literacy and illustrate that diverse thinking, learning and practicing are at the core of user interaction with technology. Although some of the interviewees did not demonstrate explicit understanding of self-driven aspects of their age identity, the findings show that, to better understand the role of socially constructed and biological dimensions of age identity in users' digital experiences and attitudes, we need to develop a micro-perspective on age identity.

In a nutshell, the study makes the following contributions to digital divide and digital inclusion scholarship:

1) It argues in favor of and empirically operationalizes Livingstone's proposal for a conceptual shift from skills to interaction in the study of

literacy in digital inclusion. Also, it argues that identity involves the ways and extent to which socio-demographic and socio-cultural factors are perceived and integrated at the psychological level and thus it suggests that we view age identity as self-driven and at the micro-level, so as to better understand the complex role of socially constructed and biological aspects of age in digital inclusion.

2) It provides qualitative evidence on the value of the notion of "interactivity" in the study of digital literacy. The interviews shed some light on the diverse and complex experiences, practices, evaluations and post-use reflections involved in users' interaction with digital technologies and services as well as with other users. Also, the interviews showed that age identity matters for users' decisions on and aspects of current and planned digital use, and provided examples of the links of age with other elements of the user identity, such as family orientation, study and career aspirations, and so on. Interestingly, the study provides further evidence that questions continuing tendencies to bracket young as digital natives and older people as digital immigrants, granting certain validity to ongoing intra-generational discourses.
3) Even more importantly, in answering the RQ, the study points to certain linkages between literacy and identity in making sense of and explaining varying degrees of digital inclusion. Specifically, the study found evidence on the interrelation between age identity and self-perception of digital literacy (e.g., the psychological impact of age leading to a mismatch between older users' perceived and factual/demonstrated digital literacy), which influences users' both ongoing and planned use of digital technologies. At the same time, the decision of some older people not to undertake training to enhance their literacy in the future is influenced by their age identity.
4) Finally, the study has been inclusive with regard to the "digital." Unlike much of the existing digital inclusion research that limits its focus on the Internet, this study acknowledged that different digital technologies can be used differently by different individuals. Along these lines, the study employed the categories "minimal" and "advanced" users, which provided the scope for identifying levels of digital inclusion in association with literacy and identity, thus casting light on the nuances and complexities inherent in digital inclusion.

In a policy context, this study makes a first step in the exploration of identity-literacy links within the realm of digital inclusion and problematizes linear and benefits-focused policy approaches to digital inclusion. On the one hand, policy makers and other stakeholders raise the importance of literacy. Literacy is a strategic goal of Ofcom, the UK media regulator (mandated in the Communications Act 2003).[3] The first international forum on media and information literacy (Fez, Morocco in 2011) declared that literacy is a fundamental human right that enhances the quality of human life. On the other hand, policy makers primarily take a "skills"/"capabilities" approach to digital inclusion. For instance, the UK government's digital inclusion strategy has focused on the need to overcome access, skills, motivation and trust challenges that deprive categories of the population of the benefits of digital inclusion. Thus, it has set ten actions in order for all citizens to be digitally included by 2020, all of which stress the importance of "capabilities" (Cabinet Office & Government Digital Service 2014). This is to say that the UK government places particular emphasis on "skills" and "capabilities" to address complex literacy levels and associated disparities that impact the degree and quality of digital inclusion. Our study demonstrates the need for policy makers to go beyond a skills-enhancement approach and to consider the links between citizens' literacy and their identity, thus taking a more granular approach to the nuances of digital inclusion and acknowledging the various possibilities involved in people's positioning in the digital realm.

Furthermore, the study offers findings that can inform media literacy training and education providers as well as policy practices in the field. For instance, practitioners and policy makers should recognize the significant implications of older people's under-assessment of digital empowerment, as questions of digital citizenship move further up the digital economy agenda and aging is an increasing area of policy concern. Public debates may need to build awareness among older people of their tendency to underestimate both themselves as digitally included and their potential within the digital economy. In addition, older users' critical discourses about the available training, and its (age-related) appropriateness can feed discussions on current training provision and bespoke training in the future. On the other hand, young users, though more confident and with higher digital aspirations, can experience a generation gap themselves as to their reluctance to adopt emerging and future digital technologies and their concerns over technology dominating the lives of minors and children. This sends a strong message to those—practitioners, training providers, educators and policy makers—who celebrate youth as the digital natives and focus their training/educating mostly on the older populations.

While ours was a small study, its qualitative findings offer original insights into the linkages between digital literacy and identity and their role in digital inclusion. Such insights can provide the ground for more in- depth and larger-scale research as well as for new policy considerations in relation to digital inclusion in the future.

Notes

1 Nonetheless, a language of "exclusion" and "divide" was not abandoned altogether. Scholars continued to refer to the digital divide as a problematic phenomenon with implications for policy making (e.g., Epstein, Nisbet and Gillespie 2011; Verdegem and Verhoest 2009).

2 Appendix 1 presents the socio-demographic and user profile of the interviewees.

3 The contents of the Act can be found at: https://www.legislation.gov.uk/ukpga/2003/21/contents

References

Angwin, Julia, and Laoura Castaneda. "The Digital Divide/High-Tech Boom a Bust for Blacks, Latinos." *San Francisco Chronicle*, May 4, 1998. www.sfgate.com/news/article/The-Digital-Divide-High-tech-boom-a-bust-for-3007911.php.

Appel, Markus. "Are Heavy Users of Computer Games and Social Media More Computer Literate?" *Computers and Education* 59, no. 4 (2012): 1339–49. doi:10.1016/j.compedu.2012.06.004.

Baron, Naomi S., and Ylva Hårdaf Segerstad. "Cross-Cultural Patterns in Mobile-Phone Use: Public Space and Reachability in Sweden, the USA and Japan." *New Media & Society* 12, no. 1 (2010): 13–34. doi: 10.1177/1461444809355111.

Berners-Lee, Tim, and Mark Fischetti. *Weaving the Web: The Past, Present and Future of the World Wide Web*. London: Orion, 1999.

Bradbrook, Gail, and Fisher, John. *Digital Equality: Reviewing Digital Inclusion Activity and Mapping the Way Forward*. London: Citizens Online, 2004.

Brandtweiner, Roman, Elisabeth Donat, and Johann Kerschbaum. "How to Become a Sophisticated User: A Two-Dimensional Approach to e-Literacy." *New Media & Society* 12, no. 5 (2011):813–33. doi: 10.1177/1461444809349577.

Cabinet Office and Government Digital Service. "Government Digital Inclusion Strategy." *Policy Paper*, London, November 10, 2014.

Carpentier, Nico. "Differentiating Between Access, Interaction and Participation." *Conjunctions: Transdisciplinary Journal of Cultural Participation* 2, no. 2 (2015): 7–28. doi: 10.7146/tjcp.v2i2.22844.

Carvin, A. "More than Just Access: Fitting Literacy and Content into the Digital Divide Equation." *Educause Review* 35, no. 6 (2000): 38–47.

Castells Manuel. *The Power of Identity*, 2nd ed. Malden, MA: Blackwell, 2010.

Clarke, Amanda, and Lorna Warren. "Hopes, Fears and Expectations About the Future: What Do Older People's Stories Tell Us About Active Ageing?" *Ageing & Society* 27, no. 4 (2007): 465–88. doi: 10.1017/S0144686X06005824.

Cuddy, Amy J. C., Michael I. Norton, and Susan T. Fiske. "This Old Stereotype: The Pervasiveness and Persistence of the Elderly Stereotype." *Journal of Social Issues* 61, no. 2 (2005): 267–85. doi: 10.1111/j.1540-4560.2005.00405.x.

Digital Inclusion Team. *The Digital Inclusion Landscape in England: Delivering Social Impact Through Information and Communications Technology*. London: Digital Inclusion Team, 2007.

DiMaggio, Paul, Eszter Hargittai, Coral Celeste, and Steven Shafer. "Digital Inequality: From Unequal Access to Differentiated Use." In *Social Inequality*, edited by K. Neckerman, 355–400. New York: Russell Sage Foundation, 2004.

Epstein, Dmitry, Erik C. Nisbet, and Tarleton Gillespie. "Who's Responsible for the Digital Divide? Public Perceptions and Policy Implications." *The Information Society* 27, no. 2 (2011): 92–104. doi: 10.1080/01972243.2011.548695.

Erumban, Abdul Azeez, and Simon B. de Jong. "Cross-Country Differences in ICT Adoption: A Consequence of Culture?" *Journal of World Business* 41, no. 4 (2006): 302–4. doi:10.1016/j.jwb.2006.08.005

Fuchs, Cristian. *Internet and Society: Social Theory in the Information Age*. New York: Routledge, 2008.

Gee, James Paul. *What Video Games Have to Teach Us About Learning and Literacy*. New York: Palgrave Macmillan, 2003.

Gui, Marco, and Gianluca Argentin. "Digital Skills of Internet Natives: Different Forms of Digital Literacy in a Random Sample of Northern Italian

High School Students." *New Media & Society* 13, no. 6 (2011): 963–80. doi: 10.1177/1461444810389751
Hargittai, Eszter. "Second Level Digital Divide: Differences in People's Online Skills." *First Monday* 7, no. 4 (2002).
Hargittai, Eszter, and Amanda Hinnant. "Digital Inequality: Differences in Young Adults' Use of the Internet." *Communication Research* 35, no. 5 (2008): 602–21. doi: 10.1177/0093650208321782.
Helsper, Ellen J. *Digital Inclusion: An Analysis of Social Disadvantage and the Information Society*. London: Department for Communities and Local Government, 2008.
Helsper, Ellen J. "A Corresponding Fields Model for the Links Between Social and Digital Exclusion." *Communication Theory* 22, no. 4 (2012): 403–26. doi:10.1111/j.1468-2885.2012.01416.x.
Helsper, Ellen J., and Rebecca Eynon. "Digital Natives: Where Is the Evidence?" *British Educational Research Journal* 36, no. 3 (2010): 503–20. doi: 10.1080/01411920902989227.
Jensen, Jens F. " 'Interactivity': Tracking a New Concept in Media and Communication Studies." *Nordicom Review* 19, no. 1 (1998): 185–204.
Katz, James E., and Philip Aspden. "Internet Dropouts in the USA: The Invisible Group." *Telecommunications Policy* 22, nos. 4–5 (1998): 327–39.
Katz, James E., and Ronald D. Rice. *Social Consequences of the Internet: Access, Involvement, and Interaction*. Cambridge, MA: MIT Press, 2002.
Kvasny, Lynette. "Cultural (Re)production of Digital Inequality in a US Community Technology Initiative." *Information, Communication and Society* 9, no. 2 (2006): 160–81. doi: 10.1080/13691180600630740.
Lee, Jae-Shin. "Interactivity: A New Approach." Paper presented at the *Association for Education in Journalism and Mass Communication Conference*, Phoenix, AZ, August 9–12, 2000.
Lenhart, Amanda, and John B. Horrigan. "Re-Visualizing the Digital Divide as a Digital Spectrum." *IT&Society* 1, no. 5 (2003): 23–39.
Livingstone, Sonia. "Media Literacy and the Challenge of New Information and Communication Technologies." *The Communication Review* 7, no. 1 (2004): 3–14. doi: 10.1080/10714420490280152.
Livingstone, Sonia. "Youthful Experts? A Critical Appraisal of Children's Emerging Internet Literacy." In *Oxford Handbook on ICTs*, edited by Robin Mansell, Chrisanthi Avgerou, Danny Quah, and Roger Silverstone, 494–513. Oxford: Oxford University Press, 2007.
Livingstone, Sonia. "Engaging with Media—a Matter of Literacy?" *Communication, Culture & Critique* 1, no. 1 (2008): 51–62. doi: 10.1111/j.1753-9137.2007.00006.x.
Livingstone, Sonia, and Ellen J. Helsper. "Gradations in Digital Inclusion: Children, Young People and the Digital Divide." *New Media & Society* 9, no. 4 (2007): 671–96. doi: 10.1177/1461444807080335.
Loos, Eugene. "Designing for Dynamic Diversity: Representing Various Senior Citizens in Digital Information Sources." *Observatorio (OBS*) Journal* 7, no. 1 (2013): 21–45.
Macnicol, John. *Age Discrimination: An Historical and Contemporary Analysis*. Cambridge: Cambridge University Press, 2006.
Manovich, Lev. *The Language of New Media*. Cambridge, MA: MIT Press, 2001.

McMillan, Sally J. "Exploring Models of Interactivity from Multiple Research Traditions: Users, Documents, and Systems." In *Handbook of New Media: Social Shaping and Consequences of ICTs*, edited by Leah A. Lievrouw and Sonia Livingstone, 163–82. London: Sage, 2002.

Millward, Peter. "The 'Grey Digital Divide': Perception, Exclusion and Barriers of Access to the Internet for Older People." *First Monday* 8, no. 7 (2003).

Min, Seong-Jae. "From the Digital Divide to the Democratic Divide: Internet Skills, Political Interest, and the Second-Level Digital Divide in Political Internet Use." *Journal of Information Technology & Politics* 7, no. 1 (2010): 22–35. doi: 10.1080/19331680903109402.

Morrisett, Lloyd. "Foreword." In *The Digital Divide. Facing a Crisis or Creating a Myth?* edited by Benjamin M. Compaine, ix–x. Cambridge, MA and London: MIT Press, 2001.

Nelson, Todd D., ed. *Ageism: Stereotyping and Prejudice Against Older Persons*. Cambridge, MA: MIT Press, 2002.

Norris, Pipa. *Digital Divide: Civic Engagement, Information Poverty, and the Internet Worldwide*. Cambridge: Cambridge University Press, 2001.

Notley, Tanya. "Young People, Online Networks, and Social Inclusion." *Journal of Computer-Mediated Communication* 14, no. 4 (2009): 1208–27. doi: 10.1111/j.1083-6101.2009.01487.x.

Olphert, Wendy, and Leela Damodaran. "People and Digital Disengagement: A Fourth Digital Divide?" *Gerontology* 59, no. 6 (2013): 564–70. doi: 10.1159/000353630.

Prensky, Marc. "Digital Natives, Digital Immigrants." *On the Horizon* 9, no. 5 (2001): 1–6.

Punie, Yves, Wainer Lusoli, Clara Centeno, Gianluca Misuraca, and David Broster, eds. *The Impact of Social Computing on the EU Information Society and Economy*. Seville: European Commission—Joint Research Centre—Institute for Prospective Technological Studies, 2009.

Rafaeli, Sheizaf, and Fay Sudweeks. "Networked Interactivity." *Journal of Computer-Mediated Communication* 2, no. 4 (1997). doi: 10.1111/j.1083-6101.1997.tb00201.x.

Reed, Jan, Margaret Cook, Glenda Cook, Pamela Inglis, and Charlotte Clarke. "Specialist Services for Older People: Issues of Negative and Positive Ageism." *Ageing & Society* 26, no. 6 (2006): 849–65. doi: 10.1017/S0144686X06004855.

Reisdorf, Bianca C. "Non-Adoption of the Internet in Great Britain and Sweden." *Information, Communication & Society* 14, no. 3 (2011): 400–20. doi: 10.1080/1369118X.2010.543141.

Rice, Ronald D. "Primary Issues in Internet Use: Access, Civic and Community Involvement, and Social Interaction and Expression." In *Handbook of New Media: Social Shaping and Consequences of ICTs*, edited by Leah A. Lievrouw and Sonia Livingstone, 109–29. London: Sage, 2002.

Robinson, John P., and Steven P. Martin. "Social Attitude Differences Between Internet Users and Non-Users." *Information, Communication & Society* 12, no. 4 (2009): 508–24. doi: 10.1080/13691180902857645.

Robinson, Laura. "Freeways, Detours, and Dead Ends: Search Journeys Among Disadvantaged Youth." *New Media & Society* 16, no. 2 (2014): 234–51. doi: 10.1177/1461444813481197.

Rogers, Everett M. *Diffusion of Innovations*, Vol. 4. New York: Free Press, 1995.
Rogers, Everett M. "The Digital Divide." *Convergence* 7, no. 4 (2001): 96–111. doi: 10.1177/135485650100700406.
Selwyn, Neil. "Reconsidering Political and Popular Understandings of the Digital Divide." *New Media & Society* 6, no. 3 (2004): 341–62. doi: 10.1177/1461444804042519.
Selwyn, Neil. "Digital Division or Digital Decision? A Study of Non-Users and Low-Users of Computers." *Poetics* 34 (2006): 273–92. doi:10.1016/j.poetic.2006.05.003.
Selwyn, Neil. "The Digital Native—Myth and Reality." *Aslib Proceedings* 61, no. 4 (2009): 364–79. doi: 10.1108/00012530910973776.
Stanley, Laura D. "Beyond Access: Psychosocial Barriers to Computer Literacy." *The Information Society* 19, no. 5 (2003): 407–16. doi: 10.1080/715720560.
Szuprowicz, Bohdan O. *Multimedia Networking*. New York: McGraw-Hill, Inc., 1995.
Tapscott, Don. *Growing up Digital: The Rise of the Net Generation*. New York: McGraw-Hill, 1998.
Thoman, Elisabeth, and Tessa Jolls. "Media Literacy Education: Lessons from the Center for Media Literacy." In *Media Literacy: Transforming Curriculum and Teaching*, edited by Gretchen Schwartz and Pamela U. Brown, 180–205. Malden, MA: National Society for the Study of Education, 2005.
Thomas, Frank, Leslie Haddon, Rosemarie Gilligan, Peter Heinzmann, and Chantal de Gournay. "Cultural Factors Shaping the Experience of ICTs: An Exploratory Review." In *International Collaborative Research: Cross-Cultural Differences and Cultures of Research*, edited by Leslie Haddon, 13–49. Brussels: COST, 2005.
Tsatsou, Panayiota. "Non-Adoption of the Internet: A Qualitative Examination. Empirical Insights and Methodological Implications." *Questions de communication* 18 (December 2010).
Tsatsou, Panayiota. "Digital Divides Revisited: What Is New About Divides and Their Research?" *Media, Culture & Society* 33, no. 2 (2011a): 317–31. doi: 10.1177/0163443710393865.
Tsatsou, Panayiota. "Why Internet Use? A Quantitative Examination of the Role of Everyday Life and Internet Policy and Regulation." *Technology in Society* 33, nos. 1–2 (2011b): 73–83. doi:10.1016/j.techsoc.2011.03.016.
Tsatsou, Panayiota. *Digital Divides in Europe: Culture, Politics and the Western-Southern Divide*. Berlin: Peter Lang, 2011c.
Van Deursen, Alexander J. A. M., van Dijk, Jan A. G. M., and Oscar Peters. "Rethinking Internet Skills: The Contribution of Gender, Age, Education, Internet Experience, and Hours Online to Medium—and Content—Related Internet Skills." *Poetics* 39, no. 2 (2011): 125–44. doi: 10.1016/j.poetic.2011.02.001.
Van Dijk, Jan A. G. M. *The Deepening Divide: Inequality in the Information Society*. Thousand Oaks, CA, London and New Delhi: Sage, 2005.
Van Dijk, Jan A. G. M., and Kenneth Hacker. "The Digital Divide as a Complex and Dynamic Phenomenon." *The Information Society* 19, no. 4 (2003): 315–26. doi: 10.1080/01972240309487.
Van Deursen, Alexander J. A. M., and Jan A. G. M. van Dijk. "Internet Skills and the Digital Divide." *New Media & Society* 13, no. 6 (2010): 893–911. doi: 10.1177/1461444810386774.

Verdegem, Peter, and Pascal Verhoest. "Profiling the Non-User: Rethinking Policy Initiatives Stimulating ICT Acceptance." *Telecommunications Policy* 33, nos. 10–11 (2009): 642–52. doi: 10.1016/j.telpol.2009.08.009.

Walton, Anthony. "Technology vs African-Americans." *Atlantic Monthly* 283, no. 1 (1999): 14–8.

Wilhelm, Anthony. "From Crystal Palaces to Silicon Valleys: Market Imperfection and the Enduring Digital Divide." In *Access Denied in the Information Age*, edited by Stephen Lax, 199–217. New York: Palgrave, 2001.

Wise, Kevin, Brian Hamman, and Kjerstin Thorson. "Moderation, Response Rate, and Message Interactivity: Features of Online Communities and Their Effects on Intent to Participate." *Journal of Computer-Mediated Communication* 12, no. 1 (2006): 24–41. doi: 10.1111/j.1083-6101.2006.00313.x.

Youngs, Gillian. *Global Political Economy in the Information Age: Power and Inequality*. London: Routledge, 2007.

Youngs, Gillian, ed. *Digital World: Connectivity, Creativity and Rights*. London: Routledge, 2013.

Zillien, Nicole, and Eszter Hargittai. "Digital Distinction: Status—Specific Types of Internet Usage." *Social Science Quarterly* 90, no. 2 (2009): 274–91. doi: 10.1111/j.1540-6237.2009.00617.x.

Appendix 16.1

Interview Participants

Older age participants				Younger age participants			
Code / Age	*User*	*Gender*	*Occupation*	*Code /Age*	*User*	*Gender*	*Occupation*
DUM1 / 58	Advanced	Male	Unemployed	YMM1 / 21	Minimal	Male	Unemployed
DMUM2 /61	Minimal	Male	Chef	YMM2 / 22	Minimal	Male	Unemployed
DUM3 / 69	Advanced	Male	Retired	YUM3 / 18	Advanced	Male	Starting Uni study
DMUF1 / 65	Minimal	Female	Retired	YUF1 / 18	Advanced	Female	Starting Uni study
DUF2 / 67	Advanced	Female	Retired	YUF2 / 18	Advanced	Female	Starting Uni study
DMUF3 / 80	Minimal	Female	Retired	YUF3 / 24	Advanced	Female	Studying PGC; law degree

Conclusion

Panayiota Tsatsou, Sherah Kurnia, and Jyoti Choudrie

This cutting-edge research book is written for researchers, students, academics, and policy makers and would not have been a feasible book project without the invaluable and timely chapter contributions of all authors. The authors have drawn from their expertise in IS, management, ICTs, new/digital media, and communications policy to provide insights into ICT-enabled services and social inclusion. The book provides numerous examples of ICT-enabled services in various sectors and their benefits to individuals, organizations, and society. ICTs have indeed become pervasive not only in the business sector but also in others sectors, including government, health, entertainment, and agriculture. Innovative ways of delivering health care, education, and other community related services have been enabled by the adoption of ICTs.

Given the importance of ICTs for improving the life and enhancing social inclusion of individuals, organizations and broader communities, equitable access to ICT-enabled services by all people is a requirement. However, adoption of ICT-enabled services is not only about the provision of infrastructure, devices, and affordable access but also relates to the competencies and capabilities in using the available services. Thus, despite the many benefits ICT-enabled services offer to society, barriers to widespread access to and effective use of the existing ICT-enabled services have been identified at various levels, ranging from individual to national level. Therefore, in order to take advantage of ICT-enabled services and enhance social inclusion through such services, concerted efforts are needed among and with various key stakeholders who are involved in developing, promoting and supplying such services.

This book contributes to ongoing debates and knowledge concerning ICT-enabled services and social inclusion. It examines the theoretical foundations of ICT-enabled services and social inclusion, presents various innovative cases of ICT-enabled services in a number of contexts, and investigates adoption, usage and management aspects related to social inclusion and usability of ICT-enabled services. As ICT-enabled services continue to spread rapidly within the society, this book provides valuable insights into the importance of ICT-enabled services for social inclusion, possible barriers to achieving social inclusion and strategies to mitigate them. Increasing the awareness of the importance of ICT-enabled services is an important first step to develop more socially beneficial and inclusive digital services.

Theoretical lenses discussed in Part 1 of this book, including the Capability Approach, the concept of connectivity as a social-technical lens, the ambient inclusion framework, and the concept of optimal level of user participation, and the offerings of useful frameworks to guide future efforts in addressing social inclusion and usability of ICT-enabled services for both research and practice. Case studies of recent and innovative applications of ICT-enabled services in a number of countries and contexts are presented in Part 2. These case studies provide insights into the drivers of development of digital services, the benefits expected and how social inclusion is addressed through such services, as well as the challenges encountered to improve access to and reach of services and the strategies employed to enhance social engagement with and usability of those services. Insights obtained from these cases studies of ICT-enabled services are invaluable for both researchers who aspire to conduct future study and practitioners who aim to develop more socially inclusive digital applications. Finally, Part 3 presents case studies that examine various aspects related to adoption and management of diverse ICT applications, aiming to enhance the current knowledge of key obstacles to effective adoption and management of such applications within the society. The chapters in this final part of the book apply different theories and concepts including stakeholder theory, theory of practices, and the concept of "interactivity" so as to better understand different types of barriers to ICT adoption and use, factors that encourage such adoption and use, as well as related strategies to manage and effectively use ICT-enabled services.

This book was developed to address the continuing need for making sense of complex socio-technical systems that are associated with the adoption, management and policy of new technological services and, thus, for unpacking the intercourse of social inclusion and usability of ICT-enabled digital services. Thus, all its chapters have aimed to offer theoretical, empirical and case study insights into the complex role of ICT-enabled services in social inclusion, showcasing the importance of the parameters of usability, patterns of adoption and management and moving away from one-sided accounts of technology features or purely socio-driven approaches in this area.

The conceptual insights the book offers show that the field of ICT-enabled services and social inclusion can employ and advance concepts and theories that derive from a range of study fields and disciplines. Thus, the "Capability Approach" and the concepts of "connectivity," "ambient inclusion," and "optimal level of user participation" that the theoretical chapters in Part 1 focus on demonstrate the conceptual richness and interdisciplinarity of this area of study. At the same time, they indicate that this study area is rather under-theorized, as it appears to borrow concepts and theoretical approaches from a range of disciplines, without original theorizations being developed or new conceptual approaches being proposed within its remit. Such under-theorization often deprives researchers of original theoretical armory and suitable conceptual frameworks that would allow them to advance empirical insights further and generate theoretically sharp and informed arguments.

Along these lines, the cornerstone concepts of usability and social inclusion must be revisited in the context of the study of ICT-enabled services, in order for researchers in this area to become better positioned to assess as well as to shed light on currently divergent and often confusing theoretical debates and presumptions regarding ICTs and social inclusion. In this regard, we acknowledge the need for advancing the theoretical work in this area and revisiting currently used theories and concepts, aiming for researchers not only to draw upon existing concepts and theories in pertinent fields of research but also to develop original theorizations, which could critically inform future work around ICT-enabled services and social inclusion.

In terms of empirical evidence on ICT-enabled services, this book confirms the existence of a large volume of empirical and case study research on a range of different ICT-enabled services. On the whole, such a significant volume of empirical study allows us to obtain an insight into both the smaller and the bigger picture of how ICTs can open up a new world of entertainment, occupation, social networking, political communication, and e-governance possibilities for diverse groups of the society. Specifically, Part 2 of the book sheds light on a range of such services, such as e-government, e-health, teleworking, e-commerce, e-activism and e-entertainment, while, both Parts 2 and 3 examine a series of actors and stakeholders who might benefit (or not) from such services (e.g., working people, professionals, farmers, activists, patients, older adults). However, we have identified a few challenges, dilemmas, and gaps in the research reporting empirical evidence that deserve further consideration.

The first challenge is to do with the extent to which heavily researched ICT-enabled services and case studies of such services do actually contribute original knowledge that can be followed up and inform new research in this area. Some of the current research seems to follow rather "secure" knowledge routes, not really pursuing to address unanswered questions, while some other research places existing knowledge at its core aiming to either challenge or confirm it. This is a challenge that relates less to the scale (small or large) of the research or the exact type of ICT-enabled services researched and more to the approach and aims that research in this area sets for itself.

Second, researchers should consider the extent to which research on ICT-enabled services moves as fast as services themselves and whether it should actually attempt to do so or not. An associated question is whether there is still a sufficiently good reason not to abandon the study of services that made their appearance a few decades ago (e.g., teleworking). On the one hand, emerging ICT-enabled services and related technologies generate new questions of research interest. On the other hand, existing and well-established services continue to matter in many contexts and for different parts of the society, while they do evolve themselves over time. Thus, researchers often find themselves trapped in a sort of dilemma, with the main body of research in this area appearing not to have achieved a kind of balance in the study of emerging and well-established ICT-enabled services. At the same time, the pace at which research attempts to keep up with technological and service changes has not received sufficient attention to date.

Third, a question we invite researchers to consider is how research insights and evidence in this area might contribute to knowledge enhancement when the same actors and stakeholders are approached through different lenses regarding their roles and the type of agency they convey. For instance, we found in this book that when terming the sample population, some members of the public are sometimes referred to as "users" and in other instances, they are termed as "consumers," without this necessarily being explained by the possibly different questions that research might wish to explore. Likewise, the same organizations and non-individual stakeholders who aim to supply and/or trade ICT-based services are sometimes labeled as organizations, in other times as professionals and some other times as corporate actors. The issue here is not to do really with labeling and language, but rather with the different assumptions and/or biases that may drive different researchers in this area.

Lastly, there appears to be a range of different focuses in the research in this area. For instance, the focus of some research appears to be on the role of demographics—especially age and profession—in the adoption and effective management of ICT-enabled services, while other research seems to be more interested in making sense of the type of relationship the user develops with technology-enabled services and associated patterns and purposes of use of those services (e.g., personal, commercial, or professional uses). However, the grounds on which the researchers set their agendas and priorities and how they position their work in the entire field of ICT-enabled services and social inclusion are questions that have hardly received any attention so far. Hence, regardless of the large volume of empirical research in this area, the choices made by researchers in terms of labeling, language and focus are not much clear or sufficiently explained. In this respect, we argue that this is a gap that needs to be addressed, something that researchers in this area must draw their attention to and reflect a lot more on in the future.

Following the earlier remarks on the existing theoretical and empirical research on ICT-enabled services and social inclusion and before we close this concluding discussion, we would like to put forward a few recommendations for future research in this area. First, we recommend the conduct of more theoretically ambitious work; namely, of work that will contribute to original concepts and theorizations and will not merely deploy existing theoretical frameworks that are borrowed from other research fields. Second, we recommend the development of a clearer understanding of research labeling, agendas and priorities in this area; thus going deeper into the geographical scope covered and the range of actors and services studied in what already comprises a significant volume of empirical research. Along these lines, we recommend the development of a better understanding of where the various research endeavors stand in this area, what contribution they make and how they can benefit future research. As ICT-enabled products and services are emerging in society and organizations rapidly, we, researchers, need to become better positioned in order to address the questions of why and what new research on ICT-enabled services and social inclusion we need and how such new research can add to existing research and contribute to knowledge enhancement in the field.

Index

About the Contributors

Adam Lodders is executive officer of the Melbourne Networked Society Institute and is responsible for administering, managing, and developing interdisciplinary research on the networked society. Working in interdisciplinary research since 2010, Adam's research interests are on law, regulation, and policy in the networked society. Adam has a juris doctor and master of public policy and management from the University of Melbourne, and a bachelor of economics and bachelor of arts (honors) from Monash University.

Anne Holland is professor of physiotherapy at La Trobe University and Alfred Health in Melbourne. Anne's research investigates new models of rehabilitation for people with chronic respiratory disease to improve treatment access, uptake, and inclusion. This includes studies of low cost home-based rehabilitation models and telerehabilitation. Anne has published over 160 peer-reviewed journal articles, and her publications have been cited more than 3700 times, including in 11 international treatment guidelines for respiratory physiotherapy, pulmonary rehabilitation, and chronic lung disease. In 2016, Anne received an award from the American Thoracic Society for her service to the field of pulmonary rehabilitation.

Antonette Mendoza is a senior lecturer in the Department of Computing and IS at the University of Melbourne. Her research expertise involves how users interact and adopt technology, how systems can be better designed to support that interaction, and once, deployed, what we can do to facilitate adoption and sustainability of technologies. She has extensive experience in software engineering, IT project management, and qualitative methods of research. She is currently collaborating with researchers on Australian Research Council (ARC) and National Health and Medical Research Council (NHMRC) projects in the health-care and homelessness environments. She is also involved in local and international collaborations with researchers on value realization of e-learning platforms and tools. Her achievements include Teaching Excellence Awards in the Melbourne School of Engineering and in the Department of Computing and Information Systems. You can find her recent publications via find an expert on the University of Melbourne webpage.

Arthur Glenn Maail is a research manager at Open Data Lab Jakarta. He obtained his PhD from the University of Melbourne in 2015. He majored in electrical engineering at the Telkom School of Engineering, Indonesia, attended a master's program in digital communication systems and technology at Chalmers University of Technology, Sweden, and received a PhD in engineering (IS) from the University of Melbourne where he did his doctoral research on the field of "Information and Communication Technology for Development." His research interests lie between technology and public policy, which include IS in developing countries, mobile technology, and data-driven public policy. Glenn's is currently working on several research projects related to implementation of open public data in developing countries which cut across different themes including smart cities and citizen participation (i.e., open cities), open public contracting data standard, and open data for budget transparency.

Avijit Sarkar is professor of operations research at the University of Redlands, School of Business. His research interests are in examining and explaining global technology adoption and utilization patterns and digital divides and GIS in business and operations research. He is co-author of the book *Global Digital Divides: Explaining Change* (Springer). His research articles have been published in journals such as *Telecommunications Policy*, *Journal of Geographical Systems*, *IIE Transactions*, *European Journal of Operational Research*, *Socio-Economic Planning Sciences* and *Computers and Operations Research*. He serves on the editorial board of the *International Journal of Business Analytics* and teaches courses in business analytics and operations research at the University of Redlands. He is a recipient of the University of Redlands Awards for outstanding research and teaching. Dr. Sarkar received his PhD and MS degrees in industrial engineering from the University at Buffalo—the State University of New York.

Bianca C. Reisdorf is an assistant professor in the Department of Media & Information and the assistant director of the Quello Center at Michigan State University. Her research interests include digital inequalities and policies, Internet use among vulnerable groups, and cross-national comparative studies that apply both qualitative and quantitative methods. She was lecturer and director of distance learning in the Department of Media and Communication at the University of Leicester in the United Kingdom before taking up her position at Michigan State University. She holds a PhD in information, communication, and the social sciences from Oxford Internet Institute, University of Oxford, United Kingdom.

Carolyn Watt is a PhD candidate (2016–2019) in the Faculty of Creative & Cultural Industries (CCI) at the University of Portsmouth. Her research focuses on social circus enhanced by digital technologies to foster social inclusion of young women. As research assistant to the CCI Enterprise & Innovation team, Watt supports the development of external bid submissions and faculty initiatives to develop research culture. Experience includes work on complex transdisciplinary projects encompassing the fields of design, science, technology, engineering, mathematics, and business (D-STEM-B). With a diverse blend of academic and performance expertise Watt has developed a wide-ranging skill set.

Christine B. Williams is a professor of political science at Bentley University. Her areas of expertise are political communication, campaigns and elections, and e-government. She serves as managing editor North America for the *Journal of Political Marketing*, and on the editorial boards of the *Journal of Information Technology and Politics*, *Journal of Public Affairs*, and the *International Journal of e-Politics*. She edited and contributed to the book *Political Marketing in Retrospective and Prospective* (Routledge 2012) and has published in a wide variety of academic journals.

Christoph F. Breidbach is a lecturer at the University of Melbourne, Department of Computing and Information Systems. His research addresses questions related to digital service innovation and transformation, and has received multiple awards, including Best Paper of the Year by the INFORMS *Service Science* section. Dr. Breidbach's publications to date appeared in leading international journals including the *Journal of Service Research*, *Industrial Marketing Management*, *Managing Service Quality*, *The Service Industries Journal*, *Service Science*, *Marketing Theory*, and other outlets. He serves on the editorial boards of the *Journal of Service Research* and the *Journal of Service Theory and Practice*, and holds a leadership position with the SIG Services at the Association for Information Systems.

Danny Samson is professor of management at the University of Melbourne since 1988 and was head of the Department of Management in the Faculty of Economics and Commerce. He has written a dozen books, most recently "Innovation and Entrepreneurship: Value Creation" (Oxford University Press, 2016). He has published over 100 academic research articles in journals including in the *Academy of Management Executive*, *Journal of the Operational Research Society*, *European Journal of Operational Research*, *Journal of Business Research*, *International Journal of Management Science*, *Decision Support Systems*, *International Journal of Production and Operations Management*, and *Journal of Operations Management*, where he was associate editor. His honors include winning a research prize for the best research article submitted to the *European Actuarial Association*, and he was voted the Outstanding Instructor by the Executive MBA class of 1985 at the University of Illinois. He has won numerous "best paper" awards, from the American Production and Inventory Control Society (APICS) and Decision Sciences Institute (DSI).

He has extensively consulted to leading and multinational companies such as Toyota and National Australia Bank.

Darja Groselj is a researcher at the Centre for Social Informatics, Faculty of Social Sciences, University of Ljubljana, Slovenia. She holds a PhD in information, communication, and the social sciences from Oxford Internet Institute, University of Oxford, United Kingdom. Her research interests focus around the use of ICTs and social inequalities, with a special interest in how online engagement is shaped by various material and motivational factors.

David J. Yates is an associate professor of computer information systems at Bentley University. Dr. Yates's research areas include computer networking, data communications, sensor networks, embedded systems, operating systems, and computer architecture. David has also held research and academic positions at the University of Massachusetts and Boston University. In the corporate arena, he was a co-founder and vice president of software development at InfoLibria—a start-up that grew to become a leading provider of hardware and software for building content distribution and delivery networks before it was acquired. He holds a PhD and MSc from the University of Massachusetts and a BSc from Tufts University.

Deana A. Rohlinger is a professor of sociology at Florida State University. She studies mass media, political participation, and American politics. She is the author of *Abortion Politics, Mass Media, and Social Movements* in America (Cambridge University Press, 2015) as well as dozens of research articles and book chapters that analyze topics as diverse as the kinds of claims individuals made in the e-mails they sent Jeb Bush about the Terri Schiavo case to collective identity in the Red Hat Society. Rohlinger has two master's degrees (communication and social science) and received her PhD in sociology from the University of California, Irvine in 2004. She has been interviewed on a range of topics including Black Lives Matter and controversies involving Planned Parenthood, as well as written commentaries for range of media outlets including *U.S. News & World Report* and *Fortune* magazine.

Efpraxia D. Zamani is a lecturer of IS and a member of the Centre for Computing and Social Responsibility at De Montfort University, Leicester, United Kingdom. Her interests cover post-adoption behavior with technology, organizational use of IT and ICT4D, among others. She has been involved in several EU-funded and privately funded research projects in the fields of mobile communications, innovation, and online social networks. Her research work has been presented in numerous conferences, and has received distinctions (Best Paper Award in the International Conference on Information Systems—ICIS 2013, and Best Paper Award Nominee in the Hawaii International Conference on System

Sciences—HICSS 47), and has been published in journals such as the *Journal of Information Technology* and the *International Journal of Electronic Commerce*.

Gillian Youngs is professor of creative and digital economy and head of innovation and impact, Westminster School of Media, Arts and Design, University of Westminster, with a background in media, business and academia. As an applied theorist, she undertakes a range of policy and practice related work. She has been researching different aspects of political economy and the Internet for nearly 20 years and her research work has been supported by the ESRC, AHRC, and British Academy. Her publications include *Global Political Economy in the Information Age* (Routledge, 2007) and the edited collection *Digital World: Connectivity, Creativity and Rights* (Routledge, 2013).

Girish J. "Jeff" Gulati is an associate professor of political science at Bentley University who earned his PhD from the University of Virginia. Dr. Gulati's areas of expertise are on the U.S. Congress, campaigns and elections, e-government, and telecommunications policy. His recent work has appeared in *Government Information Quarterly*, *New Media & Society*, *Telecommunications Policy*, and *Social Science Computer Review*. He also is an elected member of the executive board for the Informational Technology and Politics section of the American Political Science Association and serves on the senior editorial board of the *Journal of Information Technology & Politics* and editorial board of the *Journal of Political Marketing*.

H. Patricia McKenna is the president of AmbientEase (Emergent Adaptive Solutions Everywhere), a Canadian company focused on smart cities and learning cities. Patricia works within and across diverse domains of scholarship and practice (interdisciplinarity) and collaborates in team efforts to set up international, national, regional, and local information services, research projects, startups, and other creative and future-oriented undertakings. Through the UrbanitiesLab, which she founded in 2015, Patricia explores the rapidly evolving information landscape afforded by new, emerging, and next generation technologies. She engages with educators, learners, business, and local communities around use experience and unexpected possibilities for leveraging and generating new relevancies and vibrancies in 21st-century information spaces for learning, living, and working. Patricia holds a BA from the University of New Brunswick, an MLS from McGill University and a doctorate of professional studies in information management from Syracuse University. Patricia is an educator, innovator, informatics entrepreneur, and urban researcher.

James Pick is professor of business at the University of Redlands School of Business. He is past chair of the Department of Management and Business and past assembly chair of the School of Business. He is author/co-author of over 150 journal articles, book chapters, and refereed proceedings in IS, GIS, population, urban studies, and renewable energy, and author/co-author of 13 books on these topics, the most recent ones being the research monographs, *Global Digital Divides: Explaining Change* (Springer, 2015) and *Renewable Energy: Problems and Prospects in the Coachella Valley of California.* (Springer Nature, 2017). His current research interests include spatial and socio-economic modeling of the worldwide and U.S. digital divides, and study of influences on technology use in rural villages in India, and spatial and policy analysis of renewable energy development. He is a senior associate editor of *European Journal of Information Systems*, associate editor of *Information Technology for Development*, and serves on three other journal editorial boards. He holds a BA from Northwestern University, MSEd from Northern Illinois University, and PhD from University of California Irvine.

Jessica Rosales is an adjunct instructor in the School of Business at the University of Redlands. She is an MBA and BS alumna of the University. In 2015 she earned her MBA with GIS emphasis and received the 2014 Graduate Award for Excellence in Business GIS from the Center for Business GIS and Spatial Analysis (GISAB). Ms. Rosales earned her Bachelor of Science in Business (2013) and is a recipient of the Randal Walker Ethics Award (2011). She currently collaborates with the Center for Business GIS and Spatial Analysis on research projects examining business applications of GIS with the use of social media, mobile devices, and big data. Her research and maps have been featured in a variety of conference proceedings, academic publications and a federal grant. Ms. Rosales teaches GIS applications in business at the University of Redlands and is a Certification Program Coordinator at Esri in Redlands, California.

Jyoti Choudrie holds the position of professor of IS at University of Hertfordshire. She has extensive years' experience specializing in investigating the social inclusion and adoption of ICTs on society's "marginal groups," the adoption, use and diffusion of innovative ICTs in small- to medium-size enterprises and large organizations. This is based upon the principles and mechanisms of variables taken from the theories of diffusion, adoption, usage and implementation in the social, organizational and government realms and how they can be brought to fruition using modern internet related technologies; for instance, Broadband, Smartphones and online social networks to guide and improve individuals experiences of modern technology. This was achieved due to sponsored research funding schemes Royal Academy of Engineering, Microsoft and Knowledge Transfer Partnerships, and consultancy projects with organizations such as, British Telecom and AoL. To ensure that her expertise remains in the area, she has written for established journals such as *European Journal of Information Systems*, and *Journal of Information Technology*, and published a Routledge research monograph titled "Management of Broadband Technology Innovation." She has published over 100 peer-reviewed papers, commented in trade magazines such as, *Computing and Computer Weekly*, and newspapers *Times Online* and *Guardian*. She has been invited to comment on the radio about the issues of broadband adoption and policies and the impacts of technology on older adults.

Ken Clarke has 30 years of experience in a diverse range of industries from broadband services, telecoms, military and medical technology. These involved fundamental research and development, product commercialization, and professional consultation services. He has authored over 50 journal articles, book chapters, and conference papers. He is the originator of several patents and was a co-founder of a spinoff technology company. In early 2010, he joined the University of Melbourne and is currently responsible for interdisciplinary research projects, working with academics, community groups, and industry. He is active in a wide array of networked society projects such as education, virtual reality, and innovative health services.

Kirsten Larsen is the food systems research manager at the Victorian Eco-Innovation Lab, Faculty of Architecture, Building and Planning. Her work concentrates on the complex challenges of sustainability and resilience of food systems, and the design and implementation of responses that also support healthy communities and regional economic development. Ms Larsen is the director of Eaterprises Australia, co-founder and director of the OFN. She focuses on research that supports change in policy or action. Her policy development background includes extensive experience in the Victorian Government, most recently in the Department of Premier and Cabinet on climate change, food, and innovation policy. Alongside her policy and research work, she has developed pioneering solutions to food systems problems, such as OFN, an open source software platform used globally.

Luke Williamson earned a master's in IS from the School of Information and Library Science at the University of North Carolina at Chapel Hill. His research concentration was HCI and systems analysis. Past professional roles include cryptologic linguist and support technician. He currently works in financial IT at Credit Suisse as an application support analyst.

Marianne Gloet is a research fellow in the Departments of Management and Marketing and Computing and Information Systems at the University of Melbourne. She holds a PhD from Melbourne University in the field of knowledge management and innovation, and is widely published in the areas of innovation, knowledge management, IS, and human resource management. Marianne has lived and worked in Australia, Canada, the United States, Malaysia, Hong Kong, Vietnam, and the United Arab Emirates, holding senior positions in both the academic and private sectors. Marianne also consults widely to a range of private and public organizations both in Australia and abroad.

Md Mahbubur Rahim is currently a senior lecturer at the Faculty of Information Technology, MONASH University, Australia. He has research interests in such areas as e-business, e-government, social media impact, IT-enabled sustainability practices, and ERP systems. Rahim has widely published in over 80 refereed journals, conference proceedings and edited books. His total number of Google Scholar citations stands at 1,019 with an h-index of 16. He received best paper awards at ACIS2011 and e-CASE2008 conferences. Rahim has successfully supervised research projects of four PhD students in the areas of e-commerce, ERP systems, IT government, IT security, and e-procurement. He has received a number of research grants including an ARC Discovery grant. Currently, Rahim serves on the editorial board for the *Journal of Administration & Governance*.

Mohammad Hossein Jarrahi is an assistant professor at the School of Information and Library Science at the University of North Carolina at Chapel Hill. His research focuses on new work arrangements, ICT mediation and information practices of knowledge workers. His most recent project explored the roles played by various ICTs and information infrastructures in the work practices of mobile knowledge workers, and he is currently studying how extreme forms of organizational and location independence shape the use of ICTs by digital nomads.

Panayiota Tsatsou is an associate professor at the University of Leicester. Her research lies in the broader field of digital media and specifically in the areas of digital divides/digital inclusion, Internet studies, digital research, and digital media and civic participation. Her publications include the monographs *Internet Studies: Past, Present and Future Directions* (2014, Ashgate) and *Digital Divides in Europe: Culture, Politics and the Western-Southern Divide* (2011, Peter Lang). She has conducted externally funded research on digital inclusion of minority communities, the links between identity and literacy in digital literacy, and the role of digital technologies in social research.

Patrice Braun is an adjunct professor in research and innovation with Federation University Australia. Prof. Braun is an Action Researcher with a PhD in regional network development underpinned by ICT and a master's in research on the use of the Internet for community informatics. Her global research and consultancy work focuses on sustainable regional futures and ICT-enabled development. Prof. Braun is a frequent contributor to Asia-Pacific Economic Cooperation (APEC) fora.

Prakash J. Singh is professor of operations management in the Department of Management and Marketing at the University of Melbourne. He obtained his PhD from the University of Melbourne. His research interests are in operations strategy, buyer-supplier relationships, and sustainable supply chains. His research is published in a number of leading journals including *Journal of Operations Management*, *Journal of Business Ethics*, *International Journal of Production Economics*, and *International Journal of Production Research*.

Rachelle Bosua is a senior lecturer in the Department of Computing and Information System at the University of Melbourne. She obtained her PhD in 2008 from the University of Melbourne and has research interests focus on how and to what extend individuals draw on social networks to interact and collaborate on the team level. In this regards her research focuses also on the central role of the IT artifact in this context. Her research interests are in the following areas: knowledge management, social networks and social media, mobile work, and more recently the IoT and individual privacy. Her teaching expertise includes software engineering, business process analysis and business analysis modeling and design. In addition, she co-supervises multiple PhD, MIT, and MIS students.

Ranjan Vaidya is information architect at the IT Strategy, Policy, and Planning Unit, University of Auckland, New Zealand. In 2013, he completed PhD in IS at the Department of Information Systems and Operations Management, University of Auckland, New Zealand. While doing the PhD, he was awarded the Auckland University International Doctoral Scholarship. He has published journal papers and book chapters on ICT for development and his contributions have appeared in Emerald Insights and Springer. He also serves as visiting faculty at the University of South Pacific. Prior to joining the doctoral program at the Auckland University, he has worked at the Centre for Electronic Governance, Indian Institute of Management, Ahmadabad, and Indian Institute of Forest Management. His research interests include IS implementation in developing countries, critical research in IS, and impact assessment of IS projects in developing countries.

Robyn Garnett is a speech pathologist with extensive experience in delivering early intervention services in rural Australia. She specializes in working with families of children with ASD, having worked in multidisciplinary teams within both the funded and private sector. Robyn contributed to the development of Speech Pathology Australia's position paper on telepractice and is currently conducting doctoral research into a telepractice based, parent group training approach to social-communication intervention for children with autism. In addition to her clinical work, Robyn is contracted to facilitate Hanen More Than Words and Talkability workshops for Speech Pathologists throughout Australasia.

Rodrigo Mariño is a public health dentist, a principal research fellow, and a reader at the Oral Health Cooperative Research Centre, Melbourne Dental School, the University of Melbourne. Rodrigo has a dental degree from the University of Chile, and a master's in public health from the

University of Minnesota. Rodrigo has a PhD from the University of Melbourne. Rodrigo has been the principal investigator in several e-health applications in oral health projects. Rodrigo has (co)authored a total of 140 research papers published in national and international peer-reviewed journals, book chapters, and books and monographs.

Serenity Hill has extensive experience in policy development in the Victorian State Government, most recently in relation to emergency management and climate change adaptation. Serenity grew up on a family farm in Northeast Victoria, and in 2011, her attention turned from policy and research to food distribution and opening new market access channels for farmers. With Kirsten Larsen, she founded Eaterprises Australia as a social enterprise focused on transforming the way food is distributed and exchanged, which has instigated the South East Food Hub and the OFN. The OFN is a global non-profit and open source project creating online marketplaces for source identified food.

Shanton Chang is an associate professor at the Department of Computing and Information Systems, the University of Melbourne, Australia. He is currently teaching in the master's of IS program, with a focus on industry based learning. His research areas include online behavior, information needs and social technologies within the educational, health, social, and business contexts. He has over 70 refereed publications including journal papers, book chapters and conference papers.

Shawn Gaulden is a doctoral candidate and lecturer in the Department of Sociology at Florida State University. His areas of research and publication interest focus on the use of new media by social movements and political participation through online social networks. His current research focuses on political discourse and echo chambers within social media.

Sherah Kurnia is a senior lecturer at the School of Computing and Information Systems, the University of Melbourne, Australia. She is currently teaching Enterprise Systems, Shaping the Enterprise with ICT, and Business-to-Business Electronic Commerce. Her research areas include electronic commerce, inter-organizational systems, supply chain management, sustainability, strategic IT decision making, and enterprise architecture. She has over 130 publications as journal papers, book chapters, and conference papers. She has published in *Information and Management Journal*, *Journal of Business Research*, *Journal of Strategic Information Systems*, *Information Systems*, *Communications of the Association for Information Systems*, *Supply Chain Management: An International Journal and Journal of Information Technology for Development*. She has won a number of Best Paper Awards from leading Information Systems conferences. She is currently an associate editor for *Information and Management Journal*.

Sutee Pheeraphuttranghkoon completed his doctoral studies at University of Hertfordshire in 2015. His research interests include the adoption, use, and diffusion of novel ICTs. He is an accomplished web designer, with immense knowledge of mobile devices and developing countries. He has published his work extensively at the European Conference of Information Systems and Conf-IRM conferences.

Uchenna Ojiako is presently completing her doctoral studies in the University of Hertfordshire on the topic of the adoption and usage of tablet devices by 50+ in the United Kingdom. She is using a quantitative aspect to her research and specializes in older adults and ICT research due to her keen interest that emerged in her master's programme dissertation.

Victoria Carty is an activist/scholar who has written extensively on issues of social movements that include the labor movement, the peace movement, Arab Spring, Occupy Wall Street, and immigration on the U.S/Mexico border. Her most recent book is entitled, *Social Movements and New Technology*.

Zaher Joukhadar is a software engineer, data scientist, and researcher. He is currently working as a research fellow at the University of Melbourne. Zaher has a bachelor degree in informatics engineering from the University of Aleppo and a master's degree in information technology from the University of Melbourne. Most of his research centers on machine learning and image processing. Before joining the University of Melbourne, Zaher worked as AI research team leader at IDscan Biometrics. The things Zaher loves most in life when he is not sitting in front of a computer are spending quality time with friends and family, and traveling.